Thirty Years of the Chemical Weapons Convention (CWC)

Bretislav Friedrich · Ulf Schmidt · Paul Walker
Editors

Thirty Years of the Chemical Weapons Convention (CWC)

Histories, Achievements, Challenges

Editors
Bretislav Friedrich
Fritz Haber Institute of the Max Planck
Society
Berlin, Germany

Ulf Schmidt
Centre for the Study of Health, Ethics
and Society
University of Hamburg
Hamburg, Germany

Paul Walker
Chemical Weapons Convention Coalition
and Arms Control Association
Washington, D.C., USA

ISBN 978-3-031-98853-0   ISBN 978-3-031-98854-7 (eBook)
https://doi.org/10.1007/978-3-031-98854-7

Open Access funding provided by Max Planck Society.

This Springer imprint is published by the registered company Springer Nature Switzerland AG
The registered company address is: Gewerbestrasse 11, 6330 Cham, Switzerland

If disposing of this product, please recycle the paper.

Foreword: Chemical Weapons Convention Reaches Milestone

Many post-World War II international structures are what might be called bioethics adjacent: They are not strictly part of the set of widely recognized global bioethics standards and instruments, like the Declaration of Helsinki, but they overlap with concerns routinely expressed in bioethics. One of these is the Chemical Weapons Convention, which celebrated its 30th anniversary this year with the destruction by all States Parties of their declared stockpiles of chemical weapons and dual-use chemicals. This milestone was verified by the Organisation for the Prohibition of Chemical Weapons with its roughly 450-person staff, including highly skilled inspectors who were on-site for the process.

The principal US site for the development and stockpiling of chemical weapons was Edgewood Arsenal in Maryland, with its British counterpart at Porton Down. The chemicals of interest went through a sort of evolution, roughly from the World War I-era nitrogen mustards to lethal nerve agents like sarin, to deliriants like BZ. In the early years after Hiroshima and Nagasaki, a common rationale for these weapons was that they were more humane than The Bomb, but it gradually became clear that they were not useful on the battlefield, tended to be indiscriminate in their effects, and engendered their own kinds of suffering that was still more difficult to treat than wounds from kinetic weapons. Their close association with medical applications of chemistry also created a level of unease; Fritz Haber, who developed the first chemical weapons, won the Nobel Prize in chemistry in 1918. Fittingly, a conference that I attended recognizing the 30-year milestone was held at the Fritz Haber Institute in Berlin.

Fortunately, Edgewood never experienced a death like that at Porton in 1953 of a 20-year old RAF engineer who was exposed to sarin, as historian Ulf Schmidt has documented. In the 1950s human experiments for "ABC warfare"—atomic, biological, and chemical—fell under a Department of Defense policy that was copied verbatim from the Nuremberg Code without attribution, but the integration of these rules into practice was, to put it mildly, ineffective. (I have told this story in my book Undue Risk.) Many of these experiments are now part of the public imagination, especially those involving LSD. One of the most fascinating aspects of the renewed interest in psychedelics for therapeutic purposes is the way that the arc of this story has shifted toward legitimacy after an era of prohibition and the bad odor of their association with an astonishing range of characters, from CIA chemists to the counter-culture drug guru Timothy Leary. Someone to be added to this list is the psychiatrist in charge of the Edgewood hallucinogens experiments, James S. Ketchum, who published his colorful memoir in 2006, defending his research ethics.

Even the most intensive inspection measures are no guarantee. Readers may have noted that the 30-year milestone refers to "declared stockpiles." There are well-grounded suspicions that some countries have undeclared stockpiles and chemicals that have eluded the system; Russia's use of Novichok comes to mind. Neither is the future certain: British experts Malcolm Dando and Michael Crowley have warned that increasing understanding of neural mechanisms is not only good for medical science but also may expose new routes to exploitation of the brain by states or nonstate actors. What about biological weapons, now a renewed worry with the potential application of AI to create new molecules? Unlike the Chemical Weapons Convention, the Biologic and Toxin Weapons Convention, with a grand total of three hard-working regular staff, has essentially no verification mechanism.

Despite the fact that there is much more work to be done in an atmosphere that is less conducive to progress than anytime since the late 1940s, in an era of evidently diminished confidence in global institutions the destruction of these chemical stockpiles is to be celebrated. At least it provides a reminder that the darkest paths are not inevitable.

Jonathan D. Moreno
Hastings Center
University of Pennsylvania
Philadelphia, USA
morenojd@pennmedicine.upenn.edu

Jonathan D. Moreno is the David and Lyn Silfen University Professor Emeritus at the University of Pennsylvania. He is a member of the National Academy of Medicine, a fellow the Hastings Center, a member of the Philadelphia College of Physicians, a fellow of the New York Academy of Medicine, a faculty affiliate of the Kennedy Institute of Ethics at Georgetown University, and a member of the Committee on Human Rights of the National Academies of Science, Engineering and Medicine. Among his books is Mind Wars: Brain Science and the Military in the 21st Century (2012).

Preface

In 2023, the stockpiles of chemical weapons that had been declared by the state parties to the Chemical Weapons Convention (CWC) were eliminated. This momentous achievement in disarmament was the culmination of international efforts whose roots go back to the 1899 and 1905 peace conferences in the Hague and the first formal statements of the laws of war.

The CWC was drafted at the United Nations Office in Geneva and opened for signature on January 13, 1993 in Paris. It entered into force on 29 April 1997 after 65 countries had ratified it and is now binding on most states. In addition, the prohibition on use of chemical weapons is considered customary in both international and non-international armed conflicts, applying to all states and non-state actors, including those who are not party to the Convention. The implementing body of the CWC—the Organization for the Prohibition of Chemical Weapons (OPCW) based in The Hague, Netherlands—came into being with CWC's entry into force and, with its 193 member states, represents the largest multinational organization outside the United Nations. In 2013, the OPCW was awarded the Nobel Peace Prize "for its extensive efforts to eliminate chemical weapons."

In order to mark the 30th anniversary of the CWC, we co-organized a conference *Thirty Years of Chemical Weapons Convention (CWC): Histories, Achievements, Challenges*, at Harnack House, the conference center of the Max Planck Society in Berlin-Dahlem. This meeting, which took place on October 5–6, 2023, had the aim of obtaining fresh perspectives from different stakeholders—including officials, scholars, and representatives of non-governmental organizations—on where we were regarding the implementation of the CWC. Holding such a conference was in keeping with a key objective of the OPCW, namely engaging with civil society in the widest possible way in relation to chemical safety and security. The 2023 conference was a follow-up to a 2015 conference, *One Hundred Years of Chemical Warfare: Research, Deployment, Consequences*, that marked the anniversary of the German chlorine cloud attack at Ypres on April 22, 1915. This attack amounted to the first use of a weapon of mass destruction and, as such, marks a significant turning point in world history.

October 2023 was a particularly opportune moment to hold a meeting on chemical disarmament not only because of the 30th anniversary of the signing of the CWC but also because September 30, 2023 was the projected deadline for the elimination of all declared chemical arsenals. This deadline has been met. The meeting could thus reflect on the history and achievements of the CWC over the past thirty years and explore existing and future challenges as the world realigns itself with, and prepares for, a new geopolitical and security environment amid ongoing conflicts in volatile regions such as Eastern Europe and the Middle East. Also important to note is one year later, in December 2024, the Syrian regime has been overthrown after a 13-year civil war, allowing the OPCW to resolve outstanding issues with Syria's initial declaration of chemical weapons activities and hopefully bring some closure to the ongoing Syrian file. The possible use of a new family of nerve agents, i.e. Novichok, by Russia, and allegations of use of riot control agents (RCAs) in the ongoing war in Ukraine, continues to challenge the global implementation of the CWC.

The meeting was opened by the U.S. Under Secretary for Arms Control and International Security Bonnie Jenkins and by the Deputy Federal Commissioner for Disarmament and Arms Control of the German Ministry of Foreign Affairs Susanne Riegraf. The former Director General (2010–2018) of the OPCW, Ahmed Üzümcü, delivered the keynote address. In four separate sessions, officials as well as twenty-one experts—including Ambassadors to the OPCW Joseph Manso and Thomas Schieb—shared their perspectives on various aspects of and challenges to chemical disarmament. In particular, the conference addressed the question of whether the CWC can serve as a model for the abolition of nuclear weapons.

We all came together once more on the pages of this open-access volume to record what was said at the conference on the wide range of issues covered—not only for our own reference and for those who were unable to attend but to inform constructive discussions on a subject of fundamental importance for the environment and the type of world we envisage for tomorrow.

This volume is comprised of sections on 1. *Politics and Diplomacy: Perspectives on CWC* (Bonnie Jenkins, Susanne Riegraf, Ahmet Üzümcü, Joseph Manso, Thomas Schieb); 2. *Histories of Chemical Weapons in the 20th Century* (Bretislav Friedrich, Ulf Schmidt and Jo Fox, David Peace); 3. *Participation of Civil Society in CWC Implementation* (Irene Kornelly, Homeyra Karimivahed, Deborah Klein Walker, Paul Walker, and Ulf Schmidt); 4. *CWC—A Model for the Abolition of Weapons of Mass Destruction* (Alexander Ghionis, Alexander Kelle, Ralf Trapp, Michael Crowley, Paul Walker); 5. *Chemical Weapons and Terrorism: Ongoing and Future Challenges* (Alexander Kelle and Yasmin Cürük, Lincoln Sheff); and 6. *Information Warfare: The Power of Images* (Katja Schmidt-Mai). Jonathan Moreno wrote the Foreword.

We are grateful to the Board of Directors of the Fritz-Haber-Institut der Max-Planck-Gesellschaft and the Centre for the Study of Health, Ethics, and Society of the University of Hamburg for their support of the conference *30 Years of Chemical Weapons Convention* in Harnack-Haus. We are also grateful to the Fritz-Haber-Institut for funding the open-access publication of this volume. The research and editorial contributions to this volume from the University of Hamburg stem from the project "Taming the European Leviathan: The Legacy of Post-War Medicine and the Common Good," which is supported by the European Research Council under the European Union's Horizon 2020 research and innovation program (grant agreement no. 854503). We thank all those who helped to organize the conference, in particular Daria Haberland and Sabine Fräbel (Fritz-Haber-Institut der Max-Planck-Gesellschaft), Dr James Farley (University of Hamburg), and Norbert Domke and Sandra Pranzner (Harnack-Haus). Special thanks also go to Dr Kate Docking, Dr David Peace, and Dr Will Studdert from the Centre for the Study of Health, Ethics, and Society at the University of Hamburg for their unwavering support during the editorial process of the book. We also thank Annelies Kersbergen and Banu Dhalayan for overseeing the publication process of this volume at Springer Nature.

Hamburg, Germany   Ulf Schmidt
                                                          ulf.schmidt@uni-hamburg.de

Berlin, Germany   Bretislav Friedrich
                                                 bretislav.friedrich@fhi-berlin.mpg.de

Washington, D.C., USA   Paul Walker
                                                            pwalker@globalgreen.com

Contents

Editors and Contributors

About the Editors

Bretislav Friedrich is a Research Group Leader at the Fritz-Haber-Institut and Honorary Professor at the Technische Universität Berlin. Apart from his research in molecular physics, he maintains an abiding interest in the history of science and is engaged in efforts to eliminate chemical and other weapons of mass destruction. Besides numerous research papers and several books, he co-authored and co-edited *One Hundred Years of Chemical Warfare: Research, Deployment, Consequences* (Springer Nature, 2017).

Ulf Schmidt is Senior Professor of Modern History at the University of Hamburg, founding-director of the Centre for the Study of Health, Ethics, and Society, and a Fellow of the Royal Historical Society. His interests are in the history of modern medical ethics, warfare, and policy in twentieth-century Europe and the USA. He is especially interested in the history of authoritarian regimes and modern dictatorships. He is the author, among others, of *Justice at Nuremberg: Leo Alexander and the Nazi Doctors' Trial* (Palgrave Macmillan, 2004); *Karl Brandt: The Nazi Doctor. Medicine and Power in the Third Reich* (Continuum, 2007); *Secret Science. A Century of Poison Warfare and Human Experiments* (OUP, 2015); co-editor of *Propaganda and Conflict: War, Media and the Shaping of the Twentieth Century* (Bloomsbury, 2019), and *Ethical Research: The Declaration of Helsinki, and the Past, Present, and Future of Human Experimentation* (OUP, 2020). He is Principal Investigator of a six-year ERC Synergy Grant on "Taming the European Leviathan: The Legacy of Post-War Medicine and the Common Good".

Paul Walker coordinates the Chemical Weapons Convention (CWC) Coalition, is Vice Chair of the Arms Control Association, and is a member of the US Department of State International Security Advisory Board. He has worked on chemical weapons demilitarization since undertaking the first US on-site inspection of a Russian CW stockpile in 1994 when he was a Professional Staff Member of the US House of

Representatives Armed Services Committee. He led the Green Cross International Program on Environmental Security and Sustainability for 20 years and was awarded the Right Livelihood Award in 2013 "for working tirelessly to rid the world of chemical weapons." Walker holds an M.A. from Johns Hopkins SAIS and a Ph.D. in international security from MIT; he is also a US Army Vietnam-era veteran.

Contributors

Michael Crowley Bradford University, Bradford, UK

Yasmin Cürük Institute for Peace Research and Security Policy (IFSH), University of Hamburg, Berlin, Germany

Jo Fox Faculty of Humanities and Social Sciences, Newcastle University, Newcastle, UK

Bretislav Friedrich Fritz-Haber-Institut der Max-Planck-Gesellschaft, Berlin, Germany

Alexander Ghionis The Harvard Sussex Program, The Science Policy Research Unit (SPRU), University of Sussex, Falmer, UK

Bonnie Denise Jenkins U.S. Department of State, Washington, D.C, USA

Homeyra Karimivahed The Rotary Peace Centre, University of Bradford, Bradford, West Yorkshire, UK

Alexander Kelle Institute for Peace Research and Security Policy (IFSH), University of Hamburg, Berlin, Germany

Irene Kornelly Colorado Citizens' Advisory Commission for Chemical Weapons Demilitarisation U, South Denver, CO, USA

Joseph Manso Organisation for the Prohibition of Chemical Weapons (OPCW), The Hague, The Netherlands

David Peace Centre for the Study of Health, Ethics and Society, Universität Hamburg, Hamburg, Germany

Susanne Riegraf Federal Foreign Office, Berlin, Germany

Thomas Schieb Organisation for the Prohibition of Chemical Weapons (OPCW), The Hague, The Netherlands

Katja Schmidt-Mai University of Hamburg, Hamburg, Germany; Research Associate, Institute for Liberal Arts and Sciences, Hamburg, Germany

Ulf Schmidt Centre for the Study of Health, Ethics, and Society, University of Hamburg, Hamburg, Germany

Lincoln Sheff University of Bath, Bath, UK

Ralf Trapp International Disarmament Consultant, Chessenaz, France

Ahmet Üzümcü Organisation for the Prohibition of Chemical Weapons (OPCW), The Hague, The Netherlands

Deborah Klein Walker School of Public Health, Boston University, Boston, MA, USA; American Public Health Association, Washington, DC, USA

Paul Walker Chemical Weapons Convention Coalition and Arms Control Association, Washington, DC, USA

Part I
Politics and Diplomacy: Perspectives on CWC

Chapter 1
Opening Address at the Conference "Thirty Years of Chemical Weapons Convention"

Bonnie Denise Jenkins

Hello colleagues and friends. Although I could not be with you all today in person, I did not want to miss the opportunity to join you to commemorate thirty years since the adoption of the Chemical Weapons Convention. I would like to thank the Max Planck Society for hosting this conference, and a special thanks to the organizers for inviting me to offer some thoughts today about the United States' support to the Convention and the challenges to it that remain.

It is without a doubt that the United States strongly supports the Convention. Nowhere has a Convention seen more success than in eliminating a whole category of declared weapons of mass destruction.

Most recent success is obviously the United States' completed destruction of its declared stockpiles, completing a three decades-long effort that spanned eight U.S. states and involved intimate cooperation and coordination with local communities to free the country of chemical weapons. This is a remarkable achievement that required a monumental effort from countless Americans who all believed that a world without chemical weapons is a better one. The completion of the U.S. destruction effort shows our commitment to achieving the objective and purpose of the Chemical Weapons Convention: working towards a world free of chemical weapons.

Our efforts join a list of successes for the CWC.

This includes the historic achievement that earned the OPCW—the Organization for the Prohibition of Chemical Weapons and the technical body entrusted with the responsibility of implementing the Convention—a Nobel Peace Prize in 2013 with its efforts to eliminate chemical weapons in Syria.

Before 2013, the Organization was focused mainly on the task of destroying historical chemical weapons of U.S. and Russian stockpiles. With Syria's horrific attack on its civilians at Ghouta and then the regime joining the CWC, addressing

B. D. Jenkins (✉)
U.S. Department of State, Washington, D.C, USA
e-mail: bonniedjenkins@gmail.com
URL: https://www.state.gov/biographies/bonnie-denise-jenkins/

© The Author(s) 2026
B. Friedrich et al. (eds.), *Thirty Years of the Chemical Weapons Convention (CWC)*,
https://doi.org/10.1007/978-3-031-98854-7_1

chemical weapons in Syria became a major task for the Organization. The OPCW has addressed this herculean task admirably, with persistence, thoughtfulness, and flexibility. And, States Parties have also responded to the regime's actions through the establishment and funding of the Investigation and Identification Team, showing that the international community is united in support of the norm against the use of chemical weapons.

Additional examples of the success of the Convention include the responses of States Parties to Russia's attempted assassination of the Skripals with a chemical agent in 2018. Following Russia's actions, States Parties approved the addition of two families of Novichoks to the CWC Annex on Chemicals in November of 2019. States Parties also took action to address the threat of aerosolized central nervous system-acting chemicals in December of 2021, affirming that these are understood to be inconsistent with law enforcement purposes as a "purpose not prohibited" under the Convention.

These were unprecedented successes under the CWC. However, there are still challenges to the Convention that we must face with an unflinching resolve. Chief among these challenges is the continued presence of chemical weapons programs and stockpiles.

Foremost, the Russian Federation must declare and destroy its chemical weapons program and stockpile. Following the Skripals, the Russian Federation again used chemical weapons in its attempt to assassinate Aleksey Navalny. We must continue to demand clear answers from Russia on its undeclared chemical weapons program.

Only by holding accountable the perpetrators of CW use can we deter future use. Open source reports of the use of riot control agents as a method of warfare by the Russian Federation in Ukraine should concern everyone and act as a reminder that the stakes remain high. States Parties must continue to speak up and take steps to hold violators of the CWC to account.

In addition, we must continue to press the Syrian regime to cooperate with the OPCW and return to compliance with the Convention. As Secretary Blinken noted in his August 21st statement remembering and honoring the victims and survivors of the Ghouta chemical attack, Syria has yet to fully declare and verifiably eliminate its chemical weapons program despite its international obligations under the CWC and UN Security Council Resolution 2118. The Secretary also noted Syria's refusal to take any responsibility for its vile campaign of chemical weapons use, as is evident from Syria's nine subsequent chemical weapons attacks confirmed by the OPCW's Investigation and Identification Team, also known as the IIT, and the OPCW-UN Joint Investigative Mechanism. We must continue to seek justice and accountability for those responsible for these horrific acts.

While the United States has been among the strongest supporters of accountability and destruction efforts, we also continue to urge the four remaining non-States Parties to join the CWC. The Organization is preparing for that scenario by planning and gathering technical expertise that includes retaining CW destruction experience.

The United States is pleased to support this effort. With the complete destruction of the declared chemical weapons stockpile and the release of the final reports of the

IIT, an even greater importance is placed on ensuring the knowledge and expertise of chemical weapon destruction and investigation is maintained at the OPCW.

This year, as you well know, the OPCW also held its fifth Review Conference. As part of the Review Conference, States Parties focused, in part, on ensuring that the OPCW has the tools it needs to accomplish its mission, including improvements in knowledge management, as I just mentioned.

Given the continued threat of CW use, it is also essential that States Parties to the Convention, including through the OPCW, support capacity building related to deterring, responding to, and investigating CW use. The OPCW's new ChemTech Center, which opened as a kick-off event for the Review Conference, is a facility that will allow the OPCW to do just that. I have no doubt that in time, the ChemTech Center will prove to be a great resource in support of the Convention.

During the Review Conference, I also recommended that the OPCW improve organizational governance by supporting gender and geographic diversity and inclusion, and expanding education and outreach. The United States, along with Canada, Colombia, Finland, Ireland Sweden, the UK, and Northern Ireland, hosted the first-ever Women, Peace, and Security event at the Conference of States Parties in 2022.

For many years the OPCW has run an initiative focused on increasing gender diversity and equity called the Women in Chemistry Initiative. As part of our efforts to prioritize promoting the Women, Peace and Security agenda, the United States remains committed to advancing the role of women in all their diversity at all levels and in all areas of the Organization. The contributions of women are key to the future vitality of the Organization.

As we look toward the future, the CWC and the OPCW must remain agile and continue to adapt to 21st century challenges. To this end, we must deepen our engagement with stakeholders, including academics, industry, and non-governmental organizations. The OPCW should work to increase the visibility of the Convention, and the involvement of the broader international community, by supporting broader NGO participation in OPCW annual meetings.

We are grateful to have such diverse interests in chemical weapons disarmament. Conferences like this one should continue to be organized and supported to ensure every voice is heard and every perspective is considered as we think of the way forward.

I would like to conclude my remarks by thanking everyone here today for their commitment to making sure chemicals are not used as weapons. The United States is confident in the ability of the OPCW's leadership and professional staff members to carry out the significant mandates the global community have entrusted to them, and we look forward to working with States Parties to uphold the norms established under the Convention.

Whether you are an advocate, an academic, from government or an NGO, your work to uphold the international norm against the use of chemical weapons is important to global security.

We have achieved much in the past 30 years. We eliminated chemical weapons stockpiles that no longer exist and made the world a much safer place in the process.

But our work here is not yet done. As you reflect on how far we have come, and where we go from here, I wish you fruitful discussions at this conference and beyond. Thank you.

Bonnie Denise Jenkins was the Under Secretary for Arms Control and International Security at the U.S. De-partment of State. She previously served as Special Presidential Envoy and Coordinator for Threat Reduction Programs, legal advisor to the former U.S. Arms Control and Disarmament Agency, Counsel on the 9/11 Commission, and General Counsel to the U.S. Commission to Assess the Organization of the Federal Government to Combat the Proliferation of Weapons of Mass Destruction. A retired U.S. Naval Reserve Officer, Dr. Jenkins also worked for the Ford Foundation and the Rand Corporation and founded the nonprofits Organizations in Solidarity and Women of Color Advancing Peace, Security, and Conflict Transformation.

Chapter 2
Opening Address at the Conference "Thirty Years of Chemical Weapons Convention (CWC)"

Susanne Riegraf

I am very grateful to the Fritz Haber Institute for hosting this conference today.

Fritz Haber, who is known as the father of chemical warfare, played an inglorious role as advocate for the development and first use of chemical weapons. Yet, at the same time, his research won him the Nobel prize in Chemistry and is still relevant for fertilizer production until today. I highly appreciate that the organizers of today's conference have taken up the challenge to draw on his difficult heritage as a continued vocation to support the global elimination of chemical weapons.

The 30th anniversary of the Chemical Weapons Convention is a reason to celebrate.

When the CWC was signed in 1993, it was a milestone for multilateral arms control. One can confidently say that the CWC has proven over the past 30 years that it lives up to expectations.

The CWC was the first arms control treaty that combined the ban of an entire class of weapons of mass destruction with the **obligation to destroy the huge stockpiles** of chemical weapons accumulated during the Cold War. It is a great success that this year the CWC achieved this goal. Getting rid of the apocalyptic stockpiles—over 70 thousand tons of declared chemical weapons—is a huge gain in security for humankind and a major accomplishment for the OPCW.

The CWC was also the first WMD arms control instrument that combined the global ban with a **broad verification regime**, including industry verification. This is fundamental in creating the necessary trust of all participants in universal compliance. Finally, its use of the **general-purpose criterion** in defining a chemical weapon maintains its all-encompassing character in spite of developments in technology.

S. Riegraf (✉)
Federal Foreign Office, Berlin, Germany
e-mail: or12-1@auswaertigesamt.de
URL: https://www.auswaertiges-amt.de/en

© The Author(s) 2026
B. Friedrich et al. (eds.), *Thirty Years of the Chemical Weapons Convention (CWC)*,
https://doi.org/10.1007/978-3-031-98854-7_2

Until today, the CWC therefore remains a very modern and up-to-date instrument. It is a sign of its success that it has almost achieved universality, with 193 member states to date.

While these points are all good reasons to celebrate, we do not mark the 30th anniversary of the CWC in a state of cheerfulness. On the contrary—for several years, we have seen dark clouds assembling over the horizon.

The centrepiece of the CWC, the universally accepted and undisputed global ban of chemical weapons, is **under ever-increasing pressure**.

We are today in a battle to keep it alive. While we—the international community—have won a few skirmishes, we have not yet won the battle. That battle is not an isolated phenomenon neither, but is part of a broader war waged against the international rules-based order.

The appalling, repeated **use of chemical weapons in the atrocious conflict in Syria** is the biggest reversal to the CWC's aspirations that we have seen so far. Although international pressure forced Syria to become a party to the CWC in 2013 and to renounce its chemical weapons, the Assad regime has been dodging these obligations from the start. To this day, in spite of the persistent and meticulous efforts of the OPCW's Technical Secretariat, Syria never presented a complete and consistent initial declaration. Instead, the regime has continued using chemical weapons after joining the Convention, and we have reason to fear that it still tries to continue a covert chemical weapons program in spite of its international obligations. That is utterly unacceptable.

While the **"Syria fatigue"** in parts of the international community and even some steps towards regional rehabilitation of the Assad regime are deplorable, we must never acquiesce to the outrageous situation that a state party to the CWC continues clandestinely to pursue a chemical weapons capability.

What is even more worrisome is the fact that there was not a unified outcry by the entire international community to these developments, but that a few countries, and in particular Russia, covered and shielded the Assad regime's action. Even worse, **Russia has taken active steps to discredit the OPCW's actions** to investigate and publicly put the impartiality and professionalism of its Technical Secretariat into doubt.

Fortunately, the huge majority of the international community **stood together to defend the global ban on chemical weapons**. In the case of Syria, we have been successful in creating ad hoc instruments to investigate the use of chemical weapons and in a second step to attribute these cases and name the perpetrators. This task is not yet completed, and I applaud the on-going work by the OPCW's Technical Secretariat. In a third step, we must put our legal systems to work in order to bring the perpetrators to justice.

We must maintain pressure on the Assad regime to fulfil all of its obligations, and denounce all efforts to shield the breakers of international law.

Russia's own compliance record with the CWC is more than dubious. Its persistent refusal to investigate the **attacks against Sergey Skripal and Alexey Navalny**, and the enormous efforts to lay one smoke-screen after another on the facts must lead us to assume that Russia is not acting in good faith and is not in full compliance with its obligations. That is a serious concern for our security.

Against this background, Russia's persistent use of unsubstantiated claims regarding alleged plans to use chemical weapons by **Ukraine**, a country with an exemplary non-proliferation record, is even more alarming. These claims may not just serve as attempts to justify Russia's war of aggression, but could prepare the ground for so-called false flag operations in which Russia uses toxic chemicals as a weapon and tries to put the blame on Ukraine. There are already strong indications that Russia is using non-lethal riot control agents in its military operations, which is prohibited under the Chemical Weapons Convention. Russia's reckless military attacks on and around chemical facilities are additional threats to Ukraine and its civilian population.

I am grateful that so many CWC States Parties, including Germany, have responded to Ukraine's call for assistance in protecting itself against possible chemical weapons attacks.

Nonetheless, it is important to highlight that Russia's aggressive and reckless behaviour damages the CWC by **undermining its institutions** and casting doubt on universal compliance to the Convention.

We must respond by **doing more to strengthen the CWC**.

We need a strong and capable OPCW, which can confront and counter future chemical weapons risks. We need an OPCW that is prepared to investigate alleged use, to conduct challenge inspections and to render technical assistance. Even after the completed destruction of the last declared chemical weapons stockpiles, it is of the utmost importance to retain the relevant expertise in the domain of chemical weapons verification and destruction.

We see with satisfaction that the Organisation is taking the necessary steps, in particular with the creation of the ChemTech Centre.

Although the **fifth CWC Review Conference** in May this year was not able to adopt a final document because of Russia's persistent refusal to acknowledge what happened in Syria, we observed the international community's tangible and unwavering will to strengthen the Convention. At the Review Conference, we identified a broad range of subjects that contribute to this overarching goal—it is now important to use the momentum and work constructively on issues like the modernisation of the industry verification regime.

At the same time, an exchange like the one at this conference is a valuable contribution.

Civil society has a long history of supporting the CWC. NGOs, research and academic institutions or the chemical industry—all of them are important stakeholders that provide valuable expertise and outreach, thus complementing and supporting the work of the OPCW.

I am therefore particularly thankful to the Fritz-Haber-Institute, the University of Hamburg and the CWC Coalition for organising this event and providing a forum for exchange.

Let us hope our joint efforts will make it possible to celebrate the 40th anniversary of the CWC against a more cheerful background. Our joint determination that the science of chemistry should never again be used against humans, but only to the benefit of humankind, is undoubtedly a good reason to celebrate.

Susanne Riegraf joined the German Foreign Service in 2003. In July 2018, she was appointed Head of the Division for Transatlantic Relations and G7 at the German Federal Foreign Office. Prior to that, Ms. Riegraf served as the Deputy Head of the Division for Non-Proliferation and Nuclear Disarmament from 2016 to 2018. From 2012 to 2015, she was part of the EU's negotiating team in the E3/EU+3's nuclear talks with Iran as a political Advisor and, from December 2014 onwards, team leader Iran nuclear in the European External Action Service (EEAS). Prior to that, Ms. Riegraf worked three years in the German Embassy in Tehran between 2004 and 2007, followed by assignments in the Policy Planning Staff of the Foreign Ministry between 2007 and 2009 and in the Federal Chancellery between 2009 and 2012. She holds a Master's degree in International Relations, Law and Eastern European History from the University of Tübingen. Since August 2022, she is Deputy Federal Government Commissioner for Disarmament, Non-Proliferation and Arms Control at the German Federal Foreign Office.

Chapter 3
Keynote Address

Ahmet Üzümcü

Excellencies, ladies and gentlemen,

It is a great pleasure for me to address you on the occasion of this event to celebrate the thirtieth anniversary of the CWC's signature in Paris in 1993. I wish to thank the co-organizers, Professor Friedrich, Professor Schmidt and my friend Paul Walker for inviting me to speak about the achievements of the OPCW and the future challenges.

The Organization for the Prohibition of Chemical Weapons (OPCW) is very dear to me. I served as Director General for eight years, the longest, the most demanding but at the same time the most rewarding position in my diplomatic career. During these years, the OPCW was able to meet several challenges and to adapt itself to the new circumstances.

When I joined the OPCW in July 2010, consultations on the deadline for the destruction of declared stockpiles of CWs and production facilities were underway. The deadline of ten plus five years foreseen by the CWC would be over in 2012 and there was no legal basis to extend it. Lengthy and sometimes fierce debates had finally resulted in a Conference of States Parties (CSP) decision which enabled the major possessor states to continue their destruction activities with some additional transparency measures.

Another issue that was addressed by member states was not unrelated. Some countries were arguing that after the stockpiles of CWs were destroyed, the OPCW Secretariat could be downsized and a limited number of staff could run routine activities such as industrial inspections and capacity building. I thought that as Director General I could take an initiative on this matter, while the deadline issue was handled by member countries at an open-ended Working Group.

I set up an Advisory Panel chaired by Ambassador Rolf Ekeus from Sweden and composed of diplomats and experts who already worked on the CWC related

A. Üzümcü (✉)
Organisation for the Prohibition of Chemical Weapons (OPCW), The Hague, The Netherlands
e-mail: uzumcuahmet@yahoo.com
URL: https://www.opcw.org

© The Author(s) 2026

B. Friedrich et al. (eds.), *Thirty Years of the Chemical Weapons Convention (CWC)*,
https://doi.org/10.1007/978-3-031-98854-7_3

issues. I asked them to prepare a report about the future of the OPCW. They met three times in The Hague. I received the Panel's report in July 2011, which I shared with States Parties. The report concluded that "the OPCW should remain the global repository of knowledge and expertise in the field of chemical weapons", and the priority of the Organization in the future should be "the prevention of re-emergence of CWs". Following the publication of the report, I encouraged the member countries to actively participate in informal discussions in retreats to address different aspects of the CWC regime and ways to keep the OPCW capable, apt to meet future challenges. I believe that the report, as well as the series of informal consultations, helped enhance the sense of ownership of the Organization by States Parties. There was no more discussion of downsizing.

The political will among member countries for keeping the OPCW strong and capable was invigorated by the OPCW-UN joint mission in Syria. After the sarin attack in Ghouta on 21 August 2013, which resulted in the death of over 1400 people in a few hours, we were holding our breath for a possible military operation against the Syrian regime held responsible for the use of CWs by the United States and some other western countries. This didn't happen. The US and Russian delegations negotiated a framework document in Geneva according to which the Syrian Government would accept the elimination of its CW program under international verification and join the OPCW as a member. This document became the basis of the decision adopted by the OPCW Executive Council on 27 September 2013 and the UN Security Council Resolution 2118 endorsing it on the same day. As to the implementation of the decisions, I argued that the OPCW should take the lead, with the logistical and security coordination support to be provided by the UN. Since Syria had become a member of the OPCW, our organization should assume the responsibility of such a mission as had happened in the past for other possessor states. The UN wanted to lead the mission with the support of the OPCW. In the end the then-UN Secretary General Ban ki Moon proposed to run it jointly, which I accepted. We appointed together a Special Coordinator who would report to both of us.

The OPCW-UN joint mission was strongly supported by the OPCW membership. More than thirty countries and the European Union contributed to the trust fund established for that purpose and many of them provided in-kind support, such as maritime transportation or protection. The most toxic chemicals were neutralized on an American military cargo ship—Cape Rey—with the presence of the OPCW inspectors. The OPCW staff deployed to Syria were all volunteers. After the civil war began in Syria in March 2011, we thought that the OPCW could be called upon at a certain stage since we knew that this country possessed large stocks of CWs. We sent our inspectors to training programs with a view to preparing them for deployment in a conflict zone. When we asked for volunteers in September 2013, sixty of them were ready to go. This was more than we needed. Foresight paid off.

The engagement by member states and the professionalism of the OPCW staff, as well as the UN support, made the Syria mission a tremendous success. In less than a year all declared CWs and production facilities were destroyed under the verification of the OPCW. However, gaps, inconsistencies and discrepancies continued to exist on the Syrian declaration. The efforts by the OPCW's Declaration Assessment Team

(DAT) and two rounds of consultations in The Hague, between the OPCW delegation that I led and the Syrian delegation headed by Faysal Mekdad, the current Foreign Minister produced limited progress. Based on the findings of DAT, I and later my successor reported on several occasions that the Syrian declaration was not complete or accurate. This situation has not changed since then.

The OPCW had played a significant role in investigating the allegations of use of CWs in Syria. In March 2013 the UN Secretary General called me and asked whether the OPCW would be able to support the UNSGM that he was intending to activate in order to investigate an incident reported by the Syrian Government. My response was affirmative. I didn't need to take it to the member countries since the relationship agreement with the UN made the OPCW support for such missions mandatory. I later informed the membership about the UN request and my response. However, the Syrian submission for the UNSGM was followed by two more requests for investigation by France and the United Kingdom. The Secretary General instructed the team to investigate all three allegations. The team headed by a Swedish scientist was composed of nine inspectors from the OPCW and three experts from the WHO. They were all volunteers. After a few months of negotiations between the UNODA and the Syrian authorities, the mission was finally able to deploy to Damascus in August 2013. While they were preparing to visit the sites of the reported incidents, the Ghouta attack occurred on 21 August. Secretary General instructed the team to investigate the Ghouta incident first. In spite of a sniper attack at the buffer zone and the loss of an armored vehicle, the team was able to reach Ghouta in a second attempt and collect environmental and biomedical samples. The CWC procedures were followed. The samples were split at the OPCW laboratory in The Hague and sent to two designated laboratories. The analyses proved the use of sarin. The results corroborated each other.

The allegations of use of CWs in Syria continued after it had become a member of the OPCW. We could not remain indifferent to such reports. Normally, the challenge inspection mechanism should have been invoked but no member country was willing to do it. As Director General I had no authority to activate the mechanism. Following consultations with some States Parties in Spring 2014 I decided to develop a new mechanism in order to establish the facts surrounding the allegations of use. We called it the Fact-Finding Mission (FFM). We drafted a Terms of Reference that we shared with Syria. The Syrian authorities initially dragged their feet before giving access to the OPCW team in May. We put a firewall between the joint mission and the FFM, upon the request of the UN. While the team was preparing in Damascus to go to the site of an incident, a new chemical attack was reported. On 27 May, early in the morning, the FFM team on its way to Kafr Zeta came under attack at the buffer zone, between the government-held territory and the opposition-controlled area. An armored vehicle was destroyed by a remotely exploded roadside bomb and this was followed by an ambush. Fortunately, the team members survived the attack with minor injuries. Both the government and the opposition groups denied any responsibility. I had to call back the team to The Hague. There were two options: to suspend the investigations or to pursue them from outside the Syrian territory. We chose the latter. We deployed the FFM to neighboring countries to Syria from where they had the possibility to contact the victims of CWs, the health personnel who treated them and the eye witnesses. The FFM teams interviewed them, collected biomedical and environmental samples and drew conclusions as a result of meticulous examinations. The FFM investigated more than seventy allegations and established the use of CWs in twenty instances. The FFM mandate was limited to determine the use and did not get into attribution.

The FFM did not produce the desired effect, namely halting the chemical attacks through deterrence. However, I believe that it had some impact on the users and the situation could have been much worse if this mechanism was not established. The task of identification of perpetrators was later given by the UN Security Council to the OPCW-UN Joint Investigative Mechanism (JIM) In August 2015, Russia voted in favor of this resolution. This came as a surprise to many of us, including Syrians. However, the Russian position radically changed a few months later after it became militarily involved in Syria.

JIM submitted its reports to the UN Security Council and shared them with the OPCW. The reports established that the Syrian armed forces were responsible of the use of CWs in three cases and the ISIS in one case. Russia raised doubts about the findings of the FFM and JIM. They developed false narratives about incidents, sometimes contradicting each other. They questioned the impartiality and objectivity of the OPCW staff. They lobbied other member states to gain support. The Russian disinformation campaign had a limited success but the proceedings at the OPCW had become increasingly politicized and tense.

JIM's mandate was not extended by the UNSC because of the Russian veto at the end of 2017. The FFM was continuing to work on the determination of use of CWs, but a gap emerged in regard to the attribution.

In January 2018, France launched the International Partnership against Impunity. The French initiative, though welcomed by many, could not be a substitute for an attribution mechanism. Several options, including the UNSG mechanism were considered. The UN Secretary General was reluctant to initiate it. In my public statements I suggested that the OPCW Secretariat could do the job if the Director General was given a mandate.

The Salisbury incident in March 2018 triggered a turning point for the OPCW and the CWC regime. The use of Novichok in the failed attempt of killing the former GRU agent and his daughter was attributed to Russia by the British Government. The OPCW sent a team of experts to Salisbury and independently confirmed the use of Novichok. This incident showed that more had to be done to deter further uses of CWs in Syria and elsewhere. The UK and other western countries undertook a wide and effective campaign, which culminated in the June decision of the Special Session of the Conference of States Parties. The Investigation and Identification Team (IIT) was established under the authority of the OPCW Director General. This was a significant milestone in the history of the OPCW. An international organization was mandated, without a UN resolution, not only to establish the violations but also to identify those who were responsible for them. The IIT produced three reports and identified the Syrian Armed Forces as responsible for the use of CWs in several incidents, including in Douma, in April 2018. The reports were also sent to the UN

Secretary General and International, Impartial and Independent Mechanism to assist in the Investigation and Prosecution of Persons Responsible for the Most Serious Crimes under International Law Committed in the SAR since March 2011. (IIM).

As a result of the IIT reports, the CSP decided to suspend some rights of Syria until it redressed the situation and fulfilled certain specific obligations. Some senior Syrian officials had been included on the U.S. and EU sanctions lists. However, no individual had yet been prosecuted. In the absence of an international court which could be seized for that purpose, national tribunals in western countries could perhaps prosecute and convict in absentia the Syrian officials who were responsible of the use of CWs.

In retrospect, I believe that Syria should have been subject to further scrutiny before it was allowed to become a full member of the OPCW in 2013. Its compliance could have been tested during a probationary period. Although the Convention didn't foresee it, a CSP decision or a UNSC Resolution or both could have provided the legal basis for that approach. If some possessor states are to become new members in the future, in the light of the Syrian experience, a probation might be considered.

In spite of a number of incidents in Syria, in Malaysia, in the UK and Russia over the past decade, the international norm against the use of toxic chemicals as weapons is solid. No country or individual claimed any responsibility. On the contrary, the States Parties condemned on several occasions the use of CWs anywhere, at any time, by anyone, under any circumstances. More specifically, at the meeting in Ypres in 2015 on the occasion of the Centennial Commemoration of the First Large-Scale Use of CWs, they solemnly declared that any use of CWs as such was totally unacceptable and would violate the legal norms and standards of the international community, and expressed their conviction that those responsible for the use of CWs should be held accountable. This is clearly the result of international efforts that took more than a century.

In 1899 The Hague Peace Conference prohibited the use of poison at war. Following the devastating consequences of the widespread use of CWs during the First World War, the Geneva Protocol was concluded in 1925. This legally binding document prohibited the use of chemical and biological weapons in warfare, but the development, production and stockpiling were still allowed. Prior to and during the Second World War and in early years of the Cold War, the Soviet Union and the United States manufactured large stocks of CWs of different types. Extremely lethal nerve agents were developed and weaponized by both sides, especially after the war. If these weapons were used during a war, the results would have been more disastrous than the First War.

The collapse of the Soviet Union, the use of CWs during the Iran-Iraq war and the attacks in Sardasht and Halabjah by the Saddam regime against civilian Kurds did accelerate the international efforts in search of a total ban on these weapons. It is also true that protective measures against such weapons had become more effective and the safe storage of them had become increasingly costly.

When the Conference on Disarmament in Geneva successfully concluded the CWC in 1992, it was regarded as one of the most significant peace dividends at the end of the Cold War, and a major triumph in the history of multilateralism. The ceremony of the opening of the CWC for signature in January 1993, in Paris was well attended. The whole world welcomed the global ban on a certain category of weapons of mass destruction, without any discrimination.

The Convention is a well-balanced legal document. We should be thankful to the negotiators for putting together such a comprehensive legal instrument with a robust verification annex. In addition to the verification of destruction activities, a mechanism of onsite inspections at industrial plants, randomly selected by a software was foreseen. This required the cooperation of the global chemical industry and it helped to ensure a certain order and discipline worldwide. Furthermore, the articles on national implementation, assistance for response as well as the peaceful uses of chemistry provided incentives for the engagement of member states which possessed neither CWs nor chemical industry.

The initial euphoria after the entry into force of the CWC in 1997 had slightly waned when technical problems emerged in destruction activities. Potential difficulties were clearly underestimated at the negotiating phase. The members who had smaller stocks completed the destruction in a short time. The United States and Russia, which inherited the Soviet stockpiles, had realized that the process would be time consuming, costly and technically complicated. These weapons were not designed to be destroyed, but to be used in the battlefield.

The US and some other western countries provided financial and technical support to Russia at the initial stage. The progress was still slow on both sides. Although the deadline foreseen by the Convention could not be met, the two countries were firmly committed to complete the destruction in the shortest possible time. The rest of the membership was also convinced of that. A few statements of criticism were politically motivated.

The Russian Federation finalized the process in 2017 and the US in July, this year. The OPCW inspectors were permanently present in destruction plants, including during the pandemic. Safety measures were strictly followed and no major incident occurred. Civil society played a positive role in all these endeavors.

I had the opportunity as DG, OPCW to visit four plants in the United States and an equal number in Russia. I observed the professional skills of engineers and other technical staff. Over time, sophisticated capabilities were developed, especially in the US, which helped enhance the human safety, but the cost was still high. The US program of destruction of CW stockpiles has reportedly cost over fifty billion dollars.

The achievements by Russia were, however, overshadowed by the position it took on Syria-related issues and the uses of Novichok in Salisbury and against Navalny in Russia. The disinformation activities aimed at discrediting the FFM, JIM and IIT findings tarnished the image of Russia and raised doubts about its commitment to the credibility and integrity of the CWC and beyond.

In 2012 I sent a letter to a senior American official, in charge of steering the program of the CW stockpiles' destruction. I asked whether some measures could be taken to accelerate the process and I dared to add that if this was possible the common success of the organization and its membership could be awarded by the Nobel Peace Prize. US colleagues explained at length that all necessary measures were taken to complete the destruction in the shortest possible time and reaffirmed their commitment to fulfil their obligations. I had no doubt about that. Nor did the Nobel Peace Prize Committee.

In mid-September 2013, I was on a visit to China to observe the progress in destruction of chemical weapons abandoned by Japan at the end of the Second World War. The negotiations between the US and Russian delegations in Geneva were underway. I was called by the representatives of the two countries and asked what the OPCW could deliver if it was given a mandate to run a chemical demilitarization mission in Syria. I assured them that the OPCW was fully prepared for such missions. I decided to cut short my visit and return to The Hague. While waiting for my flight at the Beijing airport, a young Chinese diplomat who was aware of what was going on said, "Sir, if the OPCW assumes the role of eliminating the Syrian CW program in Syria, and if you are successful in that mission, you will definitely win the Nobel Peace Prize". We didn't need to wait that long.

The efforts of the OPCW and its member states to eliminate the CWs in the previous years was recognized by the Nobel Committee in October 2013. The level of destruction was still eighty percent but the Committee was clearly assured about the determination of the OPCW and the commitment of possessor states to complete the job. In my lecture at the ceremony in Oslo, I paid tribute to all those who, through their dedication and resolve, contributed to this hard-won success for chemical disarmament.

The Nobel Peace Prize that I had the privilege to receive on behalf of the Organization was timely. The OPCW staff were being deployed to Syria in the midst of a civil war and the prize was a huge morale boost for them. The prize also helped forge closer ties with other stakeholders. The OPCW had concluded MOUs with the IUPAC (International Union of Pure and Applied Chemistry), the International Council of Chemical Industry Associations, the World Customs Organization, etc. The OPCW also provided a platform to develop The Hague Ethical Guidelines in 2015, with the participation of other stakeholders. The OPCW was no more a tiny, obscure organization in The Hague, as described by some media outlets when the Nobel Peace Prize was announced in October 2013.

I should add here that the Nobel prize money had been used over the past ten years to acknowledge, together with the city of The Hague, those institutions or individuals who have made outstanding contributions to the goals of the CWC. My friend Paul Walker received the OPCW-The Hague award last year on behalf of the CWC coalition.

I should here touch upon briefly on the status of the NGOs in the proceedings of the OPCW. When I arrived at the OPCW, the NGOs were not allowed to express their opinions or provide their input. In Geneva where I was representing my country at different UN agencies, the situation was totally different. A much more liberal approach was followed. I supported behind the scenes the initiative taken by some delegations to the OPCW to allow the NGO representatives to speak at the annual Conferences of States Parties. The outcome was successful. I believe that Paul Walker played a crucial role in convincing those delegations who were hesitating to accord such a right with his responsible and reassuring attitude.

The OPCW is now a mature organization with 193 member states. Only four countries are expected to become members, to render the CWC fully universal. I hope that they will do it without further delay. Egypt and Israel should be convinced that it is in their interest to join the CWC and could thereby contribute to peace and security in the Middle East. I am aware of the Egyptian position promoting the establishment of a WMD Free Zone in the Middle East, but I find it unrealistic for the foreseeable future. Some other formulas and incentives must be explored.

The first year in my office at the OPCW, I was invited to London, to speak at the Royal Society of Chemistry. I was impressed by this prestigious institution, which is more than three centuries old, and the calibre of its members. The Society had a motto, in Latin of course, "Nullius in verba", which meant to emphasize its reliance on experimental rather than on metaphysical arguments. This gave me the idea to find an appropriate motto for the OPCW. Upon my return to The Hague, we drafted a few alternatives. The motto should reflect the fundamental goal of the organization and should be forward leaning. We made a survey within the Secretariat and I consulted with some Ambassadors. We finally decided on "Working together for a world free of chemical weapons". I believe that the OPCW has made great strides to achieve this goal, but we are not there yet.

Some prospective members are suspected of possessing CWs. Once they join the CWC, the safe and verified destruction of their stockpiles will need the OPCW's technical support. And these will again be declared stockpiles. There will be a need for permanent monitoring to prevent their reemergence. Apart from states, the threat of production and use of CWs by non-state actors, particularly terrorists, will continue to pose a serious challenge. This will entail the development and implementation of preventive and response measures. The knowledge and expertise on chemical weapons accrued by the OPCW over the years will have to be retained.

In view of all this, and based on the experience we had in Syria, I thought in my last few months in The Hague that additional capabilities were required. The OPCW laboratory, even though it coordinated the analysis of more than one thousand samples during the Syria missions, was a small facility with limited equipment and other resources. Following consultations with some members, especially the host country, the Netherlands, I concluded that the construction of a Centre for Chemistry and Technology was a realistic prospect. I started the initiative a few weeks before my departure from The Hague. I am very pleased to see that the project was diligently pursued by my successor, Ambassador Arias, and the OPCW staff, and a large number of States Parties contributed to the trust fund for that purpose. The Center is now operational as of May this year. I am sure that it will help ensure that the OPCW keeps pace with all relevant developments in science and technology. It will also provide a capacity-building platform for closer international cooperation in research and training with the participation of a wide range of stakeholders.

The development of such a new capability for the OPCW is timely, since the destruction of declared stockpiles is over and a new strategic direction for the Organization is being drawn. Scientific and technological developments have gained a new momentum, thus creating new opportunities, as well as challenges. The concerns raised about emerging technologies are also valid for the CWC regime. Neverthe-less, the OPCW is fortunate to have the Scientific Advisory Board composed of 25 experts from different countries, which assumes the task of observing all relevant developments and providing recommendations to the Director General and member states.

While talking about science, I would be remiss not to speak about Fritz Haber, a prominent German scientist, who is remembered by his role in the large-scale use of CWs during the First World War. Haber developed the means of dispersing chlorine as a weapon in Ypres in April 1915, and sulfur mustard in 1917, and personally oversaw their deployment at the frontline. While the successful use of chlorine was celebrated back at home with colleagues and friends, his wife Clara, also a chemist, committed suicide. According to some researchers, she could not stand her husband's active involvement in war efforts, and for others this was due to a combination of a variety of different reasons. Anyhow, Fritz Haber was promoted in military rank and was received everywhere as a loyal patriot. There was no criticism of him at all. Moreover, in 1918 Haber was awarded the Nobel Prize for Chemistry for synthesizing ammonia from nitrogen and hydrogen. This enabled large scale synthesis of fertilizers, which helped enhance food security worldwide. None of this helped him to stay in his native country; he had to leave Germany in 1933, and died in Switzerland a year later. Because he was Jewish. If you want to know more about Fritz Haber and his first wife Clara Immerwahr, you should read the two fascinating articles written by Professor Bretislav Friedrich that he kindly shared with me a few months ago.

And now to Oppenheimer. I am sure many of you watched the movie this summer, and I would advise you to do it if you haven't done it yet. The similarities are striking. Both scientists played a leading role in the development and use of two categories of WMDs. Haber wanted to help break the stalemate in trench war, and Oppenheimer to bring an end to the war. The humanitarian consequences of the atomic bombs used in Hiroshima and Nagasaki were of course more devastating. In his own words, Oppenheimer "became death". But the nuclear technology was later used to produce energy in an environmentally-friendly manner and in medicine to save lives. Indeed, the dual use of science and technology remains a conundrum.

The dilemma faced by scientists has now been exacerbated, given the amazing pace of technological advances. The question of mitigating the risks is being addressed by several countries. This requires a collective undertaking by the international community as a whole rather than being done piecemeal.

While awaiting possible regulations, the scientific community could increase its education and outreach activities in order to promote ethical values. This applies to all disciplines in an increasingly multidisciplinary world. Eight years ago, the OPCW established an Advisory Board on Education and Outreach composed of fifteen representatives from member states. The board develops recommendations to be implemented by the Secretariat, as well as member countries in order to raise awareness about the goals of the Convention, the ethical values, the peaceful use of chemistry and the risks associated with the activities of chemistry practitioners. During my tenure as Director General of OPCW, I spoke on ethics in chemistry at international conferences. I am now an honorary member of a Working Party on

Ethics in Chemistry within the European Chemical Society, a group of academics who are promoting the peaceful use of chemistry. I believe that each of us, either policy planners, diplomats, experts or scientists should contribute to the efforts aimed at preventing the misuse of science and technology.

At the end of my statement today, I would like to quote my concluding remarks of ten years ago in Oslo: "The CWC has given us a legacy that no future disarmament effort can afford to ignore. A legacy that has at its core verification, broad stakeholder engagement, consensus born of trust and, above all, a commitment to science that actively serves the cause of peace and security. It is this legacy that we must set as the keystone in an ever-widening arch of disarmament. Only by building such an arch will we be able to bridge our security and our prosperity. Destiny has ruled that we rid the world of chemical weapons. And that we achieve this in our lifetime. This is our place in history. And this is the future we are creating. A future for which our children and grandchildren can be truly thankful."

Thank you for your attention.

Ambassador Ahmet Üzümcü is a career diplomat with vast experience in multilateral diplomacy and disarmament and non-proliferation issues. Prior to his appointment as Director-General of the OPCW in December 2009, he was Turkey's Permanent Representative to the UN Office in Geneva, where he chaired the Conference on Disarmament in March 2008. Ambassador Üzümcü completed his second term as Director-General of the OPCW in July 2018. Ambassador Üzümcü has also served as Deputy Undersecretary of State for Bilateral Political Affairs at the Ministry of Foreign Affairs(2004–2006), Turkey's Permanent Representative to NATO (2002–2004) and Ambassador to Israel(1999–2002), in addition to earlier postings to NATO, Aleppo and Vienna. He holds a Bachelor's Degree in International Relations from Ankara University and speaks English and French fluently.

Ambassador Üzümcü received the Nobel Peace Prize on behalf of the OPCW in December 2013. In December 2015, H.E. Mr. Laurent Fabius, Minister of Foreign Affairs and International Development of France, decorated Director-General Üzümcü with the Légion d'honneur (rank of officer). In 2019, he was made honorary CMG, Order of St Michael and St George, by the British Government for his services to international diplomacy and the rule of law. During his tenure as Director General, he was decorated by Federal Republic of Germany and Austria for his contributions to the success of the OPCW. He was also recognized by honorary doctorates from four universities in the UK, Argentina, Russia and Italy.

Chapter 4
Challenges to the Chemical Weapons Convention I

Joseph Manso

Many thanks to the organizers for this highly informative conference, and special thanks for inviting me to offer some thoughts regarding challenges to the Chemical Weapons convention or CWC.

It's been thirty years since the opening of the CWC for signature. The United States and the world are celebrating the achievement of the complete destruction of all chemical weapons that had been declared by States Parties under the CWC.

Today the world is a safer place because we have successfully eliminated a whole category of declared weapons of mass destruction. When the United States destroyed its last munition in July, it was the end of a three decades-long effort that spanned eight U.S. states and one U.S. territory, and required intimate cooperation and coordination with local communities. Over 30,000 metric tons of chemical weapons were destroyed by the United States, and over 73,000 metric tons of chemical weapons were destroyed worldwide, all meticulously verified by the Organization for the Prohibition of Chemical Weapons (OPCW).

The U.S. commitment to destruction also went beyond our borders as we contributed support for the chemical weapons destruction of other OPCW states parties, including U.S. $3 billion for the Russian Federation, $115 million for the Syrian Arab Republic, $53 million for Libya, and $48 million for Albania.

In this context of progress, let us also look at the remaining challenges and how we might address them.

There are two core areas of concern: political, and scientific and technological.

The core political concern is that a small number of countries are still producing and using chemical weapons. The Assad regime in Syria, enabled by Russia, and Russia itself continue to demonstrate a willingness to use the very worst of the weapons the CWC was designed to outlaw.

J. Manso (✉)
Organisation for the Prohibition of Chemical Weapons (OPCW), The Hague, The Netherlands
e-mail: JosephXManso@outlook.com
URL: https://www.opcw.org

© The Author(s) 2026   23
B. Friedrich et al. (eds.), *Thirty Years of the Chemical Weapons Convention (CWC)*,
https://doi.org/10.1007/978-3-031-98854-7_4

To illustrate this point, we need only to recall other more sober milestones that have also passed in 2023:

- August 21 marked ten years since the Assad Regime used sarin in an attack that killed over 1,400 people in Ghouta.
- September 14 marked ten years since Syria acceded to the CWC, and its declaration remains incomplete and inaccurate to this day.
- April 7 marked five years since the Assad regime used chlorine gas in an attack that killed 43 named individuals and hurt dozens more in Douma. In its report of January 27, 2023, the OPCW Investigation and Identification Team attributed that attack to the Syrian Arab Air Force.
- March 4 was the 5-year anniversary of the use of a Novichok nerve agent by Russian agents to poison the Skripals in Salisbury, also poisoning other residents. Two years later, on August 20, 2020, Russian agents would again use a Novichok nerve agent to poison Russian opposition figure Aleksey Navalny, who received medical treatment here in Berlin, and now sits in a Russian prison on politically motivated charges (NOTE: Mr. Navalny died in a remote Russian prison on February 16, 2024.)

Unfortunately, the threat of CW use isn't confined to the past. Just last month, Ukraine sent a Note Verbale to the OPCW alleging a series of instances, including as recently as September 11, involving the use by the Russian military of riot control agents against Ukrainian forces.

Further worrisome evidence was a Russian state-controlled Channel 1 broadcast of an interview with a Russian officer describing the Russian Army's use of riot control agents on the battlefield against Ukrainian armed forces. Open-source reports of the use of riot control agents by Russia's forces against Ukrainian soldiers should concern everyone, and act as a reminder that we cannot take compliance with the CWC for granted and that the stakes are high.

So, what are we to do?

It is important to recall that the flip side of these challenges is that the vast majority of states remain committed to the Chemical Weapons Convention and the norm against chemical weapons use.

The few states that challenge this norm, in particular Russia and the Assad regime in Syria, are isolated and face international condemnation and sanction for doing so. It is noteworthy that even these states feel compelled to lie about having used chemical weapons. They are attempting to cover their tracks and undermine the credibility of international institutions, including the UN and the OPCW, because of the stigma and possible repercussions of using these banned weapons.

The majority of OPCW States Parties are also willing to take steps to address chemical weapons use, including supporting the creation of investigative bodies at the OPCW, such as the Investigation and Identification Team. The OPCW has also taken measures to redress noncompliance with the Convention, such as the Conference of States Parties decision in April 2023 to suspend some of Syria's rights and privileges under the Convention.

OPCW States Parties also responded to Russia's attempted assassination of the Skripals with a chemical agent by approving the addition of two families of Novichoks to the CWC Annex on Chemicals in November 2019. Following this, States Parties took action to address the threat of aerosolized central nervous system-acting chemicals by adopting a decision in December 2021 affirming that the aerosolized use of central nervous system-acting chemicals is understood to be inconsistent with law enforcement purposes as a "purpose not prohibited" under the convention.

We must continue to demand clear answers from Russia on its undeclared chemical weapons program and maintain pressure on Syria to fully declare and destroy its remaining stockpile.

Another aspect of the political challenge is that possible chemical weapons use by non-state actors remains a persistent threat and an area where we must continue to focus cooperation and capacity building efforts.

In this area both the OPCW and States Parties both have an important role. OPCW international cooperation and assistance programs can help build the capacity of States Parties to address non-state actor threats, and the OPCW can also help share best practices through events like the table-top exercise planned this autumn in the Open-Ended Working Group on Terrorism.

OPCW programming is complementary to bilateral and regional capacity building efforts to prevent and respond to CW use by non-state actors. Last year alone, the United States provided more than $150 million in assistance to States Parties to address vulnerabilities to chemical terrorism and other chemical safety issues, in cooperation with various national and regional counterparts.

This is part of a larger positive agenda for the OPCW, which in addition to counter-terrorism includes international cooperation to strengthen customs and border controls, enhance first responder capabilities, and funding from the United States for junior professional officers from underrepresented regions.

The second set of core challenges relate to developments in science and technology.

Advances in science and technology are creating both new opportunities as well as new risks.

The Chemical Industry is changing quickly. As it grows, the number of facilities is increasing and new chemical production processes are making it possible for chemical agents to be produced in smaller facilities that are less likely to be subject to the CWC verification regime.

Dual use technologies are opening the door to new delivery systems such as drones, and putting them in the hands of more people.

Biotechnology is creating new threats related to biotoxins, which implicates both the Chemical Weapons Convention and the Biological Weapons Convention.

The OPCW's new Center for Chemistry and Technology (ChemTech Center) and the Scientific Advisory Board are important tools for addressing these emerging and already existing challenges. We can also address some of them by reviewing the CWC's Article VI industry inspection regime to ensure that inspections are focused on the highest risk facilities. We should also support the OPCW's Technical Secretariat (TS) in strengthening the OPCW's capabilities for sample analysis, and OPCW engagement with industry to raise awareness of the risks and find solutions.

The TS is a key resource for the OPCW and the States parties. Its high level of competence and commitment should be maintained and enhanced.

States Parties must also remain committed to ensuring that the OPCW is resourced to remain fit-for-purpose and capable of responding to future threats. The opening of the new ChemTech Center in 2023, with its state-of-the-art laboratory and international cooperation and assistance platforms, is a concrete demonstration of this commitment.

As we work through the second biennium budget process, the United States remains committed to providing the OPCW with adequate resources, both through the regular budget and through voluntary contributions to meet these challenges. In doing so, we must also maintain fiscal responsibility.

In sum, that are two areas of core challenges facing the organization and the states parties: political challenges and those related to advances in science and technology. There are steps the States Parties can take now to address these challenges and we must maintain the political will to do so even if it means confronting Russia and Syria. Such confrontation is not an end in itself, but is the result of repeated Russian and Syrian noncompliance with the Chemical Weapons Convention. Their production and use of chemical weapons is not acceptable to the international community.

The year 2023 has brought us closer to a world free of chemical weapons, but more work is required. Addressing these challenges is the mandate of the Chemical Weapons Convention and the OPCW. The victims of chemical weapons deserve nothing less than our firm and continued commitment to implement that mandate.

Thank you for your attention and I look forward to our discussion.

Joseph Manso was United States Permanent Representative to the Organization for the Prohibition of Chemical Weapons from 2020 to 2023. Ambassador Manso has extensive experience in multilateral affairs, and is currently a Senior Fellow at the National Defense University and a Distinguished Fellow at Arizona State University.

Chapter 5
Challenges to the Chemical Weapons Convention II

Thomas Schieb

I would like to thank the Fritz Haber Institute for hosting this event and for inviting me.

As my US colleague just said, the fact that a small number of countries—Syria and Russia—are still producing and possibly also using chemical weapons poses a particular challenge to the CWC.

In addition to continuing to demand answers from Russia and put pressure on Syria, yet another challenge is to keep as many States Parties to the CWC as possible engaged and interested in these pressing issues, and to keep fighting against "Syria fatigue", as well as "Ukraine fatigue".

As long as Russia and Syria are in breach of the Convention, there cannot be "business as usual".

Unfortunately, there seems to be a certain tendency among a number of States Parties to suggest that we stop worrying about the things we allegedly cannot change anyway and instead focus on international cooperation and capacity building, as well as on a couple of OPCW housekeeping issues like geographical representation and tenure policy.

As important as international cooperation and improving the efficiency of the OPCW are, we cannot allow this, since it would weaken the credibility of the organisation and undermine the Convention.

Clearly, Russia and Syria are trying to wear down those States Parties that are seeking the truth, while pushing their distorting narratives and disinformation campaigns. We should not let them. We have to keep the course and fight disinformation, while reaching out to as many countries as possible.

Ladies and gentlemen,

T. Schieb (✉)
Organisation for the Prohibition of Chemical Weapons (OPCW), The Hague, The Netherlands
e-mail: l-cw@denh.auswaertiges-amt.de
URL: https://www.auswaertiges-amt.de/

© The Author(s) 2026
B. Friedrich et al. (eds.), *Thirty Years of the Chemical Weapons Convention (CWC)*,
https://doi.org/10.1007/978-3-031-98854-7_5

As to the more "technical challenges" the OPCW is faced with, these are undoubtedly significant as well.

Given the enormous developments in science and technology, it is extremely challenging for the OPCW to remain up to date. Scientific disciplines are increasingly converging, and the "dual use" potential—and thereby the risks—of new developments like AI and robotics are significant. The way chemicals are designed and produced is constantly evolving at an ever-faster pace.

In order to remain fit for purpose, the OPCW will have to continuously adapt, modernise and develop its technological knowledge and scientific expertise. Given the limited capacities of the OPCW's Technical Secretariat, this will only be possible if the OPCW taps into the vast reservoir of scientific research and of academia, among other things intensifying partnerships with laboratories and scientific institutions.

Apart from exchange with representatives of the scientific community, interaction is also very important in terms of outreach and advocacy: promoting the OPCW not only as a forum of scientific exchange but also as a hub of international cooperation; reaching out to a wider public, explaining the organisation's mandate and priorities, not least in search of the best minds to come to work at the Technical Secretariat; but also correcting possible misconceptions–to make clear that the OPCW is more than a mere control agency, as the chemical industry seems to perceive it sometimes.

The OPCW having accomplished the important mission of destroying all declared chemical weapons stockpiles, the public needs to be aware that the organisation will have to continue to play an important role in the future. It needs to know why the organisation matters, why the fight against the re-emergence of chemical weapons is so important, and that the organisation's activities, including support for States Parties in implementing the CWC, contribute to a safer and more peaceful world. To get these messages across, civil society organisations should serve as multipliers.

So, it is safe to conclude that it will be even more important for the OPCW now than in the past to have a meaningful and productive interaction with civil society.

This is why Germany has initiated so-called informal consultations on civil society engagement at the OPCW. The idea is to have a broad debate, starting with a stock-taking exercise, which includes looking at how other international organisations interact with civil society. This will be followed by a discussion of options for ways to render the interaction more structured and systematic. It goes without saying that civil society needs to be involved in this process as closely as possible.

Ladies and gentlemen,

The OPCW might be a "niche organisation", as you sometimes hear, but it should not be in a silo. The OPCW should broaden its view and not shy away from following cross-cutting issues which could, albeit indirectly, be relevant to CWC implementation.

Let me mention three examples:

The OPCW needs to be aware of what is going on in the field of chemistry in general. On 30 September 2023, at the 5th International Conference on Chemicals Management (ICCM5) in Bonn, a Global Framework was adopted which sets targets to address harm from chemicals and waste. Making the management of dangerous chemicals safer and equipping States Parties accordingly are also objectives of international cooperation and assistance in the framework of the OPCW. So maybe the OPCW could help implement the Global Framework? In order to find out, the OPCW needs to be involved and needs to be present at such international events.

Unfortunately, the legacy of old chemical weapons is still with us, including large amounts of chemical weapons dumped at sea at the end of WWII. On top of large stocks of conventional munitions, up to 65,000 t of chemical warfare agents and munitions were dumped into the Baltic Sea alone. At the time, this was a seemingly easy way to dispose of these dangerous stockpiles of toxic chemicals. Today we know that sea dumped chemical weapons (SDCW) pose enormous environmental risks. Large areas of sea are already or are at risk of being polluted. However, there is no easy solution. Recovery is difficult, expensive, risky and not necessarily the best solution. We definitely need more research to be able to make the right decisions.

The CWC does not include any obligations on SDCW dumped before 1985, although States Parties can engage voluntarily. We believe that, on this basis, the OPCW has a role to play. At the request of States Parties, the OPCW could act as a platform for information exchange on scientific analysis, political responses and best practices among States Parties. The OPCW could support regional efforts, like those undertaken by the Council of the Baltic Sea States.

It can be argued that implementing the CWC helps States Parties to implement Agenda 2030. Strengthening this message, especially in the countries of the so-called Global South, could help make the OPCW more interesting for the countries in question–and maybe help to encourage them to engage more in the OPCW!

Ladies and gentlemen,

Before concluding, I would like to highlight the OPCW's important role when it comes to paving the way for bringing to justice those responsible for chemical weapons use.

In a significant number of cases, the OPCW was able to attribute responsibility for chemical weapons use in Syria. The OPCW's Investigation and Identification Team has prepared thousands of files and sent them to the International, Impartial and Independent Mechanism, the IIIM, which assists in the investigation and prosecution of persons responsible for the most serious crimes under international law committed in Syria since 2011. Hopefully, this evidence of chemical weapons use will be used in legal proceedings against the perpetrators of chemical weapons use in the not-too-distant future. It is very unfortunate that not a single perpetrator has yet been convicted.

My point is this: the OPCW's findings are very important in helping to ensure progress in terms of accountability and fighting impunity in Syria. The OPCW needs to continue its important work in this respect, too.

Before starting his assignment as Permanent Representative of Germany to the OPCW in August 2022, Ambassador **Thomas Schieb** spent four years in Belgrade, serving as Ambassador of Germany to Serbia. His previous postings abroad included New York, where he worked as Political Coordinator at the German Permanent Mission to the UN, London, Brussels, Dar es Salaam and Sarajevo. Thomas Schieb holds a Master's degree in economics (University of Göttingen).

Part II
Histories of Chemical Weapons in the 20th Century

Chapter 6
In Denial of the Horrors of Chemical Warfare: Expert Assessments of Chemical Weapons During and After World War One

Bretislav Friedrich

"Denial ain't just a river in Egypt."—attributed to Mark Twain

Abstract This chapter traces the reasoning behind the widespread advocacy of chemical weapons during and after World War One by military officials and scientific experts—with Fritz Haber and John Burdon Sanderson Haldane prominent among the latter. The fact that an estimated 1% of military deaths and 6% of military injuries in the war were due to chemical weapons inspired the notion that chemical weapons were, in fact, humane. The universal abhorrence of chemical weapons as manifestly inhumane is more recent, as is their classification as weapons of mass destruction. That favorable views of chemical warfare were once prevalent is barely comprehensible today and, unless contextualized within the mode of thinking of experts at the time, appears *absurd*. The elimination in 2023 of the declared chemical arsenals by all 192 parties (representing 98% of the world's population) to the 1993 Chemical Weapons Convention (CWC) reflects the current attitude toward chemical weapons worldwide. However, chemical weapons have meanwhile been superseded by what Haldane considered unfeasible in 1925, namely nuclear weapons. Consequently, chemical weapons had been relegated to the sidelines as "poor man's atomic bomb." What Haldane did get right, however, was that, if feasible, nuclear weapons could upend organized life on the planet. We conclude by noting that our consensual coexistence with stockpiles of nuclear weapons should be judged as *beyond absurd*.

B. Friedrich (✉)
Fritz-Haber-Institut der Max-Planck-Gesellschaft, Berlin, Germany
e-mail: bretislav.friedrich@fhi-berlin.mpg.de
URL: https://www.fhi.mpg.de/209437/Friedrich

B. Friedrich et al. (eds.), *Thirty Years of the Chemical Weapons Convention (CWC)*,
https://doi.org/10.1007/978-3-031-98854-7_6

Keywords The Hague Conventions · Trench Warfare · Fritz Haber · John B.S. Haldane · Chemical Weapons Convention · Treaty on the Prohibition of Nuclear Weapons

6.1   Introduction

The era of peace and prosperity that much of Europe had enjoyed since the Franco-Prussian war of 1871 came to an abrupt end with the outbreak of World War One. Its first salvos were echoed by verbal exchanges between the academics of the warring parties.[1] This "war of the spirits" took a lethal form once the scientific communities became engaged in developing and promoting new weapons systems, in breach of the ethos of the *République des Lettres*—and, eventually, of international law. Their involvement in developing and deploying chemical weapons is among the best examples of the breach of both (Haber 1986, 18–18; 293–294, Friedrich 2015; James and Friedrich 2019, 25–44).

The war "to end all wars" didn't go as envisioned by the German military planners in their 1905 *Schlieffen-Plan* (Haffner 1964; Mommsen 2011). Within a month of the first exchange of fire between German and French troops on August 3, 1914, trench warfare produced a strategic stalemate on the Western Front following the week-long Battle of the Marne (September 5–12, 1914).[2] Ending in Allied victory, the battle set the stage for the immovable trench warfare of the next four years of World War One.

6.2   Chemical Weapons as a Means to Break the Stalemate of Trench Warfare

6.2.1   German Chemical Warfare

On the German side, the lingering idea of using chemicals to incapacitate enemy troops had been rekindled quickly after the Battle of the Marne, as the chief of the Supreme Army Command (*Oberste Heeresleitung*, OHL), Erich von Falkenhayn, established a committee to assess the suitability and availability of chemicals as

---

[1] Perhaps the most notorious example is the manifesto "An die Kulturwelt" published on October 4, 1914 and signed by 93 leading German academics. It was reprinted in English under the title "To the Civilized World" in 1919 by *North American Review* 210: 284–287.

[2] A counter-offensive by the French Army and the British Expeditionary Force against the invading German Army.

weapons (Szöllösi-Janze 1998, 321–332 and references cited therein). After unsuccessful tests of various irritants on the front,[3] von Falkenhayn decided to abandon the "smelly stuff" and make use of lethal chemicals (substances that "incapacitate permanently," as he put it) with the purpose of breaking the stalemate of trench warfare (Martinetz 1996, 18). To this end, *Geheimrat*[4] Fritz Haber, who had meanwhile become a member of von Falkenhayn's chemical warfare committee, proposed the use of chlorine[5] as a chemical weapon.

Heavier than air and thus suitable for striking enemy troops inside their trenches, chlorine held the promise of not only incapacitating enemy combatants but also disabling their "conventional" weapons by corrosion. According to Harold Hartley (1925)[6]:

> [Haber] said that owing to the small area affected by each shell it would be necessary to fire a large number simultaneously to produce any technical effect. He suggested the use of a large number of mortars for this purpose, an idea realized later in the Livens Projector, but he was told that it was impossible to produce them in time to be of use. The employment of gas from cylinders was then proposed in order to get a high concentration and to cover a wide area, the object of the General Staff in using gas being to drive us from our trenches and get back to open warfare ... The suggestion of a cylinder attack with chlorine was adopted and Haber was placed in charge of the technical preparation.

An attempt to test chlorine cloud as a chemical weapon at the proving ground in Wahn near Köln in January 1915 was aborted due to accidents that occurred while handling the gas[7] and a decision was made by the OHL to test a chlorine cloud directly in battle on the Western front. Although viewed with skepticism and mistrust by most of the military, including von Falkenhayn himself, the idea of deploying a chlorine cloud found support from generals Berthold von Deimling and Emil Ilse, who oversaw German operations in Flanders and who set their eyes on the Ypres Salient, which, according to the *Schlieffen-Plan*, lay on the German Army's route into France (James and Friedrich 2019, 26). The battlefield test, which amounted to

---

[3] The brainchild of the fledgling chemical warfare committee was the ineffective sneezing powder (*Ni-Stoff*) that was used amid the hostilities at Neuve-Chapelle already during the same month that the committee was formed. In response to the *Ni-Stoff* fiasco, Gerhard Tappen of the OHL turned for help to his brother Hans Tappen, a chemist trained by Emil Fischer. Hans Tappen proposed to fill artillery shells with the lachrymator xylyl bromide, hence dubbed *T-Stoff*, that was first tested in January 1915. It was promptly used at the Eastern front, near Łódź, in cold weather with little or no effect due to its low vapor pressure at low ambient temperatures (James and Friedrich 2011, 25–26).

[4] In translation, Privy Councilor—a title bestowed on distinguished citizens whose expert opinion was sought by the government. Haber became a Privy Councilor upon assuming the directorship of the Kaiser Wilhelm Institute for Physical Chemistry and Electrochemistry in Berlin in 1911.

[5] Chlorine was produced on a large scale by Germany's dyestuff industry.

[6] Sir Harold Brewer Hartley was a British physical chemist and, in his capacity as Assistant Director of Gas Services, a War Office counterpart of Fritz Haber. See also below.

[7] Fritz Haber himself was once caught up in a small chlorine cloud (Haber 1924, 88).

the first use of a weapon of mass destruction in history, took place there on April 22, 1915.[8]

Much of the high-ranking German military took a more scrupulous—or chival-rous—approach than generals von Deimling and Ilse and at first openly detested chemical weapons. Thus, they furnished a curious substitute for adherence to the spirit, if not letter, of the Hague Conventions of 1899 and 1907, which limited the use of poisonous substances in warfare.[9]

How well justified militarily was the mistrust of chemical weapons? Did they do what they were supposed to, or was their deployment a failure? Military historian Edward Spiers provided recently the following answer in his eloquently entitled essay, "The Gas War, 1915–1918: If not a War Winner, Hardly a Failure" (Spiers 2017):

> Contemporary claims[10] that gas warfare proved "a failure" during the First World War would have baffled wartime adversaries, who invested heavily in the research, development, and production of gas warfare. If poison gas, like other conventional weapons, never broke the stalemate of the trenches, it evolved into a weapon of harassment that compounded the effects of conventional weapons and degraded the effectiveness of enemy forces compelled to wear gas masks for protracted periods of time. The introduction of mustard gas in July 1917 greatly increased the number of gas casualties and set the scene for a steady increase in the use of chemical weapons during the later stages of the war. Like the tank and aircraft, gas was not strategically decisive, but continuing investment in this form of warfare underscored its potential utility.

Artillery shells filled with chemical agents grew from a negligible proportion in 1915 to about 50% of the German, 35% of the French, 25% of the British, and 20% of the American ammunition expenditure by the Armistice (Spiers 2017), see Table 6.1. While contemporary statistical assessments have to be read with caution, we have some idea about the proportion of chemical warfare casualties in relation to those killed or wounded by other means. According to Augustine Prentiss (1937, 649), about 90,000 soldiers were killed and 1.3 million injured by chemical weapons in World War One, as compared to a total of 8.5 million military personnel killed and

---

[8] The first chlorine cloud attack on enemy positions (French and British) took place on that day at 1700 GMT (18:00 CET) in the perimeter of the Langemarck village near Ypres, when the prevalent wind finally turned in the northerly direction. Seven previous attempts had to be aborted because of unfavorable wind. The attack released 150 tons of chlorine gas from 1600 big and 4130 small cylinders placed at a distance of about 1 m from each other, covering about 6 km of the front. The chlorine gas concentration achieved was about 0.5% at a distance of 50–100 m from the cylinders (James and Friedrich 2011, 28).

[9] The Hague Convention of 1907, Article 23: "In addition to the prohibitions provided by special Conventions, it is especially forbidden: (a) To employ poison or poisoned weapons; (e) To employ arms, projectiles, or material calculated to cause unnecessary suffering." The separate Hague Declaration Concerning Asphyxiating Gases stipulated "abstaining from the use of projectiles, the sole object of which is the diffusion of asphyxiating or deleterious gases."

[10] These would include the authoritative account by Ludwig (Lutz) Haber, who answered this very question with a resounding "Yes" (Haber 1986, 259–284).

Table 6.1   Production of chemical weapons by country, in metric tons (Haber 1986, 170)

Country	Chlorine	Phosgen	Di-phosgen	Mustard gas	Chlor-picrin	HCN	Total
Germany	58,100	18,100	11,600	7,600	4,100	–	99,500
Britain	20,800	1,400	–	500	8,000	400	31,100
France	12,500	15,700	–	2,000	500	7,700	38,400
USA	2,400	1,400	–	900	2,500	–	7,200
Total	93,800	36,600	11,600	11,000	15,100	8,100	176,200

21 million wounded, mainly by artillery fire (Brittanica 2024). Thus, an estimated 1% of military deaths and 6% of military injuries were due to chemical weapons.[11]

Strangely enough, and difficult to understand from today's perspective, the above casualty numbers inspired the notion that chemical weapons were, in fact, humane. For instance, the U.S. Assistant Secretary of War and Director of Munitions, Benedict Crowell, noted (1919, 396):

> The methods of manufacturing toxic gases, the use of such gases, and the tactics connected with their use were new developments of this war; yet during the year 1918 from 20 to 30 per cent of all American battle casualties were due to gas, showing that toxic gas is one of the most powerful implements of war. The records show, however, that when armies were supplied with masks and other defensive appliances, only about 3 or 4 per cent of the gas casualties were fatal. This indicates that gas can be made not only one of the most effective implements of war, but one of the most humane.

The universal abhorrence of chemical weapons as manifestly inhumane is more recent and so is their classification as weapons of mass destruction. While the latter is a concept of the nuclear age, one would hope that the former is not, although there was already considerable international debate in expert and public circles about the inhumane nature of chemical weapons at the time. In any case, at the time of their use in World War One, prominent munitions and war experts, as well as scientists of the Central Powers and the Entente alike, kept referring to chemical weapons as "humane." Some continued to do so well into the 1940s (Haber 1986, 285–319).

Apart from being the unquestioned driving force behind the centrally-directed development of chemical warfare in Germany, Fritz Haber was a prominent advocate of chemical warfare's "humane" character (see below). Moreover, Haber not only followed his maxim "In peace for mankind, in war for the fatherland"[12] personally, but also applied it to his entire institute and promptly redirected its resources

---

[11] As Matthew Meselson pointed out to me, there is no reliable estimate of the number of deaths from the first chlorine cloud attack. Volume I of The Problem of Chemical and Biological Warfare (Stockholm Internation Peace Research Institute—SIPRI, 1971) states that Rudolf Hanslian in 1934 withdrew his much quoted 1927 estimate of 5,000, writing that for propaganda purposes the Entente had quintupled their casualty figures.

[12] Usually quoted in this abbreviated form. For a full quote in German, see (F. Haber 1920). Haber was referring to the attitude of Archimedes. See also (Hoffmann 2014).

toward projects relevant to the war.[13] The Institute became a prototypical example of Big Science in the context of a military–industrial–academic complex, not only with respect to its sheer size but, above all, with respect to the complexity of its organizational structure and interdisciplinarity of its research methods (James et al. 2011, 30). In the words of the historian Fritz Stern, Haber's Institute during the First World War became "a kind of forerunner of the Manhattan Project" (Stern 1999, 119). The Hartley Report and its rendition in the book *The Poisonous Cloud* by Fritz Haber's son Ludwig (Lutz) F. Haber[14] provides the details of the operation, including on the research and development of protective measures such as gas masks.

Albert Einstein's pacifist views contrasted sharply with those of his friend Haber. As he would put it later: "Warfare cannot be humanized. It can only be abolished" (Rowe and Schulmann 2007, 224). Strangely enough, there is no record of Einstein's criticism of Haber's World War One efforts, although Einstein occupied an office at Haber's institute at the time and must have been aware of what the preoccupation of the institute had become.

The development of the catalytic process for ammonia synthesis on an industrial scale (the Haber–Bosch process) was already one realization of Germany's desire to manufacture domestic substitutes for key imported goods. Similar intentions led Haber to his wartime partnership with the Raw Materials Department of the War Ministry under Walther Rathenau, which eventually led him to research chemical means for waging war. As Johnson pointedly summed up the progression: "the logic of *Ersatz* [substitute] led to the problems of munitions, and eventually to poison gas" (Johnson 1990, 133).

The high explosives for the artillery shells were produced by the chemical industries of the warring nations.[15] Hence the characterization of World War One as the "chemists' war," although chemical warfare surely added much weight to this term. We note that the development and acquisition of the Haber–Bosch technology by Germany just in time for World War One was key to sustaining her supply of nitrates (via oxidation of ammonia) and thereby her war effort: without it, the embargoed supplies of Chilean saltpeter would have run out within months and the Great War would have indeed been as brief as anticipated by the *Schlieffen-Plan*, except that it would have ended not in Germany's speedy victory but rather her swift defeat. Although Germany lost the war eventually, its chemical industry did not. In fact, its massive build-up, funded in large part by state loans, could never have happened in

---

[13] The conversion to military research projects proceeded surprisingly smoothly and without noticeable resistance (Johnson 1990, 133).

[14] Lutz Haber's personal interest in the topic of chemical warfare was fueled not just by his family lineage but also by his acquaintance and friendship with Harold Hartley, whose confidant—and in a sense heir of his extensive collection of materials connected with chemical warfare in WWI—Lutz Haber had become. Sir Harold Hartley was Fritz Haber's counterpart at the British War Office during WWI who, after the war, was in charge of inspecting German research and production facilities related to chemical warfare and banned by the Versailles Treaty.

[15] The ammonia synthesis may serve as an apt reminder of the Janus-face of modern science: On one side, "bread from air," on the other, "gunpowder from air".

a peacetime economy. Moreover, the loans were actually never paid back—because of hyperinflation that descended upon Germany after the war (Friedrich 2019).

6.2.2 The Use of Chemical Weapons by the Entente

In contrast to the rich historiography available on the German, British, and U.S. chemical warfare build-up following the German chlorine cloud attack at Ypres (Heller 1984; Haber 1986; Schmidt 2017; Zoghlami 2021), studies documenting the French response are scarce (Lepick 1998, 2017, 70). An exception are the published minutes of a lecture that was given in March 1920 by Professor Charles Moureu from the Collège de France (Moureu 1920). In his 2017 account, Olivier Lepick noted (Lepick 2017, 69):

> Although France had been experimenting with chemical weapons when Germany launched its first lethal chemical offensive in the spring of 1915 in Langemark [Ypres], the German initiative came as a huge tactical surprise to the country. Soon after the initial shock and the controversy that ensued on whether Germany had violated the laws of war that day, French authorities rapidly decided, without real political debate, to retaliate in kind.

This was well in keeping with the public sentiment in France as expressed in *Le Temps* (Lepick 2017, 72):

> … no government shall not respond to such barbarian initiatives without endangering its own troops. In this perspective, the French government intends, in the strict limits of its military needs, to use all necessary means that appear appropriate to stop the German military authorities from continuing to commit such horrible murders […].

The "experimenting with chemical weapons" by the French military mentioned above by Lepick started way before the war and consisted of exploring the effects of ethyl bromoacetate [$BrCH_2CO_2CH_2CH_3$], a highly toxic chemical incorrectly classified as lachrymator. Clarence West of the U.S. Chemical Warfare Service provided the following testimonial about its testing and deployment (West, 1919):

> Before the war suffocating cartridges were shot from the cartridge-throwing rifle of 26 mm. These cartridges were charged with ethyl bromoacetate, a slightly suffocating and non-toxic lachrymator. They were intended for attack on the flanking works of permanent fortifications, flanking casements or caponiers, into which they tried to make these cartridges penetrate by the narrow slits of the loopholes. The men who were serving the machine guns or the cannon of the flanking works would have been bothered by the vapor from the ethyl bromoacetate, and the assailant would have profited by their disturbance to get past the obstacle presented by the fortification. The employment of these devices, not entailing death, did not contravene the Hague conventions.

> The only memorable operations in the course of which these devices were used before the war was the attack on the Bonnet [sic] gang at Choisy-le-roi.[16] In the war of the trenches there has been an abuse in the employment of these suffocating cartridges; an abuse because the small quantity of liquid that they contain, about 19 cubic centimeters, can produce no effect on a terrain without cover.

---

[16] Bonnot Gang (*La Bande à Bonnot*) was a French anarchist group.

In his talk in 1923 for an investigative committee of the German *Reichstag*, Fritz Haber made a key correction to West's account (Haber 1924, 81–82):

> Hydrogen cyanide [HCN], known throughout the world as one of the strongest poisons, and the first French gas warfare agent, ethyl bromoacetate, whose harmlessness all representatives of the Entente emphasize in the strongest terms, are very close to each other in terms of lethal dose. This paradoxical fact has to do with that the likelihood of poisoning is not related to the toxicity in a simple way. The inhalation of hydrocyanic acid is not bothersome in any way. One cannot die more pleasantly. Ethyl bromoacetate, however, torments the nose and throat unbearably, long before we take it up in a lethal quantity. This is the difference that makes the probability of poisoning by hydrocyanic acid much greater than the corresponding probability for ethyl bromoacetate. It is not the difference in the lethal dose, but the fact that you run away from ethyl bromoacetate long before you inhale the lethal amount, which gives ethyl bromoacetate the false reputation of harmlessness.

And he added (Haber 1924, 83):

> On January 7, 1915 General Joffre requested this weapon[17] for the front, and on February 21 of that year, the French Ministry of War issued service regulations for the use of gas rifle- and hand-grenades to the troops. In March 1915, these weapons were then used against our troops on the Western Front.

Although ineffective because of the low concentrations achieved, the intended purpose of using ethyl bromo- or chloroacetate was the same as that of the German chlorine cloud: to force the enemy out of his trench positions by exposing him to an asphyxiating agent (F. Haber 1924, 87).

Within a day of the German chlorine cloud attack at Ypres, the Commander of the British Expeditionary Force, John French, requested from the London government that "immediate steps be taken in retaliation to supply similar means of the most effective kind for the use of our own troops" (Schmidt 2015, 26). By September 1915, the decision was made by the British Government to establish a facility for the research, development, and testing of chemical weapons known as the Royal Engineers Experimental Station in Porton Down. The facility became operational in March 1916, investigating agents such as chlorine, phosgene, and mustard gas, as well as defensive measures against them. As noted by Ulf Schmidt (2015, 29):

> British scientists, many of them Fellows of the Royal Society (FRS), felt duty-bound and honoured to place their expertise at the service of the realm, and had little, if any, moral objections in developing new weapons of mass destruction.

This attitude was fostered by the public sentiment as expressed in *The Times* on April 29, 1915:

> The willful and systematic attempt to choke and poison our soldiers can have but one effect upon the British people and upon all the non-German people of the earth. It will deepen our indignation and our resolution, and it will fill all races with a new horror of the German name.

---

[17] For a lack of bromine, ethyl bromoacetate was replaced with equally toxic ethyl chloroacetate.

This description of the horrors of chemical warfare, aimed at justifying the retaliatory measures of the Entente, contrasted sharply with that provided for the German public. For instance, the *Kölnische Zeitung* claimed that (Schmidt 2015, 25).

> the letting loose of smoke clouds, which, in a gentle wind, move quite slowly towards the enemy, is not only permissible by international law, but is an extraordinarily mild method of war.

In reality, as described in (Schmidt 2015, 22)

> Often the experience of being gassed led to real and imagined clinical symptoms for years to come. The possibility of being killed by asphyxiating gases triggered deep-seated emotional responses and occasional nervous breakdowns which psychiatrists classified as "gas neurosis;" in other cases, soldiers exposed to blistering agents were classed as suffering from "gas hysteria," since the substances could cause conjunctivitis and temporary blindness (Harrison 2010,106–109). Not unlike soldiers suffering from shell shock, victims of gas war exhibited a range of respiratory and other medical symptoms, including 'functional photophobia, aphonia, hysterical cough, vomiting and palpitations' (Jones et al. 2008, 1420). Eyewitnesses recalled that "gas shock was as frequent as shellshock" (Shephard 2000, 64).

The U.S. Chemical Warfare Service oversaw the production and deployment of U.S. chemical weapons, see Table 6.1. In November 1918, The Cleveland Plant started producing the blister agent lewisite, which was, however, never used in World War One.

6.3   Favorable Views on Chemical Warfare During and in the Aftermath of World War One

The attempts to outlaw chemical warfare that found their expression at the 1922 Washington Arms Conference Treaty[18] and the 1925 Geneva Protocol[19] (Walker 2017, 380–381) were taking place against the background of outright advocacy for the use of chemical weapons in future armed conflicts. Among the most prominent advocates of the opposing view—in favor of chemical warfare—were Fritz Haber and the British polymath John Burdon Sanderson Haldane.

6.3.1   Fritz Haber's Views

Fritz Haber, Fig. 6.1, advertised the use of chemical weapons as an important milestone in the "art of war"—and saw its psychological effect as crucial (Haber 1924, 36):

---

[18] Rejected by France.

[19] Ratified by 65 states with France acting as its depositary.

All modern weapons, although seemingly aimed at causing the death of the adversary, in reality owe their success to the vigor with which they temporarily shatter the adversary's psychological strength.

Haber also pointed out that key to success in chemical warfare are "intellectual imponderables" of the troops (F. Haber 1924, 39):

A strict selection separates the troops that are capable of maintaining gas-discipline and who fulfill their combat task from the martially inferior mass of those who crumble and abandon their posts.

Furthermore, Haber emphasized that the variability of the effects of chemical weapons presents ever-new demands on the "moral resistance" of the troops, as opposed to artillery shelling, which is always the same and people get eventually used to it (Haber 1924, 37). Thus, Haber viewed chemical weapons as a strategic means to

Fig. 6.1 Fritz Jakob Haber (1868–1934). The breadth of Haber's intellectual interests was astounding—ranging from fundamental physics to physical chemistry to physiology. For his 1909 discovery of the catalytic synthesis of ammonia from its elements he received the 1918 Nobel Prize in Chemistry. Likewise, the scope of Haber's organizational activities was immense and included the founding of the Kaiser Wilhelm Institute of Physical Chemistry and Electrochemistry, as well as the co-founding of the forerunner of the Deutsche Forschungsgemeinshaft and of the Japan Institute. Neither his scientific merits nor his unbridled patriotism sufficed to stave off his loss of status and position once the Nazis rose to power. Haber's Jewish origin, as well as his democratic attitudes, were a thorn in their flesh. During the last months of his life, Haber espoused liberal Zionism and made preparations for joining Chaim Weizmann's project of building Jewish academic institutions in Palestine (Stern 2011, Bielik and Friedrich 2020).

break the stalemate of trench warfare by forcing the adversary to surrender, shorten the war, and thereby preclude the slaughter of millions by artillery and machine gun fire. It was in this sense that he referred to chemical weapons as "humane."

Haber's enthusiasm for chemical weapons had an important caveat, described by Haber's biographer Margit Szöllösi-Janze in this way (Szöllösi-Janze 1998, 332):

> Was [Haber] a fanatic of gas-warfare? One must not overlook that he saw the significance of the gas weapon in its stupefying character and assessed the inherent dynamics of its use differently than the military. In his opinion chemical weapons could be effective and lead to a shortening of the war only if the war came to an end before the adversary would be able to develop suitable protective measures and even more dangerous chemical agents. For this reason, [Haber] voiced his opposition at a meeting in 1917 with [General Erich] Ludendorff [deputy of von Falkenhayn's successor as Chief of OHL, Paul von Hindenburg] to the use of mustard gas [that represented an escalation of the chemical war], which struck him as sensible only if Germany could take advantage of its head start and win the war within a year.

And indeed, France began producing mustard gas in July 1918. Here's how Harold Hartley described the path of the Allies to mustard gas (Haber 1986, 192):

> [Hartley] was awakened early on that 13 July [1917] and was informed that the Germans had fired a new type of shell [at Ypres] that made a "plop"-like sound when it burst […] The next day [Hartley] had located some unexploded shells with yellow cross markings. They were defused, taken to [the General Headquarters], opened, the contents analyzed […] and the findings compared with the entry in *Beilstein*. By 16 July [the British] knew what the stuff was, and later analyses added little to this knowledge.

Haber's correspondence reveals his fascination with the new tactical possibilities opened up by gas warfare and the room for "scientific imagination" that such technological means for conducting war had offered (Szöllösi-Janze 1998, 327). He used (and perhaps coined) the metaphor that conventional warfare was like playing checkers whereas warfare enhanced by chemical weapons was like playing chess. Haber never regretted his involvement in chemical warfare.

As for legal issues with chemical warfare, Haber put the blame for any transgressions against international law squarely on Erich von Falkenhayn. He did so in a testimony—which took the form of lectures—delivered to an investigative committee of the *Reichstag* in 1920–1923. However, in this testimony, Haber did not shy away from playing a legalistic shell game when he argued that German gas attacks were carried out either without the use of shells (like the chlorine cloud attacks) or with shells loaded, in addition to poison gas, with explosives (whereas the Hague conventions prohibited the use of shells or grenades filled solely with poisonous substances).

After the war, OHL's—and Haber's—cavalier view of the Hague Conventions was validated by a German parliamentary committee, which concluded (Stoltzenberg 2004, 152):

> Neither the German nor the French governments, nor as far as is known, any other power participating in the war or a neutral one raised any objections against the modes of action in the gas war. From this it can be concluded that both sides viewed the Hague Conventions of 29 July 1899 and 18 October 1907 as obsolete and by silent agreement regarded them

as annulled.[20] Even when accepting this assumption, it remains a fact that the first obvious transgression of an international agreement was on the French side, whereas Germany only followed and thereby merely took a countermeasure as accepted in international law.

6.3.2   John Haldane's Views

As noted by Lutz Haber (Haber 1986, 293)

William Pope[21] condemned all anti-chemical warfare talk as mischievous and called on those who disagreed with him to explain why they wanted to abolish a "very humane, although new instrument of warfare." His [Cambridge] colleague J.B.S. Haldane also examined the subject wittily and without false sentimentality in *Callinicus.*

Indeed, Haldane's *Callinicus. A Defence of Chemical Warfare*[22] (Haldane 1925) is a tractate on the future of warfare that develops with iron logic an apologia for chemical warfare from these two premises: (a) future wars are inevitable; (b) the belligerents will use any means available to them in order to win. Written during his readership in biochemistry at Cambridge, Haldane, Fig. 6.2, notes that (Haldane 1925, 4):

We have not yet made a scientific study of the causes of war, and, until we do, may expect more wars.

Then he goes on and states (ibid., 15)

Of course, if we could utilize the forces which we now know to exist inside the atom, we should have such capacities for destruction that I do not know of any agency other than divine intervention which would save humanity from complete and peremptory annihilation.

He then follows up with an incorrect prediction of the unfeasibility of unleashing nuclear energy. This further cements his conviction that future warfare should be chemical. His main argument is based on the relatively low death toll of the chemical weapon of the day, mustard gas, compared with artillery shells (ibid., 26–27):

---

[20] This seems to be a retrospective reading by the German committee to make the narrative fit.

[21] Sir William Jackson Pope was a British chemist. During World War One he served on the Board of Invention and Research for the Admiralty and on the Chemical Warfare Committee at the Ministry of Munitions, where one of his contributions was a modified method for preparing mustard gas, see below.

[22] Callinicus of Heliopolis (Baalbeck in today's Lebanon) was a 7th Century Byzantine architect and chemist. This is how he was characterized in (Haldane 1925, 6): "It is true that the appropriately named Syrian [of Jewish descent] Callinicus [meaning victorious] had prolonged the life of the Eastern Roman Empire for another 750 years and saved a large part of Christendom from Mahommedan domination by his invention of 'Greek fire,' an inflammable liquid [of secret composition, likely consisting of a mixture of resin, petroleum, calcium, and sulfur] which was, however, later superseded by gun-powder".

Fig. 6.2 John Burdon Sanderson Haldane (1892–1964). Trained as a mathematician at Oxford, Haldane worked principally in the fields of physiology and genetics. He tackled questions related to the origin of life and was the originator of the "primordial soup" theory. His 1957 theory of evolutionary rates is still of relevance today. He was known as "the man who knew almost everything" (Monk 2020) and characterized as "the cleverest man I ever knew" by Peter Medawar (Medawar, Preface to Clark 1984). His father, John Scott Haldane (1860–1936), a physiologist, was dispatched to Ypres after the first German chlorine cloud attack and co-authored an influential report for Prime Minister Herbert Asquith about his findings. Photograph by Olive Edis, 1925. All rights reserved © National Portrait Gallery, London

The most interesting thing, however, about mustard gas is that, though it caused 150,000 casualties in the British Army alone, less than 4,000 of these (or 1 in 40) died, while only about 700 (or 1 in every 200) became permanently unfit. Yet the Washington Conference[23] has solemnly agreed that the signatory powers are not to use this substance against one another, though, of course, they will use such humane weapons as bayonets, shells, and incendiary bombs.

Further down, he even links the use of mustard gas by the U.K. to the realization of the above by the British leadership and its hope of establishing a war of movement—rather than immovable trench warfare—by chemical means (ibid., 49):

It was not, therefore, until the Germans had demonstrated upon the persons of some tens of thousands of British soldiers (we had 14,000 casualties, though with only 400 deaths, during the first three weeks of the mustard gas war) that there was something to be said for a weapon that was not primarily designed to kill, that we began to use it.

---

[23] That is, the conference leading to the Washington Arms Conference Treaty.

The reluctance to embrace chemical warfare by the British establishment Haldane blames on "intellectual torpor" comparable to that that led to the demise of the Roman and Spanish Empires. He exemplifies this torpor by referring to opposition to animal testing at Porton Down (ibid., 74).

> the physiologists at the experimental ground at Porton, in Hampshire, had considerable difficulty in working with a good many soldiers because the latter objected so strongly to experiments on animals, and did not conceal their contempt for people who performed them. And yet these soldiers would have had no hesitation in shelling the horses of hostile gun-teams, and the vast majority of them were in the habit of shooting animals for sport.

Finally, Haldane underpins his case in favor of chemical warfare by a comparison of the ferocity of the Second Punic War and the German–French struggle in World War One and concludes that (ibid., 80).

> Modern transport and hygiene made its scale possible; the weapons used merely served to prolong it.

For (ibid., 50)

> It seems, then, that mustard gas would enable an army to gain ground with far less killed on either side than the methods used in the late war, and would tend to establish a war of movement leading to a fairly rapid decision, as in the campaigns of the past.

We add that Haldane was not only wrong in his assessment of the unfeasibility of developing nuclear weapons, but also of the innovation potential of chemical weapons: The nerve agents discovered as part of insecticide research by Gerhard Schrader at IG Farben in 1936 exceeded by orders of magnitude the toxicity of any other gaseous compounds and would later be weaponized with the aim to kill rather than to incapacitate (Schmidt 2015, 74–99). On the other hand, U.S. research in the Edgewood Chemical Center during the 1950s-1970s on novel psychoactive agents was driven by a vision of "war without death" (Johnson, 2024).

Both the Washington Arms Conference Treaty and the Geneva Protocol were in part driven by concerns about the use of chemical weapons against civilians and included provisions for their protection (Haber 1986, 248–249; 250–253). Fritz Haber never anticipated the use of chemical weapons against civilians. As Daniel Charles put it (Charles 2005, p. 174), "In this respect, Fritz Haber's imagination remained trapped in the nineteenth century." In contrast, J.B.S. Haldane took the use of chemical weapons against civilians into account and noted (Haldane 1925, 38–61):

> I would suggest that it is more likely to-day that poisonous gas will be used against British soldiers or civilians in future wars … A shut room on a first or second floor would be nearly proof against gas released in the neighbourhood if it had not got a lighted fire to drag contaminated air from outside into it. Moreover, civilians could, and would, rapidly evacuate an area which has been heavily soaked with mustard gas, whereas soldiers have to stay on at the risk of their lives … But let us tell our civilian population before and not after they are attacked with blistering gases that the blisters produced are considerably less dangerous than measles.

6.4   Conclusion

While J.B.S. Haldane was never involved in chemical warfare in practical terms, Fritz Haber was—even after World War One: In 1919, under Haber's tutelage, Germany launched a secret program to continue the development and production of chemical weapons. In order to avoid inspections stipulated under the Versailles Treaty, the program had been moved to third countries, one of them the Soviet Union. Hugo Stoltzenberg, the father of Haber's biographer Dietrich Stoltzenberg, was in charge as Haber's proxy (Stoltzenberg 1994, 330–350). Haber's involvement came to an end only in 1933, when the Nazis rose to power in Germany and he fell out of grace. The chemical weapons production lines in Germany were converted, in part, to accommodate the manufacture of pest control agents, legal under Versailles (Szöllösi-Janze 1998, 447–480). The necessary research and development was provided by Haber and his Kaiser Wilhelm Institute. Among the pest control agents then developed was also "Zyklon B," later used in the Nazi extermination camps to poison millions of people, mainly Jews, among them several members of Haber's family (Szöllösi-Janze 1998, 462–464).

After the Nazi takeover, Haber was forced to emigrate, see, e.g., (Bielik and Friedrich 2020). As noted by Lutz Haber (Haber 1986, 312):

> It is known that [Fritz Haber] would have preferred to go to Switzerland for language and health reasons, but he received no invitation. The former enemies now rallied: in the summer of 1933 Haber was in London, spoke to [Frederik George] Donnan, who alerted [Harold] Hartley, who contacted [William] Pope, and so the old-boy network worked again in unexpected ways. Pope offered laboratory space and other facilities and in September wrote about arrangements for Haber, his sister, his secretary, and a studentship for Dr. [Josef J.] Weiss, his last assistant. And so Haber left Germany for good and was welcomed in Cambridge by the Vice-Chancellor [of the University of Cambridge].

Thus, the combination of mutual personal respect[24] with converging views on the future of warfare in general and chemical warfare in particular transcended any national animosities these members of the "old-boy club" may have felt when they were facing off with one another from the opposing sides of the Western Front.

That such favorable views of chemical warfare were once prevalent is barely known today and, unless contextualized within the mode of thinking of experts and military officials at the time, appears absurd. The elimination in 2023 of the declared chemical arsenals by all 192 parties (representing 98% of the world's population) to the 1993 Chemical Weapons Convention (CWC) reflects the current attitude toward chemical weapons worldwide (Walker 2017, 379).

However, chemical weapons have meanwhile been superseded by what J.B.S. Haldane considered unfeasible in 1925, namely by nuclear weapons. Consequently, chemical weapons had been relegated to the sidelines as "poor man's atomic bomb" (United Nations 2022). What Haldane did get right, though, was that, if feasible, nuclear weapons could upend organized life on the planet.

---

[24] For instance, Hartley was referring to Fritz Haber as "the great Haber" (Haber 1986).

Our consensual coexistence with stockpiles of nuclear weapons should be judged as beyond absurd.[25] As John Polanyi aptly remarked (Polanyi 2017):

> Today millions flee in search of a safe place, while at every compass bearing lie stocks of nuclear weapons. These weapons testify to the power of scientific thinking, but also to the risk of human folly. For it is folly to rely on this most terrible form of war, as offering the surest path to peace. As scientists we should warn of the dangers inherent in such a paradox.

Beatrice Fihn of ICAN (International Campaign to Abolish Nuclear Weapons[26]) in her Nobel Peace Prize lecture put it thus (Fihn 2017):

> The story of nuclear weapons will have an ending, and it is up to us what that ending will be. Will it be the end of nuclear weapons, or will it be the end of us? One of these things will happen.

Can the CWC and its enforcement by the Organization for the Prohibition of Chemical Weapons serve as a model for bringing an end to nuclear weapons? It is our hope it can. For it must.

Acknowledgements I'm grateful to Dr. Paul Walker (Arms Control Association) and Prof. Ulf Schmidt (Universität Hamburg) for discussions concerning disarmament and for their feedback on this chapter.

---

[25] The 2017 Treaty on the Prohibition of Nuclear Weapons has so far 93 signatories and 70 state parties, none of which is a nuclear power https://en.wikisource.org/wiki/Treaty_on_the_Prohibition_of_Nuclear_Weapons. The equivalent destructive power of "first-strike" nuclear weapons (Our World Data 2024) currently stands at about 250 kg of TNT per human inhabitant of the planet (2 Gt total). A regional nuclear winter (Robock et al. 2023) would be induced by exploding just one Mt, i.e., 0.05% of the "first strike" stockpile. Nuclear Winter plays havoc with the concept of Mutually Assured Destruction (MAD), as the use of nuclear weapons by only one side would lead to nuclear winter, i.e., Self Assured Destruction (SAD) (Robock and Toon 2016).

[26] Apart from the *International Campaign to Abolish Nuclear Weapons*, the Nobel Peace Prize for efforts to eliminate nuclear weapons has been awarded to: *International Physicians for the Prevention of Nuclear War* (in 1985), the *Pugwash Conferences and Joseph Rotblat* (in 1995), and *Nihon Hidankyo* (in 2024).

References

Bielik T, Bretislav F (2020) Far apart and close together: Fritz Haber and Chaim Weizmann. Isr J Chem 60(10–11):1061–1076

Charles D (2005) Master mind: the rise and fall of Fritz Haber, the nobel laureate who launched the age of chemical warfare. HarperCollins, New York

Clark R (1984) JBS: the life and work of J. B. S. Haldane. Oxford University Press, Oxford

Crowell B (1919) America's munitions 1917–1918: report of Benedict Crowell, the assistant secretary of war. Director of Munitions. Government Printing Office, Washington

Encyclopedia Britannica (2024). https://www.britannica.com/event/World-War-I/Killed-wounded-and-missing

Fihn B (2017) Nobel lecture given by the nobel peace prize laureate 2017. International Campaign to Abolish Nuclear Weapons (ICAN), delivered by Beatrice Fihn and Setsuko Thurlow, Oslo, 10 December 2017. https://nobelprize.org/prizes/peace/2017/ican/lecture

Friedrich B (2015) Fritz Haber und der Gaskrieg. Physik in Unserer Zeit 46:118–125

Friedrich B (2019) Fritz Haber at one hundred fifty: evolving views of and on a German Jewish patriot. Bunsen Magazin 21(3):130–144

Haber F (1920) Die chemische Industrie und der Krieg. Die Chemische Industrie 43:250–352

Haber F (1924) Fünf Vorträge aus den Jahren 1920–1923. Springer, Berlin

Haber LF (1986) The Poisonous cloud. In: Chemical warfare in the first world war. Clarendon Press, Oxford

Haffner S (1964) Die sieben Todsünden des Deutschen Reiches im Ersten Weltkrieg. Lübbe-Verlag, Bergish Gladbach

Haldane J (1925) Callinicus. A defence of chemical warfare. E. P. Dutton & Company, New York

Harrison M (2010) The medical war. In: British military medicine in the first world war. Oxford, Oxford University Press

Hartley H (1925) Report on the German chemical warfare organisation and policy, 1915–1918 (PRO/WO/33/1072)

Heller CE (1984) Chemical warfare in world war i: the American experience, 1917–1918. Leavenworth Papers. Combat Studies Institute, U.S. Army Command and General Staff College, Fort Leavenworth, Kansas, USA

Hoffmann D (2014) ...im Frieden der Menschheit, im Kriege dem Vaterland. Universität und Wissenschaft im Ersten Weltkrieg. In: Die Berliner Universität im Ersten Weltkrieg, H. Henemann, Berlin

James J, Steinhauser T, Hoffmann D, Friedrich B (2011) One hundred years at the intersection of chemistry and physics. The Fritz Haber Institute of the Max Planck Society 1911–2011. De Gruyter, Berlin

Johnson J (1990) The Kaiser's chemists. Science and modernization in imperial Germany. UNC Press, Chapel Hill

Johnson J (2024) War without death? In: Science and the state: governmental research in war and peace during the twentieth century. Bielefeld University Press, Bielefeld

Jones E, Everitt B, Ironside I, Wessely S (2008) Psychological effects of chemical weapons: a follow-up study of first world war veterans. Psychol Med 38(10):1419–1426

Lepick O (1998) La Grande Guerre Chimique, 1914–1918. Presses Universitaires de France, Paris

Lepick O (2017) France's political and military reaction in the aftermath of the first German Chemical offensive in April 1915: the road to retaliation in kind. In: One hundred years of chemical warfare: research, deployment, consequences. Springer, Heidelberg

Martinetz D (1996) Der Gaskrieg 1914/18. Entwicklung, Herstellung und Einsatz chemischer Kampfstoffe. Das Zusammenwirken von militärischer Führung, Wissenschaft und Industrie. Bernhard & Graefe Verlag. Bad Neuenahr—Ahrweiler

Mommsen W (2011) Die Urkatastrophe Deutschlands. Der Erste Weltkrieg 1914–1918. Klett-Cotta, Stuttgart

Monk R (2020) JBS Haldane: the man who knew almost everything. The New Statesman, 4 November

Moureu C (1920) Chimie de guerre, les gaz de combat. Librairie de l'enseignement technique, Paris

Our World Data (2024). https://ourworldindata.org/grapher/estimated-megatons-of-nuclear-weapons-deliverable-in-first-strike

Polanyi J (2017) Tributes to and reminiscences of Michael Polanyi. Bunsen Magazin 19:117–119

Prentiss A (1937) Chemicals in war: a treatise on chemical warfare. McGraw-Hill, New York

Robock A, Xia L, Harrison C, Coupe J, Toon OB, Bardeen CG (2023) Opinion: how fear of nuclear winter has helped save the world, so far. Atmos Chem Phys 23(12):6691–6701

Robock A, Toon OB (2016) Self-assured destruction. The climate impacts of nuclear war. Bullet Atom Sci 68(5):66–74

Rowe DE, Schulmann R (2007) Einstein on politics. Princeton University Press, Princeton

Schmidt U (2015) Secret science. In: A century of poison warfare and human experiments. Oxford, Oxford University Press

Schmidt U (2017) Preparing for poison warfare: the ethics and politics of Britain's chemical warfare programme, 1915–1945. In: Symposium "100 Years of Chemical Warfare", pp 1–28. Springer, Berlin

Shephard B (2000) A war of nerves. In: Soldiers and psychiatrists, 1914–1994. Jonathan Cape, London

Spiers E (2017) The gas war 1915–1918: if not a war-winner, hardly a failure. In: Symposium "100 Years of Chemical Warfare", pp 153–169. Springer, Berlin

Stern F (1999) Einstein's German world. Princeton University Press, Princeton

Stern F (2011) Fritz Haber: flawed greatness of person and country. Angew Chem Int Ed 51(1):50–56

Stoltzenberg D (1994) Fritz Haber. Chemiker, Nobelpreisträger, Deutscher, Jude. Wiley-VCH, Weinheim

Stoltzenberg D (2004) Fritz Haber. Chemist, Nobel Laureate, German, Jew. Chemical Heritage Press, Philadelphia

Szöllösi-Janze M (1998) Fritz Haber 1868–1934. Eine Biographie. C.H. Beck, München

United Nations (2022). 'Poor Man's atomic bomb' made of dual-use biological, chemical material replaces nuclear weapon for non-state actors. First Committee Told. https://press.un.org/en/2022/gadis3693.doc.htm

Walker P (2017) A century of chemical warfare: building a world free of chemical weapons. In: Symposium "100 Years of Chemical Warfare", pp 379–401. Springer, Berlin

West C (1919) The history of poison gases. Science 49(1270):412–417

Zoghlami H (2021) Franco-British responses to chemical warfare 1915–8, with special reference to the medical services, casualty statistics and the threat to civilians. Med Hist 65(2):101–120

Bretislav Friedrich is a Research Group Leader at the Fritz-Haber-Institut and Honorary Professor at the Technische Universität Berlin. Apart from his research in molecular physics, he maintains an abiding interest in the history of science and is engaged in efforts to eliminate chemical and other weapons of mass destruction. Besides numerous research papers and several books, he co-authored and co-edited *One Hundred Years of Chemical Warfare: Research, Deployment, Consequences* (Springer Nature, 2017).

Chapter 7
Coping with Crisis: Rumours, Science and Poison Warfare in Inter-War Europe

Jo Fox and Ulf Schmidt

Abstract This chapter uses the 1930 Meuse Valley incident, when 63 people were killed in Belgium due to toxic industrial gases mixing with a dense fog, to explore historical and methodological questions about rumour, fear, science, and war during times of crisis. Exploring the intersection between modernity, scientific understanding, and human behaviour, the chapter demonstrates how rumours created different systems of knowledge and scientific inquiry. The chapter highlights how rumour construction in the Meuse Valley illuminates the significant psychological impact of chemical warfare on populations, the continuity of pre-existing belief systems in the face of scientific analysis, and the persistence of international mistrust. The chapter also raises questions about the parameters for trusting knowledge, and how parameters can change in times of crisis or war. These questions are particularly pertinent in an age of post-truth politics, the attempts of national governments to control social media, and the regulation of information and 'fake news'.

Keywords Rumours · Science · Knowledge · Trust · Mistrust · War · Fear · Modernity

7.1 Community in Crisis

On 1 December 1930, the 9,000 inhabitants of the Meuse Valley in Belgium, an over-populated and heavily industrialised area characterised by steelworks, zinc smelters, glass manufacturers, fertiliser and high explosive plants, found their entire valley engulfed in a thick, grey fog. The weather had been unusual. Anticyclonic conditions with high atmospheric pressure, together with a temperature inversion at about 70 or 80 m above ground, and a steady but modest easterly wind blowing from the

J. Fox
Faculty of Humanities and Social Sciences, Newcastle University, Newcastle, UK

U. Schmidt (✉)
Centre for the Study of Health, Ethics, and Society, Universität Hamburg, Hamburg, Germany
e-mail: ulf.schmidt@uni-hamburg.de

© The Author(s) 2026
B. Friedrich et al. (eds.), *Thirty Years of the Chemical Weapons Convention (CWC)*,
https://doi.org/10.1007/978-3-031-98854-7_7

town of Liége into the valley to Engis and Huy, had created a situation where toxic particles and gases emitted from factories were unable to rise, and—after mixing with the fog—accumulated under an 'atmospheric ceiling' in the two-kilometre-wide corridor formed by the valley. As the day wore on, there were no signs of the fog disappearing. The inhabitants of the Meuse Valley were not unduly concerned. Such meteorological phenomena were not uncommon in this part of the country. Fog lasting for more than three days had occurred four times over the last thirty years (in 1901, 1911, 1917 and 1919 respectively). The memory of these events among the generation of Belgians who had grown up before and had lived through the First World War was still vivid. So, too, was their memory of poison warfare after the first attack with chlorine gas near the town of Ypres in April 1915 had initiated a Europe-wide chemical arms race.[1]

At first, the people of the Meuse Valley went about their daily business, giving little to no attention to the impenetrable fog that had descended into the valley. After three days, however, fatalities began to occur in and around the towns of Liége and Engis. Within forty-eight hours, between 4 and 5 December, more than 60 people died in horrific circumstances; men and women coughing and choking, fighting for breath, suffering from severe asthma-like symptoms, accompanied by 'signs of pulmonary oedema such as cyanosis, rapid breathing, and even frothy sputum' (Nemery 2001, p. 704). Almost all fatal cases suffered from acute circulatory failure; most were elderly suffering from pre-existing heart and lung problems but younger people were also affected. Some of the livestock, cattle especially, as well as birds and rats, also died (Roholm 1969, p. 64).

Rumours quickly spread from house to house, first by word of mouth, then by self-appointed experts and media representatives, that the valley was under attack from 'poison fog', killing peasants and factory workers through asphyxiation.[2] Local citizens were quick to blame large quantities of German chemical munitions buried in and around Liége during the First World War, suggesting that leaking poison gas munitions, containing, in all likelihood, mustard gas and phosgene, were responsible for the fatalities. The confluence of a highly unusual meteorological phenomenon with recent memories of modern gas warfare, and its associated human suffering through asphyxiation, created the conditions for the emergence of rumours in which fact and fiction merged, revealing the lasting psychological legacies of the First World War and the persistence of pre-existing belief systems in the face of scientific analysis and probable explanation.

---

[1] For a comprehensive history of poison warfare and human experiments in the twentieth and twenty-first centuries see Schmidt (2015); here esp. pp. 28f, p. 72f; Schmidt (2017), pp. 1–28; Harris and Paxman (2002), p. 21.

[2] 'Poison Fog' (1930).

7.2 Conceptualising Rumour

This paper takes the Meuse Valley incident as a starting point to raise broader historical and methodological questions associated with rumour, fear, science and war at times of crisis. Examining the interplay between modernity, scientific understanding and human behaviour, the paper uses the phenomenon of rumour construction and dissemination to reveal how populations make sense of the complex world around them and the socio-political context in which they live. Rumour-mongering is often characterised as a pathological condition, indicative of a sick body politic; instead, this paper positions rumour as an intrinsically human behaviour that reveals deeper, often private and undetected, trends in popular opinion. Countering rumour formation is as much about understanding what the population is prepared to believe as manufacturing the conditions in which rational thought and scientific explanation prevail.

The rumours circulating in the Meuse Valley in 1930 attest to the profound psychological impact of chemical warfare, the persistence of international mistrust, and the alienating effects of modernity. Numerous studies of nineteenth and twentieth century Europe point to the sublimation of the individual to the collective experience of the mass, especially in times of crisis. The emergence of modern European nation states in the nineteenth century accelerated processes of centralisation, bureaucratisation, and standardisation. Government agencies and corporations created new working environments in which individuals and communities perceived themselves as powerless. As modernity engendered feelings of being reduced to a cog in a soulless bureaucratic and technological machine, rumours functioned as a means to re-assert agency. Rumours, as the historian James Scott has pointed out, became the 'weapons of the weak' or a safety valve to cope with the disassociation of the individual in modern life and escape from the constraints of modern society. Rumours also often revealed the popular fascination with the fantastic and surreal, expressing unconscious fears and anxieties of the unknown.

The particular circumstances of the First World War gave rise to significant discussion about the deleterious effects of rumour in the immediate post-war period. Most commentaries focussed on state-sponsored psychological warfare and the manipulation of populations in pursuit of war aims, rather than rumour formation and circulation as intrinsically human behaviour. Marc Bloch's 1921 essay, 'Reflections of a historian on false news during the war', published in *Revue de Synthèse Historique*, is a rare exception (Bloch 1921, pp. 13–35). Bloch's perspective on rumour formation was formed out of his observations whilst serving near Braisne in Northern France. In his article, Bloch recounted the capture of a German scout. Misunderstandings as to the soldier's origins led to the assumption that he was a spy, and rumours began to swirl among the troops. That the rumour caught hold was in part due to the destabilising effects of the information environment created by war and the destabilisation of trusted sources of news. Bloch recalled that, quoting a humourist:

"the prevailing opinion in the trenches was that anything could be true except what was allowed to be printed." Hence—given this deficiency of the newspapers, to which was added

the uncertainty of postal relations on the front-line, irregular and considered to be monitored—there was a prodigious renewal of oral tradition, the ancient mother of legends and
myths. By a bold stroke that the most audacious of experimenters would never have dared
to dream of, censorship, abolishing the centuries gone by, brought the frontline soldier back
to the means of information and to the state of mind of the old ages, before the newspaper,
before the printed news sheet, before the book. (Bloch 1921, p. 32 Eng. Trans).

Here, Bloch witnessed a tension between the controlled infosphere and human
scepticism and the desire to know, where given knowledge might be challenged
by the inherent 'wisdom'—or sense-making—of the crowd. With a lack of trusted
information came a regression to what the sociologist Tamotsu Shibutani has termed
'improvised news', 'where individuals or groups construe a meaningful interpretation ... by pooling intellectual resources' (Shibutani 1966, pp. 46–49). This process
functions as a means of predicting new developments in a fast-moving, change-
able and unstable situation so that we may prepare for all possibilities. Shibutani
defined rumour-mongering as an act of 'collective problem solving' and as 'part and
parcel of [our] efforts ... to come to terms with the exigencies of life', especially at
times of 'sustained collective tension' (Shibutani 1966, p. 62). It serves as a mechanism for reasserting individual control when individuals perceive themselves to be
powerless and when events appear unpredictable. The hyper-modern information
environment combined with volatile and unpredictable circumstances, then, forces
the individual not to seek solace in scientific or accepted knowledge, but rather to
retreat to more primitive forms of sense-making, to create, in Bloch's words, 'zones
of myth formation' (Bloch 1921, p. 33).[3]

The rumour about the alleged German spy also circulated largely due to the
widespread pre-existing belief among the French soldiers that the Germans had an
established spy network. As Bloch noted, the false story about the identity and intentions of the German captive spread quickly because it resonated with pre-existing
beliefs, prejudices, and emotions (Bloch 1921, pp. 29–33) More often than not,
rumour functions as a mean of deeper emotional expression, particularly emotions–
such as fear, anxiety, despondency and despair. The creation and transmission of
rumour might be understood as the search for stability, certainty and explanations, a
means of seeking out the 'truth', for answers, seeking to take control of a situation by
checking out stories that could not be or would not be verified by the authorities. This
pursuit is of considerable importance, since it potentially reveals the psychological
needs of the population. It is unsurprising that rumour tends to peak at times of crisis
and fundamental instability, just as it is unsurprising that those conditions produced
a critical mistrust of official sources, generated conspiracy theories, and exposed
vulnerabilities. But, as Bloch understood, this does not mean that the spreading
of rumours should be defined as pathological conduct. Far from it. Rumour is an
inherent human behaviour. As Bloch notes, 'false news, in all the multiplicity of
its forms–simple gossip, deceptions, legends–has filled the life of humanity' (Bloch

---

[3] For further discussion see, for example, Deutscher (1967), Portelli (1999), Turner and Fine (2004),
Coast and Fox (2015), Welch (2019).

1921, p. 15). In many ways, understanding societies as complex emotional communities potentially provides a better guide to behaviours than the often superficial, polarised, and most importantly socially sanctioned readings provided by public opinion sources. What people are prepared to believe is perhaps just as historically significant as what actually happened, and this gives rise to a series of critical questions identified by Bloch: 'How do they [false stories] arise? From what elements do they draw their substance? How do they propagate, gaining in magnitude as they pass from mouth to mouth or from writing to writing? No question more than these deserves to excite anyone who likes to think about history.'

Here rumour and misinformation are transformed not into a challenge for the historian, but an opportunity. Passing on misinformation is not irrational: it is motivated by very human needs, whether as a form of exhibitionism (the need to assert status, being in the know), a means of entertaining another and gaining popularity, to bond with others, to prepare (individually or collectively) for the unexpected, for reassurance and emotional support, or to externalise fears, wishes and hostilities. With this in mind, false news becomes, in Bloch's words, 'a mirror in which the collective consciousness surveys its own features', a mechanism for understanding individual and collective motivation (Bloch 1921, p. 31). When this approach is applied to the rumours of the Meuse Valley, it is possible to witness the creation of a distinct 'zone of myth formation' that is separate from scientific or rational explanation, but at the same time informed by individual and collective experiences and widely-communicated but ultimately unrelated events which facilitated the spreading of rumours, as the chemical warfare incident in Hamburg discussed below shows. That the responses and behaviours in relation to these phenomena are transhistorical (similar processes may be witnessed, for example, during the COVID-19 pandemic) speaks to the importance of understanding the process of rumour formation and dissemination, and the possibilities for countering mis- and disinformation in scientific communications.

7.3 A Canvass for Rumours

Far from being a simple imaginative creation, the type of rumours circulating in the Meuse Valley were not only based on previous wartime experiences, but also on a widely-reported case of a chemical warfare disaster in the German city of Hamburg. In 1928, the accidental release of eleven metric tons of phosgene from a storage tank led to three hundred reported causalities and ten deaths within a couple of days.[4] The scale of the accident attracted international media attention. On 21 May, various newspapers reported that Hamburg and neighbouring towns had been struck by 'Poison Gas Terror' after warm weather had caused the bursting of phosgene

---

[4] 'Poison Gas Terror', Nottingham Evening Post, 21 May 1928, issue 15,569, p. 6. The accident followed hot on the heels of a earlier incident in which poison gas had escaped from the German Dye Trust, causing two deaths and considerable soul-searching by the German authorities; 'Poison Gas Cloud', Western Gazette, 25 May 1928, p. 6.

gas barrels at the Stolzenberg chemical factory in the district of Veddel. Respirators and ammonia dispatched from Berlin and distributed among the affected population arrived too late to avert large numbers of casualties.[5] Having been mobilised to assist in the removal of casualties, most of whom were suffering from breathing difficulties, Reichswehr troops and ambulance crews witnessed 'scenes of panic' as hundreds of 'screaming' inhabitants were 'fleeing in all haste' from the toxic fumes.[6] Mirroring the wartime coverage of atrocities in occupied France and Belgium, British news outlets competed over the most colourful and emotive language to describe the incident. Whereas *The Times* called Hamburg a 'plague-stricken' city in which the 'nervous population' was 'haunted by fear', the *Daily Mail* told readers that the poison gas was 'still mowing down its victims'. The *Western Gazette* reported people 'staggering' out of their home and onto the street in a 'state of panic', 'collapsing on the pavement' and 'screaming in terror'.[7] Reported casualty numbers varied greatly. The *Evening Telegraph* and *Post* reported that eleven people had died in 'gas terror', while ninety people received hospital treatment.[8] A day later, the *Aberdeen Press and Journal* suggested that '1,000,000 persons' were in peril and that 'thousands' were fleeing in total panic from 'deadly clouds'.[9] Only a change in wind direction had apparently saved the lives of thousands of people. Few offered an explanation for the gas cloud: the cause of the disaster appeared at first to be a 'mystery', creating an environment that encouraged speculation and community 'sense-making'.[10]

Indicative of deeper public fears, connections were soon made between the 'poison gas cloud' that had brought death and sickness to all living beings in its path and the imminent threat of aerial bombardment, and how chemical and biological weapons might affect large cities and civilians in future wars.[11] The war of the future, the *Nottingham Evening Post* remarked in response to the near-catastrophe in Hamburg, would not be a war of 'battling armies' under the command of aging generals but one in which the battlefields would be whole cities, and in which the 'mechanical, wireless-driven, aeroplane would fly over the land, leaving wholesale destruction and death in its wake'.[12] A vivid description invited not only a psychological imagining

---

[5] 'Poison Gas Terror', Nottingham Evening Post, 21 May 1928, issue 15,569, p. 6.

[6] Gas Terror in German City', Evening Telegraph and Post, 21 May 1928, p. 5.

[7] 'The Poison Gas Disaster', Times, 23 May 1928, p. 16; 'Poison Gas Claims More Victims', Daily Mail Atlantic Edition, 24 May 1928, p. 7; 'Poison Gas Cloud', Western Gazette, 25 May 1928, p. 6.

[8] 'Gas Terror in German City', Evening Telegraph and Post, 21 May 1928, p. 5.

[9] 'Poison Gas Clouds Sweeps Over Hamburg', Aberdeen Press and Journal', 22 May 1928, p. 7.

[10] 'Poison Gas Cloud', Western Gazette, 25 May 1928, p. 6; 'Poison Gas Tragedy', Western Daily Press, 25 May 1928, p. 12.

[11] 'Poison Gas', The Citizen, 23 May 1928, p. 4; 'A Poison Gas Disaster', Times, 22 May 1928, p.16; 'German Lesson in Death Gas', Daily Mail Atlantic Edition, 24 May 1928, p. 7; see also Grayzel, S.R., At Home and under Fire. Air Raids and Culture in Britain from the Great War to the Blitz (Cambridge, CUP, 2012); Jones, E., 'Civilian Morale During the Second World War: Responses to Air Raids Re-examined', Social History of Medicine, 17 (2004), 463–479; Shapira, M., The War Inside. Psychoanalysis, Total War and the Making of the Democratic Self in Postwar Britain (Cambridge, CUP, 2013), pp. 32–42.

[12] 'Poison Gas Horrors', Nottingham Evening Post, 29 May 1928, p. 2.

of the narrowly-avoided plight of the residents of Hamburg but the fate that might befall civilians on the future home front:

> Breathing that becomes more difficult every minute. The sudden lack of air. And then, gradually, a feeling of suffocation. You begin to gasp for breath; you grow afraid when breath seems to fail, gasp yet more frantically, become terror stricken, and then look round at your comrades. They, too, are gasping for breath. What is it all about? There is no apparent reason for this suffocating feeling. Nothing can be seen except what is normal. As you look your throat grows more parched than ever, your tongue is dry, and your mouth burns as though a candle were alight in it. Then you writhe in agony and finally die.[13]

Aerial warfare, readers were told, would also involve the use of biological weapons carrying 'millions of germs', causing 'mortal disease' and untold suffering on an unimaginable scale.[14] The 'untameable horror' such scenarios of future military conflict evoked among communities created fertile ground in which any notion of toxic clouds became intertwined with a sense of impending doom. Rumours, in this context, were vocalisations of deeper, often private, fears over the specific horrors of a future war, the articulation of which serves as the release of a safety valve on unspoken concerns now aired in a public forum.[15] Unsurprisingly, pacifist and militarist movements, according to *The Times*, used the incident as 'propaganda' for their respective causes, with the former predicting that the disaster might be a 'blessing in disguise' if the manufacture of poison gas were prohibited, and the latter calling for greater military preparedness.[16] However, many agreed that the best hope for humankind to step back from the abyss seemed to be to cultivate a 'peace spirit' between the family of nations.

The public speculation was intensified by the entry of 'expert testimony' that seemed to confirm worst fears. The Earl of Halsbury, former Assistant Inspector of High Explosives, authored a widely-publicised dystopian novel, *1944*, which painted a grim picture of future wars, arguing that peoples' fear of poison gas attacks was 'no idle dream'. In 1917, Halsbury outlined the first comprehensive plan for strategic bombing which subsequently informed Allied military operations during the Second World War.[17] By joining the chorus of those calling for improved civil defence measures 'against gas attacks by hostile aircraft', and by drawing attention to the need to study weather conditions to bolster of military defence, Halsbury exacerbated fears that next time death was likely to come from the air in the form of poisonous clouds contaminating humans, livestock and the environment.[18]

Moreover, media reporting about the tragedy amplified unresolved issues about Germany's war-time reparations and disarmament and, more broadly, about Germany's legal obligations resulting from the Treaty of Versailles and subsequent

---

[13] 'Poison Gas Horrors', Nottingham Evening Post, 29 May 1928, p. 2.

[14] 'Poison Gas Horrors', Nottingham Evening Post, 29 May 1928, p. 2.

[15] 'Poison Gas', The Citizen, 23 May 1928, p. 4.

[16] 'Poison Gases', Times, 24 May 1928, p. 17.

[17] See also Royal Airforce Museum, Department of Research and Information Services, AC73/2, Giffard, Hardinge Goulburn (1880–1943), 2nd Earl of Halsbury, papers, 1915–1928.

[18] 'The Poison Gas Disaster', Times, 23 May 1928, p. 16.

agreements. Journalists questioned why and how a German factory had been in possession of large quantities of phosgene, a substance used both in the manufacture of dyes and as a chemical weapon, given that Germany was prohibited from manufacturing such agents under article 171 of Versailles.[19] The case, critics observed, raised complex issues of dual use which highlighted potential shortcomings in international arms control. Both France and Britain had requested an explanation from the German government as to why the gas had escaped. The proposal by Britain that the League of Nations should look into the matter was flatly rejected by the German authorities, who saw this as a purely 'German affair'. Germany's robust response stemmed from a complex legal situation; although the country was prohibited from manufacturing phosgene for 'war purposes', there seemed to be no limit for exporting the agent for industrial and agricultural purposes up to the production limit of the three licensed phosgene factories. This meant that Germany was permitted to produce up to nine tons of phosgene per day. Any existing stockpiles of the gas could thus not easily be challenged by the League of Nations.

While there was no suggestion that Germany had acted in 'bad faith' in order to prepare for another military confrontation, the notion that the phosgene gas had been a 'remainder of war-time supplies' was met with disbelief after Hugo Stolzenberg, the owner of the Hamburg-based chemical factory where the accident had occurred, refused to disclose the origins of the stockpiles and got entangled in inconsistencies about the date he had acquired the gas, fuelling an atmosphere rife with suspicion and speculation.[20] The socialist newspaper *Vorwärts*, for one, tapping into already apprehensive public sentiment about Germany's war-time conduct, alleged that the Stolzenberg factory had commissioned the production of poison gas and gas shells at a factory in Troitsk, near Moscow, on behalf of the Reichswehr during the military occupation of the Ruhr Valley between 1923 and 1925. The debate eventually reached conspiratorial dimensions after commentators, without explaining the legal context, alluded to a significant discrepancy between the amount of phosgene Germany was permitted to export annually and permitted quantities in the three designated factories.[21]

The incident sufficiently alarmed Allied countries, since it revealed not only the destructive power of chemical weapons in any future war, especially in large cities and other urban settings, but, more worryingly, Germany's ability to conduct clandestine military preparations in breach of international laws and treaties.[22] Debates about the Hamburg disaster contributed to an emerging European consensus which foresaw unannounced aerial gas warfare as a deciding factor in future conflicts (Evers

---

[19] 'Gas Terror in German City', Evening Telegraph and Post, 21 May 1928; 'Poison Gas Cloud Sweeps City', Courier and Advertiser, 22 May 1928, p. 3.

[20] 'A Poison Gas Disaster', Times, 22 May 1928, p.16; 'Death Gas Sweeps Hamburg. Phosgene', Daily Mail, 22 May 1928, p. 13; 'Poison Gas Deaths', Aberdeen Press and Journal, 23 May 1928, p. 7; 'Poison Gas Tragedy', Western Daily Press, 25 May 1928, p. 12; 'Hamburg Gas Disaster', Times, 26 May 1928, p. 11.

[21] 'The Poison Gas Disaster', Times, 23 May 1928, p. 16; 'Poison Gas Claims More Victims', Daily Mail Atlantic Edition, 24 May 1928, p. 7.

[22] 'The Hamburg Disaster', Derby Daily Telegraph, 22 May 1928, p. 4.

2014, p. 269ff). By the late 1920s, the horrors of gas warfare and its associated imageries had become firmly established in the collective European memory, and vivid descriptions of new types of aerial warfare continued to fuel the popular imagination. Previously unimaginable scenarios of gas attacks on urban communities, painted in evocative terms by literary figures such as Walter Benjamin and Kurt Tucholsky, created a greater awareness of risk and uncertainty in society. As early as 1925, Walter Benjamin wrote in the *Vossische Zeitung* that any future war would be a "gas war" with a "ghostly frontline" in which the distinction between combatants and non-combatants no longer existed. Such a war would entail "an absurd degree of risk" and be based on "fear and annihilation" (Schmidt 2015, p. 72f). The prospect of air attacks with chemical weapons changed not only the image of warfare, but also its associated temporality and spatiality in the popular imagination. Death from the air could come at any time and in any place. Prime Minister Stanley Baldwin's famous phrase that "the bomber will always get through", delivered in 1932 to the House of Commons, left European societies in no doubt that, however well the country was prepared, airpower would prohibit civilians and combatants from entertaining any sense of security in their homeland. Any notion of, or reference to, gas war turned into an emblematic shorthand to depict the ugly face of modernity, in which people lived in constant fear of sudden death. Places of safety no longer existed in the minds of most. By the end of the 1920s, the Golden Age of security and optimism so vividly described in Stefan Zweig's *The World of Yesterday*, at least as far as the middle and upper middle classes were concerned, had given way to an emerging "risk society", to use Ulrich Beck's phrase, in which the rapid pace of scientific and technological development led to an ever-growing spectrum of neuroses and anxieties (Zweig 2009, p. 23ff). It is perhaps for these reasons that the causes of the Meuse and Hamburg incidents were quickly positioned within a greater geo-political context and in sensationalist terms, rather than subjected to calm forensic and scientific enquiry. Here, it is more revealing to witness what people were prepared to believe than what was actually true. Shibutani has noted the distinction between accuracy and credibility: 'men act on the basis of their beliefs, which are not necessarily demonstrated truths' (Shibutani 1966, p. 7).

7.4 Reasoned Rumours–Irrational Science

Both the Hamburg disaster and the Meuse case exhibited a number of significant similarities. For one, the people exposed to phosgene in Hamburg suffered from similar symptoms to the people of the Meuse Valley, ranging from headache, dizziness, nausea, vomiting, and irritant cough, followed by pulmonary symptoms (Hegler 1928). In both cases, the number of casualties suffering from breathing difficulties and asphyxiation were in the hundreds. Two days after the fog had lifted, Queen Elisabeth, visiting the stricken valley in her motorcar in a show of solidarity and to stem the rising tide of rumours, received a rousing welcome by cheering crowds. In response to the concerns raised by people, who repeatedly told her that "fog alone could not have done that", and to shift the public discourse on a more evidence-based level, she expressed the wish that the "mystery" be investigated by a scientific commission under the chairmanship of Dr Nolf, her medical advisor.[23]

Such considered appeals failed to take root, in part undermined by the space given to self-appointed experts to speculate on the causes. In such a volatile public atmosphere, the deployment of speculative rather than evidence-based scientific explanations helped to exacerbate the spreading of rumours. One line of thought and its associated rumours, advanced by none other than Colonel Joaquin Enrique Zanetti, a professor of chemistry at Columbia University, New York City, who served with the United States Chemical Warfare Service from 1917 to 1945, suggested that toxic sand—containing 'germs brought from the near east'—had produced 'muddy rains' which had been blown from the Sahara Desert and then been inhaled to deadly effect by the Belgian people. Another theory suggested that toxic gases had been released from polluted soil. Professor Victor E Levine from the Department of Biochemistry at Creighton University School of Medicine, Omaha, Nebraska, United States, on the other hand, demonstrating some stark resemblance to the type of astrological explanations given by medieval physicians for the emergence of bubonic plague in the fourteenth century, suggested that the 'deadly gases' had originated from the 'tail of a dissipated comet'.[24] Levine was far from being alone in creating vivid images and associations to major health disasters in times long past. In Britain, J.B.S. Haldane, reader in biochemistry at Cambridge University, and son of John S. Haldane (1860–1936), one of the experts who had identified the chlorine gas used by the German military in 1915, suggested that Belgium was 'suffering from a return of the medieval "Black Death"'. Defending his comments, Haldane responded by saying that he 'did not allude to the bubonic plague' when talking about the Belgian fog but had instead referred to 'pneumonic plague' by which he meant 'an acute respiratory infection attacking the lungs'.[25] As the debate was spiralling out of control, the Belgian government, determined to calm nerves and reassure its citizens, felt the need to announce that the deaths had been due 'solely to the cold fog', attempting to scotch the most persistent of rumours: that 'war gases buried by the retreating German armies had escaped'.[26] It was this rumour which appeared and reappeared in national and international newspapers. By mid-December 1930, the US magazine *Time* went as far as raising the spectre of an elaborate military conspiracy hatched by the Belgian and French governments by suggesting that the French military had only recently produced an 'enormous cloud' near the city of Lille, some eighty miles from the Meuse Valley, 'a cheap, harmless artificial fog made from chalk, sulphuric acid and tar products' which could be employed to 'hide the movement of troops

---

[23] 'Belgian Queen & the Fog of Death', 9 December 1930; see also 'The Fog Scourge: Inquiry into the Belgian Mystery', Lancashire Evening Post, 9 December 1930; 'The Death Fog: Queen of Belgians Visits Stricken Area', Hull Daily Mail, 8 December 1930; 'Poison Gas Theory Supported in Fog Horror', Sheffield Independent, 8 December 1930.

[24] 'Poison Fog' (1930).

[25] 'Poison Fog' (1930).

[26] 'Poison Fog' (1930).

in war time'.[27] *Time* ominously asked its readership whether this had all just been a coincidence, suturing unconnected events together in an attempt to impose a meta-narrative that could, in some senses, appear more comforting that a series of random, indiscriminate events affecting our lives.

It is possible to conclude then that the mainstream media acted as handmaidens to collective mania and speculation over the incidents in Hamburg and Meuse, perhaps as a mechanism to drive sales through sensational coverage. Yet, reframed as 'improvised news', according to Shibutani, this process is better situated as a 'communicative pattern that develops when men who are involved together in a situation in which something out of the ordinary has happened pool their intellectual resources in an effort to orient themselves' or in order to 'define a situation' (Shibutani 1966, p. 9). Rumour continues to be a form of 'improvised news' until the content is verified or refuted, sitting in that intermediate space—possibly with deleterious consequences—until resolution. Such a process is witnessed in relation to the Hamburg and Meuse cases. What concerned most experts for the next seventy years were the conditions and factors responsible for causing the disaster: air pollution, the wind and climatic conditions, the size and type of particles, the noxious agents, and so on. As the sciences of toxicology, chemistry, meteorology, and human physiology were developing, expert opinion—even if not in agreement—became a much-valued source for assessing the tragic events. Dismissed as unreliable, unreasonable and at times fanciful, the role of 'rumour', and its not-so-distant relative 'fear', a reasonable human reaction, has generally been overlooked. However, a closer look at the content and beliefs which underpinned these rumours—and the fears resulting from them—not only offers unique insight into greater levels of 'rationality' and 'reasoning' in the construction and dissemination of rumours, but also suggests that seemingly irrational connections made by the inhabitants from observations and past memories may have been closer to the truth than previously assumed.

7.5 A Type of Nerve Gas Poisoning

Opinion is still divided as to what caused the deaths in the Meuse Valley. In 1930, an expert commission under the chairmanship of Jean Firket, professor of pathological anatomy and forensic medicine at the University of Liége, was tasked with investigating why there had been a sudden ten-fold rise in mortality in the area. The commission concluded that the average age of those killed by the 'toxic fog' had been about 62 years, the youngest a 20-year-old woman, the oldest 89 years of age. Compared to male victims, female victims were older, with an average age of between 56 and 70 years. Most deaths had occurred in and around the town of Engis. The Belgian investigation was not as comprehensive and methodologically sound as John Snow's epidemiological study into the causes of the London cholera of 1854, but it established beyond doubt, by way of exclusion, that 'air pollution

---

[27] 'Poison Fog' (1930).

could kill', a phenomenon which had not hitherto received extensive attention. The commission identified sulphur compounds such as sulphur dioxide (SO_2), emitted by the coal burning industry and by private households, in conjunction with sulfuric acid (H_2SO_4) produced through oxidation reactions, as the main culprits for causing the deaths in the valley. It also established that fine soot particles released by the factories, having absorbed the compounds, had functioned as a means by which the toxic agents could penetrate 'deeply into the respiratory tract', thus affecting anyone with respiratory problems particularly severely (Nemery 2001, p. 707).

Equally of interest, yet dismissed by the expert commission, was the possibility that hydrofluoric acid (HF) and fluorides released from fertiliser plants had created a toxic cocktail for the valley's inhabitants. This explanatory model was initially advanced by leading lung specialists from Holland before being taken up in 1937 by the toxicologist Kaj Roholm from Copenhagen, who argued that the victims of the Meuse Valley had 'suffered from acute intoxication by gaseous fluoride compounds which emanated from certain factories in that area' (Roholm 1969, p. 62; also Roholm 1937, p. 126). As late as 2001, scientists argued in *The Lancet* that Roholm's explanatory model 'could not be completely ruled out as a contributory factor', especially in the vicinity of the fertiliser plants (Nemery 2001, p. 706).

The available material seems to suggest, however, that none of the scientists and scholars writing about the Meuse Valley events since the 1930s have ever considered the scientific evidence in conjunction with contemporary rumours about chemical warfare agents. A closer look at this connection shows that the rumours extant at the time may have contained information about the cholinesterase-inhibiting ability—and therefore chemical warfare potential—of fluorides, which was only fully recognised by Allied scientists after the post-war discovery of German nerve agents (Schmidt 2015, pp. 74–99). Information about the toxicity of fluorides and their mode of action was extremely limited at the time of the disaster; the commission only referred to 'a deplorable lacuna in the literature'. In 1947, Lord Edgar D. Adrian, Baron Adrian of Cambridge, distinguished professor of physiology at Cambridge University, who had close links to Britain's top-secret chemical warfare establishment at Porton Down, made it plain that the first observations of the cholinesterase-inhibiting action of fluorophosphonates had been made in 1941, five years after Roholm had published his findings in the *Journal for Industrial Hygiene and Toxicology*. Yet, Adrian and his colleagues had dismissed the chemical warfare potential of these compounds (Adrian et al. 1946, 1947; Hodgkin 1977, p. 44). In other words, the fast-acting and highly toxic German nerve agents, discovered by the advancing Allied forces in April 1945, were believed to be part of 'a family' of agents which Allied scientists had examined but rejected as a warfare agent.[28]

The type of fluorides emitted in the Meuse Valley in 1930s, and the way in which they reacted with the dense fog over a period of several days, can be ascertained with a reasonable degree of accuracy. It is therefore not only possible but also highly likely that gaseous fluorides would have acted as a cholinesterase-inhibiting, and thus highly

---

[28] TNA, WO32/21200, Brunswick (DSWV) to Assistant Chief of the Imperial General Staff ACIGS (W), 14 April 1945.

toxic, agents which may well have killed—perhaps in conjunction with sulphur compounds—those inhabitants living and working in and around the factories in the Meuse Valley. The existing evidence certainly seems to point in that direction: the fluorine compounds released would, in all probability, have been hydrogen fluoride (HF), silicon tetrafluoride (SiF4) and its aqueous solutions, compounds which are not only particularly toxic and potentially lethal in minute concentrations, but also readily absorbed through the skin and the mucous membranes.[29] Data collected from clinical and autopsy findings indicate that death had occurred from a systemic failure rather than from a local effect on the respiratory organs. Roholm concluded that 'acute cardiac failure and rapid death pointed to a severe general effect and to prompt absorption of the toxic agent' (Roholm 1969, p. 68).

We know that the air during the days the fog took hold of the Meuse Valley contained toxic levels of fluorine. Fifty-six of the sixty fatalities occurred in the vicinity of fluoride-emanating factories, especially in the eastern half of the valley. No fatalities happened in or around factories releasing compounds other than fluorine. Once the fog had lifted, no further deaths occurred. Those who, during those cold December days, climbed up the hills surrounding the valley, and managed to get above the fog, remained free from clinical symptoms. Whereas explanations such as the role of the fog alone, a lack of oxygen, or carbon monoxide could be ruled out given the clinical symptoms observed, there is the possibility that the deaths were caused by lethal levels of sulphur compounds, as suggested by the Belgian expert commission. Another possibility which needs to be considered is that the active agents that caused the deaths of the people in the valley could have been gaseous fluorides (Roholm 1969, p. 70). The clinical symptoms reported by family physicians, and the fact that adrenalin-based drugs—themselves used as an antidote for nerve gas poisoning—were effectively used to relieve the symptoms, are additional indications that the chemical compounds may well have been a precursor of modern nerve agents, today known as a weapon of mass destruction. The rumours circulating in Belgium and elsewhere about buried German chemical munitions were not based on fact but they were also not entirely untrue. Indeed, as extraordinary as it may sound, in this case the rumours are likely to have anticipated a family of warfare agents which was unknown to the international scientific community until they were discovered and developed by German chemical warfare scientists in the mid-1930s.

---

[29] The humidity present in the fog would have led to a 'partial hydrolysis of silicon tetrafluoride to hydrogen fluoride and fluosilicic acid, two extremely active compounds, especially when in atomized form. They produce mists like sulfuric acid and hydrochloric acid, but are much more toxic'; Roholm 1969, p. 66; see also Roholm 1937.

7.6 Disbelieving True Rumours

As with the broader impacts of interwar debates on propaganda, the effects of the interwar rumours regarding Meuse and Hamburg lasted into the next war. Fears expressed over future aerial bombardment and biological and chemical warfare took on new significance with the renewed hostilities, and belligerents aggressively exploited these fears in psychological warfare campaigns. Psywar operatives, such as Britain's Psychological Warfare Executive (PWE) with its 'sibs' (rumour) unit, saturated the info-sphere with rumours, some fantastic, some more pointed, in order to provoke that 'horrible prickly sensation that comes to a man in the blackness of an empty house'.[30] Indeed, Ronald Turnbull, an SOE (Special Operations Executive) operative in Scandinavia, was convinced of the power of rumour to destabilise morale in profound ways. These 'skilfully directed shafts of propaganda', he wrote to Ivone Kirkpatrick at the Ministry of Information in 1940, were potentially 'worth many divisions and many squadrons of aircraft'.[31]

Rumours of poison gas were a particularly rich seam for psy-warriors to exploit, drawing on the fears created, for example, by the sensationalised descriptions of the effects of gas in incidents such as those in Meuse and Hamburg, and on deep-seated memories of Ypres. Two groups of poison gas rumour were particularly prominent. The first related to secret advanced technologies, designed to unnerve populations with generalised statements about new scientific discoveries that would give adversaries the upper hand in the conflict. On 19 May 1942, for example, PWE communicated via operatives in Germany that, 'Goebbels has instructions to prevent the British from using gas by all means possible. Hence all these frightening statements and rumours. Apparently the British are in a very much better position for using gas than we are'.[32] Rumours were often grouped together in the hope that they would collide and gain momentum. On the same day, PWE approved a rumour for distribution that Lord Cherwell, the British physicist and Churchill's chief scientific adviser 'has had to take two months sick leave following an unexpectedly successful experiment with the new British blue gas' and, in a separate rumour, that there was 'a great row going on behind the scenes in the House of Commons because Churchill has refused to give JBS Haldane, who invented the new British blue gas, any decoration because of his political history. He deserved the George Cross for what he did experimenting in the gas chamber'.[33] The second group was designed to provoke fear about the unseen, insidious effects of gas warfare, where victims of an attack were not safe in shelters or homes or that they could be secretly contaminated with disastrous effects. Another two connected rumours, released on 27 May, told the German population that, 'the RAF have ready a new gas bomb which penetrates 100 feet and

---

[30] TNA, FO898/70, Memorandum from Ronald Turnbull to Kirkpatrick, Director of Foreign Publicity, Ministry of Information, 23 June 1940.

[31] TNA, FO898/70, Memorandum from Ronald Turnbull to Kirkpatrick, Director of Foreign Publicity, Ministry of Information, 23 June 1940.

[32] Q/703 (Sib number assigned by PWE), 19 May 1942, in Richards (2010), p. 182.

[33] Q/709 and Q/713 (Sib numbers assigned by PWE), 19 May 1942, in Richards (2010), pp. 182–3.

then explodes. The gas works up through the loosened earth and goes through the cracks in the walls of deep shelters' and that, 'the English have an odourless gas which works through the clothes and produces only a slight feeling of warmth, but the victim is then sensitised to this, and if he is exposed even to a very mild dose of the gas a second time, he gets an attack of asthma. The effect lasts for six months'.[34]

That the British were undertaking such a rumour saturation campaign themselves, including with specific reference to poison gas, peaking in the spring of 1942, is not insignificant. The military and special operations units were fully aware of the scale of the global disinformation campaign, introducing a new level of scepticism about intelligence received. Indeed, at home, rumours suggesting the existence of fast-acting deadly new gases 'against which the respirators are useless' had been circulating among the British population throughout the Second World War. As early as 1940, Haldane, who in the mid-1920s had claimed chemical warfare agents to be a more "humane" weapon of war, had made it clear that he did "not believe [any of] them" (Haldane 1940, p. 163). Rumours relating to chemical agents were nonetheless monitored—and sometimes spread—by Home Office officials to counter potential enemy propaganda, but their truthfulness had rarely, if ever, been tested, nor had the intelligence behind them been passed on to the relevant agencies (Haldane 1940, p. 163). Rumours had a life of their own, detached from the seemingly hard-nosed military intelligence-gathering that informed Britain's chemical warfare corps. In those cases where vital intelligence based on oral testimony and rumours was shared with British agencies, no further action had been taken. The saturation of rumours in the midst of a global war created a heightened sense of disbelief among experts and the general public of military intelligence, and raised the evidential bar for reliable, and by implication actionable, information. Validating rumours now required conformation not only through additional sources of information, but also ideally through material evidence and objects, for example textual records, photographs, and chemical substances. The saturation of rumours in the modern public sphere and their aggressive use as a weapon of modern warfare, in other words, led to a reassertion of material objectivity.

The existence of new, highly toxic nerve agents is generally believed to have been one of the best-kept secrets of the Second World War, and unbeknown to Allied intelligence until British ground forces discovered ten truckloads of the strangely coloured shells in a railway yard at Espelkamp, north of Lübbecke, in April 1945 (Schmidt 2015, p. 158f). The Allied discovery of nerve agents, so the story goes, was based on the physical discovery of the actual shells and on the chemical examination of the liquids contained therein at Porton Down, Britain's chemical and biological warfare facility since the First World War. It was quickly established that inhalation of the substance led to a constriction of the pupils and a tightness in the chest. After injecting the purified agent into cats, a Porton scientist reported that the substance acted as an inhibitor of cholinesterase: "The symptoms are tremors, convulsions and finally death from respiratory failure. Death is usually rapid, and takes place in a few

---

[34] Q/765 and Q767 (Sib numbers assigned by PWE), 27 May 1942, in Richards (2010), pp. 183–4.

minutes". As far as possible means of protection was concerned, the response was: "No information yet" (Schmidt 2015, p. 158f).

Archival documents, however, reveal a significantly more complex picture in which intelligence information about newly-developed German warfare agents was received in 1942/43, at the peak of the PWE poison gas rumour campaign. The information received was neither sufficiently believed nor acted upon. In March 1942, the Chiefs of Staff Committee first discussed rumours that the Nazi regime might have developed a "new German CW agent", and attached the "greatest importance" to developing effective methods of detection. Given the level of secrecy attached to the "new agent", the commanders in chief of the three service departments were instructed to give the information "the most limited circulation", which was compatible with preparations for possible chemical warfare.[35] The file contains limited information about the exact nature of the "new agent", except that it seemed to be "more dangerous" to the eyes and lungs and undetectable by smell or conventional methods. Much of the technical information may have been removed from the file at a later date for reasons of national security, but it may also have not been recorded in the first place. It seems certain, though, that the relevant passages did not attract the attention of scholars studying the history of chemical warfare agents in the post-war period.

Rumours about new German warfare agents, if not backed up by detailed contextual or chemical data, quickly faded into oblivion within the war-time bureaucracy; British officials were able to glean no knowledge about the type of the newly discovered gas bomb from captured documents, and German prisoners of war had referred to a new substance called "Trilon". Evidence obtained from an unidentified German army officer captured in North Africa in May 1943 revealed that the German military had developed a new colourless warfare agent with "astounding properties".[36] His testimony provided Allied intelligence with considerable detail about the likely composition and mode of action of German nerve agents prior to their actual physical discovery in 1945, but the information was neither believed and studied nor acted upon. The officer, a former chemical scientist trained at the Technical High School in Charlottenburg, Berlin, showed himself extremely knowledgeable of the research work undertaken by the *Waffenamt Prüfwesen* at the Spandau Citadel. Under the leadership of Wolfgang Wirth, and in total secrecy, scientists from the Institute for Pharmacology and Toxicology at the Military Medical Academy conducted animal and human experiments to advance existing physiological knowledge of nerve gas poisoning and develop possible antidotes.[37] Most importantly, the officer described the symptoms of the agent he knew only by the name of "Trilon 83", the German

---

[35] TNA, WO193/723, Possible New German C.W. Agent, 30 March 1942.

[36] TNA, WO193/723, Report on Information obtained from P/W M(Am)15, 28 March Bn., captured 11 May 1943; Chemical Warfare, S.I.R 14, 3 July 1943.

[37] BIOS, Final Report, No. 138, Item No. 8, 'Interrogation of German CW Medical Personnel', August/September 1945; also Tucker 2006, pp. 36ff.

code-name for Tabun or GA. He told Allied investigators that the smallest concentration of the agent was sufficient to cause, within a "few minutes", the characteristic reactions:

Strong pupillary contractions to a pin-head and asthma-like difficulties of breathing. In any heavier concentration death occurs in about a quarter of an hour. It cannot be classed with any of the other war gasses as it is a *nerve poison* (emphasis added), producing vasoconstriction and the pupillary contractions.[38]

In retrospect, it might seem astonishing that the information supplied did not raise serious alarm bells in Allied military and intelligence circles. Here was a German scientist who had been intimately involved in developing a new family of chemical warfare agents, but the information was treated as little more than an unconfirmed rumour. The officer was unable to name the exact chemical formula, which he described as an "alkyl [probably methyl or ethyl] derivative of monoflourphosphoric acid", although the correct formula later turned out to be $C5H11O2N2P$, or ethyl dimethylamine cyanophosphate, possibly because access to this type of information was highly classified. Still, given the level of detail he was able to provide, it must have been apparent that he had been directly involved in researching the substance and had himself suffered from its effects. Experiments with these agents were "extremely dangerous", he told interrogators, because they showed "no perceptible threshold of irritation" compared to conventional chemical warfare agents. In one incident, he said, a chemist had lost his life despite attempts to reanimate his respiratory system through injections of lobeline. Treatment through atropine was considered "too strong" and believed to cause eye damage. By the time the officer had left the research department at the Spandau Citadel, tests with "T.83 Trilon" had been completed and the mass production of the nerve agent had commenced in Germany's chemical weapons factory in Dyhernfurth, code-named 'Hochwerk'—a small town northwest of Breslau, in the eastern province of Silesia—where the German conglomerate IG Farben had established the first ever plant for the manufacture of Tabun after the outbreak of war. "A large quantity [of the nerve poison] must be ready", he told Allied officials in 1943. Remarkably, on reviewing the report, military officials felt that the matter did not warrant further attention given the lack of "corroborating evidence". Despite being classed as "reliable", the report was not passed on to the relevant military agencies such as Porton Down, or indeed to the groups of Cambridge scientists around Edgar D. Adrian, who since 1941 had studied a whole range of highly toxic, cholinesterase-inhibiting substances on behalf of the Ministry of Supply (Schmidt 2015, p. 159f). In retrospect, it looked like an intelligence failure of monumental proportions, but it was more than that.[39]

The increase in rumours may have allowed for a more critical approach in questioning the veracity and origin of information, especially within the complex

---

[38] TNA, WO193/723, Report on Information obtained from P/W M(Am)15, 28 March Bn., captured 11 May 1943; Chemical Warfare, S.I.R. 14, 3 July 1943.

[39] TNA, WO193/723, Report on Information obtained from P/W M(Am)15, 28 March Bn., captured 11 May 1943; Chemical Warfare, S.I.R. 14, 3 July 1943; Tucker 2006, p. 55.

machinery of government, but it also created greater expectations of the verification of intelligence information which, by its very nature, often relied on hearsay and oral testimony. Highly important intelligence not supported by corroborating or material evidence, for example through the chemical formula or the substance itself, thus failed to be classed as *reliable* information, and no longer warranted the status of actionable information. In response to the saturation of rumours in wartime, the bar as to what was classed as 'objective' scientific information was raised to such an extent that otherwise *actionable* knowledge was no longer seen as credible. True rumours, in other words, could easily be disbelieved within the fluid environment of armed conflict. This became particularly apparent in the case of a new family of chemical warfare agents, the nerve agents, which German scientists had developed in the mid-1930s.

7.7 Summary

Rumours created specific linkages between different systems of knowledge and scientific inquiry. Under certain conditions, rumours can blur the boundaries between trained judgement and mechanical objectivity, and become a challenge to existing knowledge and expertise to such an extent that individual actors and institutions respond outside known norms in perhaps unproductive ways, for example by dismissing rumours as uncorroborated, non-credible information or as a pathological condition. Rumours, in such a scenario, are dismissed as something 'subversive' which needs to be tackled and overcome, rather than as a form of 'improvised news' to which one might respond in an investigative mode. While rumour can precede knowledge, as we have seen in the case of the dissemination of, in all probability, gaseous fluorides in the Meuse Valley disaster, it can also create a climate in which it functions as a barrier to follow up investigative leads, as we have seen in the wartime rumours about German nerve agents.

But rumours can also be an expression of individuals and groups mistrusting official sources of information, supplied by government agencies and established media outlets, thereby generating the need for a community-led 'improvised news' which counters censored or curated news sources from the centre, such as the public scepticism of Queen Elisabeth's appeal to reason in Meuse. This raises a series of broader questions: what are the parameters for trusting knowledge, and how do these parameters change in a crisis or wartime situation? Conversely, at what point do rumours function as 'improvised news' when trusted knowledge cannot be substantiated or is easily accessible? This is particularly pertinent in an age of post-truth politics and national governments' attempts to regulate social media, and control the flow of disinformation, especially in crisis situation. The fast and almost instantaneous movement of rumour in a global media environment reflects in many ways a return to early twentieth century dilemmas and methods of information management. Forms of established communications, as well as official and legal guidance, appeal to socially responsible citizens and communities; they appeal to the rational elements of human behaviour in emotive, uncertain times. However, whereas the authorities previously targeted individuals and groups in their campaigns, nation states and supra-national entities today aim to ensure greater corporate responsibility in the wake of globalisation. Corporations flouting the rules, like individuals before them, are being portrayed as placing themselves outside the community and the greater common good.

As this paper has shown, the study of rumour has to be applied within specific contexts, since rumours are not transhistorical but historically contingent, even though human behaviour and reactions to them may well be. We can detect certain patterns that repeat over time, and other aspects which are historically specific. By combing both longitudinal studies with microhistories, the challenge for future historians will be to determine what some of these patterns are. While acknowledging the importance of the threat of prosecution to regulate individual and corporate behaviour, many continue to prefer to exercise the moral pressure of the immediate social circle and local community rather than pursuing matters through national and international courts, frequently citing the distinctions between liberal democratic and authoritarian government (Gluckman 1963, p. 309). Such evidence supports Gluckman's assertion that rumour functions as a mechanism for articulating and cementing community values and prevalent morals, enacted from below rather than enforced from above.

References

Adrian ED, Feldberg W, Kilby BA (1946) F: inhibiting action of fluorophosphonates on cholinesterase', (Letters to Editor). Nature 158(4018):625

Adrian ED, Feldberg W, Kilby BA (1947) The cholinesterase inhibiting action of flurophosphonates'. Br J Pharmacol 2(1):56–58

Bloch M (1921) Réflexions d'un Historien sur les Fausses Nouvelles de la Guerre. Revue De Synthèse Historique 7:13–35

Coast D, Fox J (2015) Rumour and politics. History Compass 13(5):222–234

Deutscher I (1967) Improvised news: a sociological study of rumor. By Tamotsu Shibutani. Book review. Soc Forces 46(2):298–299

Evers K (2014) Risking gas warfare: imperceptible death and the future of war in weimar culture and literature. Ger Rev: Lit, Cult, Theory 89(3):269–284

Gluckman M (1963) Gossip and scandal. Curr Anthropol 4(3):307–316

Haldane JBS (1940) Science in peace and war. Lawrence and Wishart, London

Harris R, Paxman J (2002) A higher form of killing. The secret history of chemical and biological warfare, 1st edn. Arrow Books, London

Hegler C (1928) On the mass poisoning by phosgene in Hamburg. Clin Obs 54(37):1551–1553

Hodgkin A (1979) Edgar Douglas Adrian, baron Adrian of Cambridge. 30 November-1889-4 August 1977. Biogr Mem Fellows R Soc 25(25):1–73

Mysterious Fog Brings Death to 58 Belgians, The Cornell Daily Sun (1930)

Nembery B, Hoet PHM, Nenmar A (2001) The meuse valley fog of 1930: an air pollution disaster. The Lancet 357(9257):704–708

Poison Fog (1930) Time 16(24):21

Portelli A (1999) The death of Luigi Trastulli and other stories. State University of New York Press, Albany

Richards L (2010) The whispers of war. Underground propaganda rumour mongering in the second world war. Psy War, Peacehaven

Roholm K (1937) The fog disaster in the Meuse Valley, 1930. A flourine intoxication. J Ind Hyg Toxicol 19:126–137

Roholm K (1969) The fog disaster in the Meuse Valley, 1930. A flourine intoxication, Flouride 2(1):62–70. Republished J Ind Hyg Toxicol 19:126–137

Schmidt U (2015) Secret science. A Century of Poison Warfare and Human Experiments. Oxford University Press, Oxford

Schmidt U (2017) Preparing for poison warfare: the ethics and politics of Britain's chemical warfare programme, 1915–1945. In: Symposium "100 years of chemical warfare". Springer, Berlin, pp 1–28

Shibutani T (1966) Improvised news: a sociological study of rumor. The Bobbs-Merrill Company, Indianapolis

Turner P, Fine GA (2004) Whispers on the color line: rumour and race in America. University of California Press, Berkley and Los Angeles

Welch D (2019) Epilogue "we are all propagandists now": propaganda in the twenty-first century'. In: Propaganda and conflict. War, media and shaping the twentieth century. Bloomsbury Academic London, pp 328−342

Zweig S (2009) The world of yesterday. Memoirs of a European. Pushkin Press, London

Jo Fox began her career at Durham University as a lecturer in Modern European History. She became Durham History Department's first female professor in 2010 and its first female Head of Department in 2016. She was appointed as the first female Director of the Institute of Historical Research at the School of Advanced Study in 2017. Jo Fox is a specialist in the history of propaganda, rumour and mis- and dis-information. She is the author of two major monographs, and has published widely in leading journals (such as the *Journal of Modern History*). Her most recent research has focused on 'The Political Warfare Executive, Covert Propaganda and British Culture' with James Smith, Pat Waugh and Guy Woodward (Durham) (funded by the Leverhulme Trust), and she was Principal Investigator on an AHRC COVID-19 rapid response project on 'COVID-19 rumours in historical context.' She also regularly assists museums and archives in their public programmes and exhibitions. Jo has previously served as the Honorary Communications Director of the Royal Historical Society. She is a National Teaching Fellow (2007), a Fellow of the Royal Historical Society (FRHistS), and a Fellow of the Royal Society for Manufactures and the Arts (FRSA). From 2019-2023, she served as Chair of the Humanities and Social Science Early Research Awards panels at the Wellcome Trust.

Ulf Schmidt is Senior Professor of Modern History at the University of Hamburg, founding-director of the Centre for the Study of Health, Ethics, and Society, and a Fellow of the Royal Historical Society. His interests are in the history of modern medical ethics, warfare, and policy in twentieth-century Europe and the United States. He is especially interested in the history of authoritarian regimes and modern dictatorships. He is the author, among others, of *Secret Science. A Century of Poison Warfare and Human Experiments* (OUP, 2015) and is the co-editor of *Propaganda and Conflict: War, Media and the Shaping of the Twentieth Century* (Bloomsbury, 2019). He is Principal Investigator of a six-year ERC Synergy Grant on "Taming the European Leviathan: The Legacy of Post-War Medicine and the Common Good".

Chapter 8
'Woe to You for Being a Grandchild': Mutations and the Ethical Case Against WMDs Among Post-war British Geneticists

David Peace

Abstract This chapter explores the intersection between genetic research and ethical opposition to Weapons of Mass Destruction (WMDs) among post-war geneticists in the United Kingdom (UK). Focussing on the mutagenic effects of radiation and chemical exposure, it highlights how studies on Drosophila impacted growing concerns among members of the genetics discipline about the cross-generational impact of WMDs. The chapter examines the role of key historical actors, such as Lionel Penrose, John B. S. Haldane, and Charlotte Auerbach, in not only shaping research within the discipline of human genetics but also in opposing international nuclear weapons testing. It examines the nature of this resistance to WMDs based on genetics research in Parliamentary debates, the formation of the Genetical Society's committee on radiation effects, and critiques of the Medical Research Council's 1956 White Paper on nuclear hazards to human health. The chapter emphasises how post-war genetics research informed the ethical stance against WMDs, driven predominantly by concerns about the responsibilities towards the population health of future generations. It also examines the role of eugenics within these debates, and how opposition towards WMDs provides a lens to examine the post-war disciplinary shifts within genetics towards attempts to safeguard health across generations.

Keywords Genetic mutations · Post-war Britain · WMDs · Geneticists · Radiation · Mustard gas · Carcinogenesis · Mutagenesis · Reproductive health · Population health · Nuclear weapons testing · Human genetics · Environmental radiation

D. Peace (✉)
Centre for the Study of Health, Ethics and Society, Universität Hamburg, Hamburg, Germany
e-mail: david.peace@uni-hamburg.de

© The Author(s) 2026

B. Friedrich et al. (eds.), *Thirty Years of the Chemical Weapons Convention (CWC)*,
https://doi.org/10.1007/978-3-031-98854-7_8

8.1   Introduction: Gene Mutations and the Ethics of WMDs

In the 1920s, there was a growing consensus among geneticists that gene mutations formed the primary basis of evolution. However, inspired by the lack of experimental means to induce mutations under laboratory conditions, the American geneticist Hermann J. Muller developed a series of experiments to not only explore the nature of 'sudden hereditary changes' but to artificially produce them.[1] In these experiments, Muller used Drosophila, commonly known as fruit flies, due to their relatively simple genetic structure. Unlike humans, who have 46 chromosomes, fruit flies have only 8 large-sized chromosomes, including three pairs of autosomes and one pair of sex chromosomes. This simplicity in their chromosomal structure made them ideal candidates to test how X-rays could induce mutations through the breaking of chromosomes exposed to radiation and their attachment to another—also known as 'translocation'. Focussing on germ cells—those cells responsible for passing genes from one generation to the next—he examined the impact of radiation exposure in causing significant sudden changes in these cells and the effects of these changes on offspring (Muller 1929, 1930, 1941). These tests revealed that 'sex-linked' mutations in mated flies could be observed in offspring, as sex-cells were shown to be particularly susceptible to artificially induced mutations, leading to various unique expressions in later generations, such as 'snipped wings' and 'bloated wings' (Muller 1930).

Spurred by Muller's research, between winter 1940 and the following spring, Charlotte Auerbach, a geneticist at the University of Edinburgh's Institute of Animal Genetics, conducted a series of large-scale tests on the effects of chemicals on inducing mutations. Auerbach had been contacted by Alfred J. Clark from the University's Pharmacology Department, alongside her colleague John M. Robson, to arrange a study on the effects of mustard gas on gene mutation. This research was based on a contract Clark had with the Chemical Defence Establishment of the War Office to investigate the long-term consequences of the compound on human cells (Beale 1995, p. 27). The choice of mustard gas was not random. As early as the First World War, mustard gas had been shown to produce similar lesions 'strikingly like an X-ray burn of the skin.'[2] In Auerbach and Robson's experiments, male Drosophila were exposed to mustard gas and tested for sex-linked lethals—those mutations that lead to death among offspring if they are inherited (Auerbach 1944). From her trials, 7.3% lethals were obtained in more than 1,000 chromosomes, as compared with 0.2% in control groups. In later tests, even higher mutation rates were obtained, in some instances up to 24%. In addition to the cross-generational effects of exposure to mustard gas mutations, the experiments also demonstrated a marked impact on reproductive health, for example where doses of mustard gas that produced more than 20% sex-linked lethals effectively sterilized all treated males.

---

[1] Muller (1927). For a comprehensive overview of the 'decisive contributions' made by Muller to scientific research on induced mutations and the genetic hazards caused by radiation exposure see Pontecorvo (1968). For biographical details given by one of Muller's students see Carlson (1981).

[2] Skin Lesions Produced by Mustard Gas (1918).

Due to the potential for weaponisation of the chemicals used in these experiments, Auerbach and Robson were denied permission by the War Office to publish their findings until after the war (Auerbach 1945; Auerbach and Robson 1946). In the decade that followed, similar trials on chemically induced mutations conducted by Iosif Rapoport at the Institute of Experimental Biology of the Academy of Sciences, Moscow, Milislav Demerec at Cold Spring Harbor Laboratory, New York, and Raymond Latarjet, at the *Laboratoire Pastuer de l'Institut du Radium*, Paris, confirmed their findings.[3] These early trials collectively lent themselves to a growing post-war concern about the reproductive and hereditary health effects of induced mutations. Yet it was the dropping of atomic bombs on the cities of Hiroshima and Nagasaki in August 1945 that altered perceptions among geneticists about their research priorities regarding the reproductive and hereditary effects of induced mutations on human health.[4] This shift in research priorities was reflected in the US government's founding of the Biology Division at Oak Ridge National Laboratory in 1947 to facilitate research on mutation studies with aims that included exploring the ability of cells to recover from damage caused by radiation exposure.[5] Oak Ridge had by the early-to-mid 1950s begun to demonstrate under laboratory conditions the genetic effects of chemical and radiation-induced mutations in germ-cells carried across generations in numerous animal trials, suggesting that mutations most likely affected humans in a similar capacity.[6]

Muller, nearly three decades after his initial research on induced mutations in Drosophila, articulated his fears about the health effects on future generations caused by the use of these new weapons. The suffering caused by these weapons to human health and welfare, he argued, would not be confined to a single generation. If people were able to survive the initial use of such weapons and reproductively 're-establish' themselves numerically over the course of centuries, the toll among later generations of premature death and 'failure to reproduce' would be colossal (Muller 1955, 63). Deaths across successive generations caused by the transmission of lethal mutations would be 'several times as numerous as the deaths that had occurred' among the population who had originally been bombed.

As this chapter will demonstrate, the hereditary link between artificially-induced germ-cell mutations in humans and their impact on population health led many

---

[3] Demarini (2020), Demerec (1947), (Demerec 1948). For a comprehensive overview of the work of Iosif Abramovich on chemical mutagens, in addition to his resistance against Lysenkoism in Soviet Union, see Eiges (2013); Yanovskiy and Socol (2023); Demerec and Latarjet (1946); Latarjet (1948).

[4] An early expression of the fears of induced mutations and their impact on human health was expressed by a future colleague of Auerbach's at the University of Edinburgh, the developmental biologist Conrad Hal Waddington. Fearing the mass deployment of X-Ray treatment, Waddington believed that without further trials their use may lead to 'an increased incidence of the pathological condition in the population' and led to a 'serious deterioration of … human material.' Waddington (1938).

[5] For an extensive overview of the post-war founding of the Biology Division at Oak Ridge and the development of its research in mutations see Rader (2006).

[6] Research on induced mutagenesis at Oak Ridge included studies in collaboration with Mary L. Alexander at the University of Texas' Department of Zoology. These studies particularly focused on the mutational damage to germ cells (1954), (1955).

geneticists to reflect upon the cross-generational implications of using WMDs. These reflections extended not only to targeted peoples but also to their offspring and to the generations belonging to an imagined irradiated future caused by weapons testing, removed from the geneticists of the post-war period in some instances by many centuries. By focussing on a small network of influential geneticists in the UK connected with Charlotte Auerbach, such as Lionel Penrose and John B. S. Haldane, this chapter will illustrate how geneticists in post-war Britain came to oppose the use of WMDs. There is an extensive existing historical literature on the nature of international scientific activism and opposition to WMD testing and use in the context of the Cold War, particularly in examining the roots of organisations such as the Pugwash Conferences on Science and World Affairs (Nickerson 2013; Kraft (2018). This literature has tended to emphasise the role of the scientific community in the formulation of political action, such as influencing the modern environmental movement, and the transnational character of action against WMDs, including broader histories of institutions and international treaty conventions such as the OPCW and its role in preventing the use of potentially mutagenic chemical warfare agents (Burkett 2013; Vucinic et al. 2015; Richmond 2017; Kraft 2018; Sachse 2018). This chapter will focus on how opposition to WMDs on the grounds of their mutagenic effects played a role in shaping the post-war discipline of human genetics and the integration of its emerging research into its ethical opposition to these weapons. It will demonstrate how their opposition was rooted not only in concerns about safeguarding human health and welfare, but also integrated research on the cross-generational impact of induced mutations, particularly in the context of the UK's nuclear weapons tests of the early 1950s. It will examine the debates among geneticists during this period regarding induced mutations, radiation exposure, and genetic diseases and 'birth defects', and show how these debates played a crucial role in establishing the research objectives and disciplinary aims of human genetics in post-war Britain. The chapter will use the debates among geneticists involved in the Medical Research Council's (MRC) 1956 White Paper on 'The Hazards to Man of Nuclear and Allied Radiations' to highlight how the discipline began to reorient its priorities towards safeguarding health across generations, beyond the eugenic aspirations that had influenced many of the geneticists of that generation in the preceding decades (Peace 2021, 2024).

8.2 Parliamentary Debates on Nuclear Testing and Population Health

On the morning of 3rd October 1952, on the small archipelago of the Monte Bello islands off the north-western coast of Australia, Britain conducted its first nuclear weapons test, code-named *Hurricane*, becoming the world's third atomic power. There were many hopes and aspirations attached to the success of the project. As Lorna Arnold has highlighted, with the abrupt ending of the UK's atomic partnership with the US in 1946, following the US Atomic Energy Act, and the successful demonstration of the Soviet's atomic capabilities in 1949, Britain's post-war military security and hoped-for status as a major global power depended on the success of an independent atomic programme (Arnold 2001, p. 34). Though the resources and manpower available paled in comparison to the testing conducted by the US at Bikini Atoll in the Pacific, the detonation at Monte Bello was 'almost universally acclaimed as a triumph', demonstrating British technical and scientific prowess (Arnold 2001, p. 34; Arnold and Smith 2006, p. 29). However, not all voices were eager to cheer for Britain's new-found ability to reduce a city to a ruin with a single bomb.

In winter 1954, Frederick Lee, a British Labour politician, challenged the Prime Minister Winston Churchill in the House of Commons whether he would order a 'detailed investigation' of the effects of 'radioactive particles and gases' on human health and the climate.[7] In his reply, Churchill appears to have been more minded to the 1952 Great Smog of London rather than the long-term effects of nuclear fallout, insisting that 'climatic effects are not susceptible to experimental tests,' and that 'fog and rain … cannot be caused by radioactive particles or gases.' Concluding that effects of radiation on health were being 'continuously carried on,' the Prime Minister concluded that 'no additional detailed investigation … seems called for.'[8] Yet, the opposition did not allow Churchill off the hook so easily. Labour MP Somerville Hastings, a surgeon who had contributed to the post-war Labour Party's health service policies, pressed further, asking whether the Prime Minister was aware that 'pneumoconiosis, silicosis, and X-ray cancer' could develop even years after the detonation of a bomb. Churchill replied that he was not only 'out of [his] depth', but that he did not see how Hasting's comments were of 'relevance' to the matter.[9] The dismissive reply by the Prime Minister revealed not only a cavalier attitude towards legitimate concerns about the use and testing of these weapons, but also the stark lack of knowledge about their impact on human health and welfare by the very politicians charged with forming policies about their potential use in a global conflict.

In March 1955, Labour politicians were able to again bring the question of the health impact of nuclear weapons into debate within the House of Commons. The motion for debate was brought forward by Labour MP Edith Summerskill, a physician, who urged the Government to consider the long-term effects of continuing nuclear tests, emphasising the dangers facing humanity due to ongoing radioactive contamination of the world's atmosphere, particularly highlighting its potential impact on future generations.[10] Summerskill stressed the necessity for an international conference involving participants from the USA, USSR, UK, and France to assess the risks posed by their collective weapons tests on the future of humanity. She argued that the issue transcended not only party politics but also national and ideological boundaries, underscoring the responsibility placed on all leaders to address the cumulative global impact of radiation exposure. The grounds which she cited

---

[7] Hansard (18 November 1954). Vol. 533, cols. 559–560.

[8] Ibid., col. 560.

[9] Ibid., col. 560.

[10] Hansard (22 March 1955). Vol. 538, cols. 1881–1882.

for her motion of debate were the potential genetic implications of exposure to radiation, noting work conducted by the American geneticist Alfred Sturtevant at the California Institute of Technology that warned of the potential cross-generational impact of radiation induced mutations in germ cells (Sturtevant 1955).[11] As such, Summerskill called on the Government to take seriously the prospect of international scientific collaboration to prevent the potentially deleterious genetic impact of continued nuclear testing on future generations.

The response from the Government, represented by the Minister of Health, Iain Macleod, was not so receptive to the suggestions within Summerskill's motion. Macleod acknowledge the gravity of the issue presented by the Opposition, but he quickly shifted in his response to emphasise the complexity and long-term nature of the problem, suggesting that a full resolution would take generations, possibly centuries. He mentioned that the Medical Research Council (MRC) had been studying the issue of induced radiation mutations for many years, and that the findings so far had indicated that the increase in radioactivity due to nuclear tests was very small and not an immediate threat to public health, arguing that the total radiation exposure from nuclear tests was much less than that of a single chest X-ray. In response to Summerskill's suggestion of the need for a joint international response to the potential health dangers posed by extensive weapons testing, Macleod rejected the idea outright, stating that the Government was happy with the arrangements it had with its partners in the US and Canada, and that involving the Soviets would be a non-starter due to political and ideological differences. However, he attempted to reassure Summerskill that the Government took the health implications of these tests seriously and that extensive research by the MRC would be supported to monitor the effects of nuclear bombs on health and genetics.

Yet, Macleod's assertions that there was no immediate threat to public health did not correspond with the controversial debate within the field of human genetics that had been caused by the question of induced mutations and radiation exposure. Two years prior to the Monte Bello tests, in 1950, Muller published an extensive essay for the American Journal of Human Genetics highlighting the issue. Far from Macleod's comments in the House of Commons, Muller emphasised the long-term genetic risks of radiation exposure, warning that even a moderate increase in mutation rates could have significant cross-generational effects, highlighting that damage caused by radiation exposure could accumulate over successive generations, resulting in widespread health issues that might not be immediately apparent yet would manifest significantly over decades and centuries (Muller 1950). In the UK, this debate over the genetic impact of radiation exposure would not remain confined to parliamentary discussion. A week after the debate motioned by Summerskill, the Prime Minister was challenged again by the Labour MP Arthur Henderson on whether he would agree 'to publish a White Paper containing all available facts and information …

---

[11] Ibid., col. 1886.

on the effects of continuing radioactive contamination of the world's atmosphere.'[12] Churchill replied that he had already invited the MRC to compile a report on the 'medical aspects of nuclear radiation', which would include 'genetic aspects'.

8.3 The Genetical Society's Response to the UK's WMD Testing

On the same day that Churchill announced in the House of Commons the future publication of the MRC's White Paper on the genetic effects of radiation exposure, a committee of the Genetical Society met to formulate a response. This was an important step for the discipline of genetics across the UK into the debate on health and weapons testing. Founded in 1919, the Society had been instrumental in organising the field of genetics as a discipline in the UK by providing an institutional platform to share findings and identify the major areas of research focus (Crew 1969). The Society's members included some of the most prominent names within British genetics, such as William Bateson, often regarded as one of the founders of the field of modern genetics and one of the key figures in founding the Genetical Society. Others included Julian Huxley, Ronald A. Fisher, and John B. S. Haldane, each of whom were influential in what some had come to term the 'modern synthesis', which integrated Darwinian ideas of natural selection with Gregor Mendel's principles of inheritance (Huxley 1942). Fundamental to this synthesis was the idea that random genetic mutations create variations within populations, and that natural selection acts upon this variation and leads to evolutionary changes over significant periods of time, generation after generation (Mayr 1982, pp. 738–743).[13]

On Tuesday 29 March 1955, the Society appointed a Committee on the Genetical Effects of Radiation who decided to prepare an 'authoritative factual report'.[14] Though the committee was small, its members included some of the then most influential geneticists working in the UK. Among them was Auerbach, University of Edinburgh, and Cyril Darlington, Professor of Botany at Oxford University, whose recent research included studies on the induction of cancer by chemical agents via genetic mutation, and the chemical basis of chromosomes (Darlington 1948, 1955). The physicist Louis Gray from the British Empire Cancer Campaign Research Unit in Radiobiology at Mount Vernon Hospital, London, who had recently conducted trials on the possible uses of radiotherapy in cancer treatment (Gray 1953; Thomlinson and Gray 1955). Also on the committee was John Thoday, a botanist at the

---

[12] Hansard (29 March 1955). Vol. 539, col. 197; for a contemporary report detailing the Parliamentary voting and international context that led to Churchill's announcement, see Biological Hazards of Radiation (1955).

[13] Mayr (1982).

[14] Letter from L. S. Penrose to C. Auerbach, d. 1 April 1955. PENROSE/2/35/6, Wellcome Collection.

Strangeways Laboratory, Cambridge, who had conducted experiments on chromosomal aberrations caused by X-rays (Thoday 1946, 1954, 1947). Finally, Lionel Penrose, perhaps one of the most influential figures within British medical genetics, who as Galton Professor of Eugenics at University College London had prior to 1945 led innovations in studies of mutation rates in human populations and the supposed hereditary basis of "mental deficiencies".[15] Alongside Auerbach, Penrose led much of the efforts to form the committee and shape its agenda.

The committee agreed that any statement from the Genetical Society should not be published until after the Government had issued the MRC's White Paper, which was expected by them to be released in the early summer.[16] During their initial meeting they reviewed a number of memoranda that had been invited from members of the Genetical Society. Among those circulated was a short memo by Penrose that outlined his initial views on the genetic effects of radiation.[17] His memo highlighted several reported incidences where microcephaly was present in children whose mothers had undergone pelvic therapeutic irradiation during pregnancy. However, he queried the significance of this association based upon two reasons. Firstly, that in many of these cases X-ray treatment had occurred at the fifth month or later, a period that Penrose considered too late to influence the development of microcephaly. This belief was based upon a survey of clinical observations conducted in 1929 that concluded that although therapeutic pelvic irradiation during pregnancy could damage a growing embryo and lead to congenital defects, the use of pelvic irradiation posed no such danger (Murphy and Goldstein 1929a, 1929b). It would not be until the later-half of the 1950s that the first surveys would be published linking direct foetal irradiation with increased risk of leukaemia and other potential severe injuries.[18] However, based upon the 1929 survey, Penrose believed that although the authors were unable to demonstrate any significant results that connected childhood health conditions with direct pelvic irradiation on their parents, he believed that if 'pre-conception irradiation produced a mutation in the ovary, it would not follow that the effects would appear in the first generation' but in successive generations after.[19] This potential intergenerational harm caused by exposure to radiation would come to be a defining part of Penrose' opposition to the use and testing of WMDs, in addition to his calls for the meticulous recording of all pregnancies occurring after 'post-radiological treatment' to assess the impact of radiation exposure on 'foetal deformities'.

---

[15] Penrose' studies prior to 1945 on the hereditability of "mental deficiencies", though sometimes critical, coincided with the debates about eugenics and its potential applications during the interwar period. For Penrose's engagement with eugenic thought during this period see, Penrose (1938), (1938b), (1944).

[16] Letter from L. S. Penrose to C. Auerbach, d. 1 April 1955. PENROSE/2/35/6, Wellcome Collection.

[17] Views of Professor Penrose, Galton Laboratory–University College (undated). PENROSE/2/35/5, Wellcome Collection.

[18] Stewart (1958). For full details of the historical surveying of the epidemiological link between leukaemia and direct foetal irradiation see Greene (2017).

[19] Views of Professor Penrose (undated), p. 2.

Based upon the review of the memos submitted, such as that by Penrose, the committee decided to draft a series of papers divided among its members. It was agreed that Auerbach would focus on the experimental production of mutations, and variations in sensitivity, Gray would draft an overview of the differences between natural background radiation and 'radiation produced artificially ... with special reference to military, medical and industrial activities', Thoday on the effects of radiation in population genetics, and Penrose on human mutation and its effects on health.[20]

In addition to these draft papers, both Penrose and Auerbach were eager to bring John Haldane onto the committee.[21] Haldane and Penrose had previously collaborated together, significantly in the early 1930s in the first attempts to calculate 'mutation rates' within human populations and their effect on health. Their early work in the field focussed on creating estimates of the rate at which mutations might occur among humans when compared to the rates observed in Drosophila, specifically focussing on suspected genetic conditions such as tuberous sclerosis and haemophilia (Penrose 1935; Haldane 1935). Tuberous sclerosis—characterised by the appearance of non-cancerous tumours across the body—was chosen due to its high mortality rate and lethality; that very few people affected were assumed to have children to pass the condition hereditarily. As such, based upon the collection of family histories (pedigrees) of patients with the condition, they estimated that 25% were 'sporadic', with no evidence of a prior family member with the condition, that hereditary transmission was very unlikely, and that the condition was presumably the result of a mutation (Penrose and Haldane 1935, p. 907). When based on the population of England, such a rough estimate would imply a mutation rate of 1 in 120,000 per generation to cause the condition in individuals with no prior family history of tuberous sclerosis. When compared to the highest mutation rate observed in Drosophila (1 in 300,000), the higher rate estimated by Penrose and Haldane seemed to indicate that humans were more susceptible to mutation with potentially significant implications for human health.

In addition to their efforts to bring Haldane onto the committee, Penrose asked Auerbach to 'sound out' some of her colleagues at the University of Edinburgh. She replied that the most senior member of her department, Conrad Hal Waddington 'hesitantly said–but probably emphatically thinks' that the Genetical Society's committee was redundant.[22] She replied to Penrose that her own 'personal feeling' was that the committee had 'value, especially for bringing together information.' She feared that there was a growing consensus among geneticists, as expressed in Penrose's own memo, that 'all we know about the matter is that we don't know anything.' She had heard such a sentiment also from her own colleagues at Edinburgh, and though at first she felt inclined to agree with them, on reflection she felt that 'this feeling is in great part due to the shift in emphasis in genetical research away from radiation

---

[20] Memo, 'Committee of the Genetical Effects of Radiation of the Genetical Society', d. 26 May 1955. PENROSE/2/35/5, Wellcome Collection.

[21] Letter from Penrose to Auerbach, d. 27 April 1955. PENROSE/2/35/6, Wellcome Collection.

[22] Letter from Auerbach to Penrose, d. 27 April 1955. PENROSE/2/35/6, Wellcome Collection.

genetics, and the resulting lack of knowledge of what has already been achieved.'
In this sense, she was correct, even in the previous decade, between 1945 and 1955,
there had been surveys conducted on the genetic effects of radiation in mammalian
and plant experiments, with significant discussion on the theorised impact of induced
mutations on human populations. For instance, in 1954, based upon experimental
work such as that conducted on induced mutations in Drosophila, Herman Muller
introduced the concept of 'genetic load', which he believed would emphasise the
theoretically severe implications of increased mutation rates across an entire population (Muller 1954). This concept was a new addition to the then post-war debate on the
'maximum permissible dose' of ionising radiation—measured in units of 'roentgen'
(r)—that an individual could be exposed to without seriously harmful effects (Stone
1952; Morgan 1953; National Committee on Radiation Protection 1954). Muller
highlighted that 80r of radiation can double the natural mutation rate, resulting in 12
new mutations per 10 zygotes instead of the natural 6, thereby doubling the 'genetic
load' on surviving individuals within that population. This increase would mean
that more individuals would have to suffer 'genetic death' to maintain the genetic
equilibrium, potentially posing an existential threat to the entire population's future
fitness and survival.

Based on her belief that there was a need to articulate that 'a great deal [was]
already known' about the impact of induced mutations on population genetics, Auerbach argued that the Genetical Society's committee should press to bring out their
own statement 'fairly soon and independent of the [MRC's] White Paper'.[23] Yet, a
little over a week later, on 7 May, Auerbach again wrote to Penrose in a dejected tone,
stating that Waddington believed that the Genetical Society's committee was entirely
defunct.[24] Though Penrose initially tried to fight against this, seemingly assured that
other members of the committee did not hold Waddington's view, it became clear that
the MRC did not want to risk dissent among prominent British geneticists against
its own statement.[25] A few weeks later, it became clear that the Genetical Society
would no longer issue its own independent statement, and instead Penrose, Haldane,
and Auerbach would be invited, in varying capacities, to submit their views to and
contribute to the discussions leading to the publication of the White Paper.[26] Penrose
would be invited to sit on the MRC's committee on 'Genetic Effects', and both
Auerbach and Haldane were invited to submit their reports for discussion. Though
Auerbach appears to have been generally in agreement that it was better if the data
collected by the members of the Genetical Society be incorporated into the White
Paper—so as not to form a 'rival report'—she still harboured doubts as to whether
the data would be 'incorporated in the sense in which it is meant.'[27]

---

[23] Letter from Auerbach to Penrose, d. 27 April 1955.

[24] Letter from Auerbach to Penrose, d. 7 May 1955. PENROSE/2/35/6, Wellcome Collection.

[25] Letter from Penrose to Auerbach, d. 10 May 1955. PENROSE/2/35/6, Wellcome Collection.

[26] Letter from Penrose to Auerbach, d. 23 May 1955. PENROSE/2/35/6, Wellcome Collection.

[27] Letter from Auerbach to Penrose, d. 12 May 1955. PENROSE/2/35/6, Wellcome Collection.

8.4　Genetic Mutations and the MRC's Eugenic Controversies

On 1 June 1955, the Committee of the Medical Aspects of Nuclear Radiation's Panel on Genetic Effects met for the first time, convened by Harold Himsworth, a physician and Secretary of the MRC.[28] The panel comprised of some of the UK's most influential scientists, such as Peter Medawar, then Professor of Zoology at University College London; Joseph Mitchell, a radiologist and Regius Professor of Physic at the University of Cambridge; Waddington, Professor of Animal Genetics at Edinburgh University; and Penrose, one of only a few geneticist on the panel with direct expertise in medical genetics.

This first meeting soon revealed several key disagreements among the panel on the genetic effects of radiation exposure. The discussion began with Himsworth introducing what he termed the relatively 'less difficult' topic of the relationship between radiation dose and mutation rate at low doses, and the mutation-rate doubling in humans. While there was initial consensus among the panel that high doses of radiation exhibited a 'linear' relationship with the mutation rate, the situation at low doses of radiation was more contentious. Toby Carter, a colleague of Waddington at the Institute of Animal Genetics, Edinburgh, noted that the most recent studies had shown a linear relationship down to very low doses, yet Waddington considered only doses above 100r as high enough to show clear linearity. Yet unpublished work seemed to hint that significant changes were possible even at the lowest dose, but was rejected by some on the panel, insisting that it was important 'to accept only experimental data'.[29] However, it was in the debate about the doubling dose in humans—the amount of radiation needed to double the mutation rate—where opinions began to diverge significantly. Waddington argued that the difficulty in estimating this figure due to the unknown spontaneous mutation rate—the natural occurrence of sudden hereditary changes without external influence—suggested that any figure between 3r to potentially 300r could be plausible. However, both Penrose and Mitchell proposed that a much more accurate prediction of spontaneous mutation rates in humans could be within the specific range of 50r to 100r, with 80r being the most likely figure, echoing Muller's assessment.

The spectrum of estimates regarding spontaneous mutation given by the few geneticists on the panel reflected many of the uncertainties among them that carried over into discussions on the effects of artificially increasing mutation rates in a population. Wadding believed that focussing on genes with major effects—those causing significant, easily noticeable changes or diseases—might overlook the more common minor effect genes, which cause subtler changes that can still impact overall health and fitness. Penrose disagreed with the clean division between major and minor argued for by Waddington, suggesting that many genes that may have initially been identified because of their major effects have also been shown to have important

---

[28] Committee on the Medical Aspects of Nuclear Radiation, Panel of Genetic Effects: Minutes of the First Meeting Health on Wednesday 1st June 1955. PENROSE/2/35/4, Wellcome Collection.
[29] Ibid.

minor effects when present in different genetic combinations. He emphasised the case of heterozygous carriers—individuals with one copy of a recessive gene—which could still exhibit significant health issues, such as thalassemia and sickle-cell anaemia, as examples where minor effects in carriers are widespread and impactful on health across a population. This fundamental disagreement illustrated the lines of division forming between the members of the panel and their attempts to balance the immediate and severe genetic mutations caused by artificially induced mutations with the more pervasive, and often subtle, genetic variations that can combine across generations to influence population health.

It was across the debates concerning the genetic effects of induced mutations on population health across generations where apprehensions about the future health and 'quality' of human populations intertwined with broader eugenic ideas and concerns. Perhaps this is not so surprising, as many of the geneticists who had been either on the Genetical Society's committee or the MRC's panel had been advocates of eugenics in the interwar period, with some continuing to advocate for a form of eugenic thought following the end of the Second World War. In 1952, Cyril Darlington had become a Fellow of the Eugenics Society—a voluntary organisation established in 1909 advocating for the supposed improvement of the human race through preventing the reproduction of those deemed 'unfit'—and was elected its Vice-President in 1954.[30] In the same year he was elected a Fellow of the Society, Darlington published an extensive book on human genetics that included an attempt to articulate a 'biological interpretation' of human history that argued 'civilization' was 'genetically limited'—that 'the capacity to make use of European culture is entirely different in West Africa and in India' (Darlington 1953, p. 410). Similarly, Waddington had become a Fellow of the Eugenics Society in 1948 and Medawar would later join as a Fellow in 1959. Perhaps most prominently was Penrose, who coming from a Quaker background had always struggled with, and often resisted, the relationship between eugenics and his research on genetics and mental health. His research in human and medical genetics during the debates on the hereditary causes of 'feeble-mindedness' during the 1930s through to the post-war period often highlighted the under-appreciated role of environmental factors in the cause of health conditions over simple ideas of direct heredity then prevalent among many advocates of eugenics.[31]

In an undated memo shared with the MRC panel entitled 'non-segregating traits in humans', Penrose highlighted the intersection of eugenic ideas within the theoretical difficulties in quantitatively detecting genetic changes due to an increase in mutation rates.[32] He argued in the memo that such changes would depend on the loci involved—the position on a chromosome where genes are located—in addition to

---

[30] For a full historical account of the history of the Eugenics Society prior to 1945, see Mazumdar (1991). For an account of the post-war intellectual continuity of eugenics in Britain, please see Peace (forthcoming 2026).

[31] For a detailed history of Penrose's complex relationship with the Eugenics Society and interwar research on eugenics and 'feeble-mindedness', see (Mazumdar 1991), 215–231.

[32] Penrose, L. S.: Memo, 'Notes on non-segregating traits in man, i.e., those showing continuous variation'. MRC Committee on the Medical Aspects of Nuclear Radiation, Panel of Genetic Effects (undated). PENROSE/2/35/4, Wellcome Collection.

the individual spontaneous mutation rate. According to Penrose, data on these factors was insufficient and as such largely unknown for most traits in humans. As such, the effects of fresh mutations would only be noticed when the trait they may cause is already rare in a population, making any increase in its occurrence more apparent. From a qualitative perspective, Penrose believed that there was a general lack of agreement about the nature of changes in the genetic structure of human populations through induced mutations. He attempted to highlight and summarise some of the eugenic ideas held by his contemporaries. Some believed 'civilisation' and medical advancements were allowing 'deleterious genes' to become more frequent, potentially leading to declines in physique and intelligence across generations through advanced medical institutions and technologies. Others held the view that populations were in a state of genetic equilibrium, or even experiencing continual evolutionary improvement. Irrespective of which of the two positions were taken regarding induced mutations through radiation exposure, Penrose believed that it was 'agreed by all geneticists' that there were significant eugenic considerations: 'according to the one view it would cause a more rapid decline in quality than was already in progress and, according to the opposite view, it would prevent the slight natural improvement.'[33]

Another undated memo from Waddington entitled 'the importance of genetic factors in "normal" men and animals' also stressed the eugenic concerns about the qualitative impact of radiation on future generations.[34] He outlined first that accepted principles of genetics were often premised on experiments on animals with clear hereditary 'abnormalities', such as 'unusual colours of the eyes or coat, or easily recognisable structural alterations.' He claimed that humans who 'exhibited hereditary conditions as striking as those' would be considered 'freaks'.[35] He feared that if the panel concerned itself primarily with the impact of radiation on producing 'freaks' it would suggest to the readers of their White Paper that genetics was only concerned with rare conditions, whereas in reality the report should stress that genetic factors play a crucial role in the general well-being of the entire population. He also argued the eugenic position highlighted by Penrose, that as societies became more equal, reducing environmental differences between people, such as poverty and education, the importance of hereditary differences would increase.

This was a sentiment that was reflected in the memorandum circulated to the committee by Haldane, titled 'Long Term Effects of Radiation on Human Populations'.[36] His concerns were framed within the context of the eugenic debates of the

---

[33] Ibid.

[34] Waddington, C. H.: Memo, 'The importance of genetic factors in "normal" men and animals'. MRC Committee on the Medical Aspects of Nuclear Radiation, Panel of Genetic Effects (undated). PENROSE/2/35/4, Wellcome Collection.

[35] In a note passed to Himsworth, Penrose stated that he 'absolutely reject[ed], as unscientific ... the term "circus freaks" which is most objectionable.' Penrose, L. S.: Note on MRC 55/669, CNR 55/55. PENROSE/2/35/4, Wellcome Collection.

[36] Haldane, J. B. S.: Memo, 'Long Term Effects of Radiation on Human Populations'. MRC Committee on the Medical Aspects of Nuclear Radiation, Panel of Genetic Effects (undated). PENROSE/2/35/4, Wellcome Collection.

immediate post-war period. Only a few years earlier, in 1947, Haldane had expressly articulated his concerns about the relationship between induced recessive mutations and their potential 'dysgenic effect', highlighting that over centuries each new mutation could significantly alter the health of future generations.[37] In his memo, Haldane repeated the concerns of the 1947 article and its relationship to the primary concern of the MRC panel on the effect of small doses of radiation over long periods, such as those experienced by populations exposed to the fallout of distant nuclear tests. He believed this was an important distinction from groups exposed to massive doses of radiation, such as survivors of nuclear bombs or industrial accidents within the civil uses of atomic energy. He argued that the genetic effects on a large population fall into two classes. The first consists of 'point mutation'—that damage is only caused to an individual if they inherit two copies of the same mutated gene—which was theoretically more likely within closely interbred populations with high rates of consanguineous mating. The second, includes chromosomal changes, where incidence of genetic changes increase with the intensity of the radiation dose given within a short period of time. For example, a dose of 20r experienced by an individual within the space of a few minutes would be more effective in producing chromosomal changes than the same dose of radiation spread over the course of an entire year. He also dismissed the idea entirely that radiation induced mutations could induce 'desirable' traits, arguing that any such imagined benefits would not only be incredibly rare, but they would be substantially outweighed by the overwhelmingly negative effects of other mutations on human health across a population. On this point, Haldane concluded that even small increases in mutation rates could have significant long-term effects. While many mutations may be recessive, and as such not immediately noticeable, their accumulation over successive generations could lead to a substantial lowering of fitness affecting health.

Penrose's other memos that were circulated to the panel attempted to reframe the discussion of the genetic effects of radiation on humans, particularly future generations, away from expressly eugenic concerns of future population 'fitness' and 'quality'. In his 10 May memo, originally drafted for the Genetical Society's committee, entitled 'Gene Mutation in Man', Penrose aimed to underscore the naturally occurring genetic diversity of human populations, stating that humans are a 'polymorphous species' with a vast array of genetic variations.[38] He began by first outlining that this diversity had seemingly arisen through spontaneous mutations over hundreds of thousands of years, many of which were harmful and subject to natural selection, while the rare beneficial mutations were retained over generations over the course of human evolutionary history. This hereditary basis had allowed for these variations in human populations—from blood groups to differences in hair colour and height—but also for the appearance of genetic diseases and defects. In many of these cases, Penrose argued, spontaneous errors in gene copying during cell division and natural sources of background radiation were primary factors, not

[37] Haldane (1947).

[38] Penrose, L. S.: Memo, 'Gene Mutation in Man'. MRC Committee on the Medical Aspects of Nuclear Radiation, Panel of Genetic Effects (undated). PENROSE/2/35/4, Wellcome Collection.

simple direct patterns of inheritance within unique family pedigrees as had been insisted by eugenicists. As such, the concerns about the increase of radiation exposure were even more important outside of a eugenic context. If naturally-occurring background radiation was responsible for many genetic diseases caused by spontaneous mutations, any increase in radiation exposure could proportionally increase the frequency of all mutations. As a course of future action, Penrose proposed the MRC should advocate in its White Paper for studies on the importance of detecting mutations in heterozygous carriers, proposing a comprehensive study of pregnancies in a selected population, together with serological blood surveying and family history studies across multiple generations.

Another of Penrose's memos, titled 'Long Term Effects of Nuclear Radiation on Man', again attempted to shift the focus away from eugenic concerns towards an examination of the long-term health effects of radiation. He emphasised that the problem should be focused on the collective future impact on humanity, acknowledging the distinct natural differences between population groups. He raised two questions: what effect does spontaneous mutations have in humans and how would increased mutation rates induced by radiation manifest themselves? While animal and plant genetics, then the only laboratory-based data collected, provided some good indicators to answer these questions, specific human data was needed. Based upon his own clinical observations that most mutations lead to harmful effects, such as medical conditions including tuberous sclerosis (epiloia) and retinoblastoma, which generally only appear in later generations, in some cases often spanning thousands of years. To underscore the long-term nature of the problem facing the MRC panel, Penrose argued that if Julius Caesar had used nuclear weapons in 55 BC during his invasion of the British Isles, it would only be in the twentieth century that the period of 'maximal genetical effects' would be apparent in the widespread expression of health conditions caused by induced mutations.

The long-term nature of the problem was a particular concern when the panel attempted to integrate one of the only mass human data sets available on artificially induced mutations caused by radiation exposure—the data collected after the atomic bombings of Hiroshima and Nagasaki. The data collected by American and Japanese scientists as part of the Atomic Bomb Casualty Commission had presented few definitive findings, primarily due to a lack of initial data on the natural frequency of hereditary disease in Japan prior to the dropping of atomic bombs.[39] As such, he argued the paucity of data collected on genetic effects of radiation meant that it would not be possible to draw any concrete conclusions, again emphasising the need for comprehensive, long-term, cross-generational surveys. This was a view that was generally accepted by the other panel members during a debate on 29 June 1955 about the nature of the data gathered among those exposed to radiation at Nagasaki and Hiroshima.[40] As most human genes would be recessive, the genetic effects of

---

[39] Four reports from the Atomic Bomb Casualty Commission were shared with the MRC panel, including: Neel (1953), Sutow (1955), Yamazaki (1954), Plummer (1952).

[40] Minutes, Committee on the Medical Aspects of Nuclear Radiation, Panel on Genetic Effects, 29 June 1955. PENROSE/2/35/4, Wellcome Collection.

immediate exposure to an atomic explosion would most probably be hidden during the first generation as it would require an individual to inherit two copies of the same mutated gene, one from each parent. The only conclusion Penrose believed that was allowed by Japanese data was that any additional increase in background radiation caused by the use of a nuclear weapon would likely be harmful.[41]

The long-term nature was again emphasised in the extensive report on experimental data on induced mutation submitted by Auerbach to the MRC panel.[42] Auerbach believed that even though there was no direct experimental data available for humans, it was entirely reasonable to extrapolate from mammalian studies (e.g. mice) that visible birth defects and severe genetic effects in the first generation after radiation exposure would be rare among humans. Echoing both Penrose and Haldane, she predicted that the most significant effects would spread over many future generations, with a high likelihood of subtle, long-term consequences, rather than major, immediate, noticeable impacts. As such, any absence of noticeable effects in the first generation following radiation exposure from a nuclear explosion would not imply any absence of genetic damage. Taken together, Haldane, Penrose, and Auerbach's contributions to the MRC's panel underscored not only the long-term implications of predicting the genetic effects of radiation, but the necessity of long-term, cross-generational monitoring and research. However, Auerbach's prescient concern that the memo would not be 'incorporated in the sense in which it is meant' was not far from wrong. Based on the evidence available from the minutes of the panel's meetings, her contributions received neither discussion nor comment.

8.5   The Geneticists' Critiques of the MRC White Paper

In June 1956, the MRC published their White Paper under the title 'The Hazards to Man of Nuclear and Allied Radiations'. As has been highlighted by Alison Kraft, the White Paper was in effect a 'carefully crafted manifesto for nuclear energy' that trod softly so as not to contradict the UK Government's nuclear policies—as reflected in its opening chapter that 'the future development of our civilization is closely bound up with the exploitation of nuclear energy' (Kraft 2018, p. 82). Much of the White Paper's statements on the genetical effects of radiation were also overtly cautious. For instance, with regard to the potential link between induced mutations caused by radiation and neonatal deaths, stillbirths, and 'congenital malformations', the report stated that these conditions were largely caused by 'the environment of the unborn child', such as 'illnesses and other conditions in the mother' (Medical Research Council 1956, p. 37). As such, the White Paper claimed that even though it was not possible to make 'detailed calculations … it is certain that the total effects of doubling

---

[41] Penrose: Gene Mutation in Man, PENROSE/2/35/4.

[42] Auerbach, C.: Memo, 'Spontaneous and Radiation-Induced Mutation'. PENROSE/2/35/4, Wellcome Collection.

the mutation rates in one generation would be slight.' The conclusions of the White Paper were based in part upon the data collected on populations in Hiroshima and Nagasaki, even though the geneticists invited to the MRC's panel on genetic effects expressed scepticism about the use of this data. Based upon the effects recorded by the Atomic Bomb Casualty Committee, the White Paper claimed that any effect of dropping atomic bombs on the genetic and reproductive health of the survivors and their children was 'not highly significant statistically' (Medical Research Council 1956, p. 41).

In a March 1957 pamphlet, written on behalf of the Medical Association for the Prevention of War, Penrose expressed in no uncertain terms his scepticism about the conclusions of the MRC White Paper.[43] He highlighted the fact that the primary scientific debate centred on estimating the damage caused by artificially contributing to background radiation through testing and weapons use, not disputing altogether the occurrence of radiation induced germ cell mutations. He noted that then-current knowledge suggested that genetic risks to future generations from nuclear tests conducted could lead to an estimated maximum increase of one percent in hereditary diseases. This seemingly marginal increase would translate to roughly 101 cases where there were previously 100 within a population. As such, he contended, the MRC report was technically accurate in stating that the effects of testing a nuclear weapon were, in a sense, 'insignificant', as they would be 'overshadowed' by other factors such as the 'medical and industrial uses of radiation.' However, this technical accuracy paled in comparison to the potential new cases of hereditary disease caused by nuclear testing among a 'generation of many millions', and would become vastly larger when that one percent increase is applied to the 'whole world population' across successive generations.

From a perspective of ethics and human health, Penrose argued that when considered at the level of the population, 'every single case of severe hereditary disease or of cancer in an individual is an 'abominable thing'.[44] On both scientific and ethical grounds, he stated in no uncertain terms that the 'pollution of the atmosphere by nuclear weapons is to be deplored' on the basis of both the 'short and long term biological effects' that these use of these weapons posed. This was a view that he expressed clearly in a BBC interview broadcast on 1 April 1957. In the interview he emphasised the need for the public and legislators to distinguish between two different problems regarding these new weapons: the short-term individual effects and the long-term effects for 'subsequent generations as yet unborn.'[45] When pressed by his BBC interviewer 'from [his] own point of view' whether he felt there were any 'political or military' priorities to conduct nuclear weapons tests outweighed the potential dangers of the use of these weapons, Penrose replied that in every instance

---

[43] Penrose (1957). PENROSE/2/35/1, Wellcome Collection.

[44] Ibid.

[45] Transcription of Broadcast with the BBC, d. 1 April 1957 (As Broadcast). PENROSE/2/35/6, Wellcome Collection.

they 'must be repudiated' as each test threatened the 'use of such bombs against the civil population' and would have dire consequences in the present and for future generations. It was Penrose's belief in the importance of controlling radiation sources to safeguard public health across generations that eventually spurred his advocacy for a comprehensive monitoring system to track radiation exposure levels and the formation of inspection schemes to prevent the unauthorised manufacture of nuclear weapons, a decade prior to the signing of the Treaty on the Non-Proliferation of Nuclear Weapons (NPT) in 1968 (Penrose 1958).

Penrose was not alone in voicing his opposition to the White Paper among his fellow geneticists. Shortly after its publication, Auerbach wrote a critical review of the White Paper for *Nature*, decrying that the report was 'written in a spirit of scientific humility and caution which may well leave the general reader with a feeling that it would be best not to bother much about a subject on which so little is known' (Auerbach 1956a, p. 454). Her criticism returned back to her very first apprehensions expressed in the early days of the Genetical Society's committee meetings; that the knowledge already gained in the field of radiobiology and mutation studies had been ignored. Fearing that as there was no categorical statement that 'any radiation is genetically undesirable' in the White Paper and that it may lead to apathy among the public and legislators, 'who will prefer to get its information second-hand', she suggested that a document needed to be written clearly expressing for a general audience the cross-generational impact of genetical mutations on human health and well-being.

Answering her own call-to-arms, Auerbach published a short pamphlet titled 'Genetics in the Atomic Age', which, as highlighted by Marsha Richmond, was intended as an introduction to the 'basic principles of heredity few had learned in school' (Richmond 2017, p. 363). The pamphlet, containing diagrams and illustrations of the impact of germ-cell mutations on grandchildren, again emphasised the long-term risks of induced mutations through radiation exposure, explaining that low doses may not cause immediate visible damage but may induce genetic mutations that express themselves in the descendants of exposed individuals many generations later (Auerbach 1956b). Echoing Penrose, the pamphlet highlighted that these mutations would contribute to the existing burden of already harmful genes among human populations, leading to an increased rate of genetic diseases over time. She noted that although imagined medical advancements of a distant future may theoretically be able to alleviate some genetic conditions, their accumulation over generations would ultimately lead to unnecessary suffering and deaths. She concluded the book by stressing a common ethical responsibility 'for which posterity will hold us responsible', urging that actions taken in the present may not yield immediately visible results but will be offset across future generations, necessitating international regulations on the uses of radiation. A year later, a German edition of the book was printed, with a title that made clear to the reader to whom they owed an ethical responsibility to safeguard—*weh' dir, daß du ein Enkel bist*; woe to you for being a grandchild (Auerbach 1957).

8.6   Conclusion: The Grandchildren of Monte Bello

In recent decades there has been a reassessment of—some may say reckoning with—the international history of WMD weapons testing and its impact on human health. These have included histories of testing and human experimentation, the effects of low-level radiation on population health, and the ecological effects of testing and its environmental consequences (Moreno 1999; Walker 2000; Schmidt 2015; Higuchi 2020). In May 1983, a survey conducted by the Australian Radiation Laboratory on the Monte Bello archipelago found that although radiation levels at the ground zero sites of the weapon tests were relatively low—and of no risk to tourists—there remained 'radioactive metal fragments' scattered in large numbers across the islands (Cooper 1983). A follow-up survey conducted by the same laboratory in 1990 found that there was still a 'potential hazard to health from residual radioactive contamination on the Monte Bello Islands', particularly a 'potential inhalation hazard' at the site of the fallout plume of one of the weapons tests (Cooper 1990). In addition to these studies of the environmental effects of the UK's weapons tests, there have also been epidemiological studies of Monte Bello veterans and their families potentially exposed to radiation induced mutations. In 1990, the author Sue Rabbitt Roff conducted a survey of the Monte Bello servicemen and their families in collaboration with the British Nuclear Tests Veterans Association and the New Zealand Nuclear Tests Veterans Association (Roff and Holdstock 1999). Based upon the survey of 1,041 servicemen who reported 2,261 live born children and 2,342 grandchildren, there appeared to be correlation between radiation exposure and impacts on genetic and reproductive health across generations. 12.8%, 128 of the veterans surveyed, reported having no children at all, supplemented by a further 50—5%—who reported having no children after returning from the weapons tests. Among this group, accounting for those who did not wish to have children, Roff estimated that the data suggested a 'post-test infertility level of about 15%' (Roff and Holdstock 1999, p. 23). In terms of child health, Roff's survey highlighted that conditions were reported in 893 children, approximately two in five. The conditions reported ranged from early hair loss and dental problems, such as missing or excess teeth, to epilepsy, cerebral palsy, Down's syndrome, muscular dystrophy, cystic fibrosis, and multiple sclerosis. Among the grandchildren of the Monte Bello veterans there were 484 self-reported cases (21%) of health conditions, which ranged again from dental problems to leukaemia, neuroblastomas, spina bifida, neural tube defects, and reproductive problems (Roff and Holdstock 1999, pp. 23–24).

Recent surveys have focussed on both the health impacts of radiation exposure on the veterans of nuclear tests and the genetic risks to the families of these veteran through germ-cell mutations. This research remains contentious. In the early 2000s a collaborative epidemiological survey conducted by the Radcliffe Infirmary, Oxford, and the National Radiological Protection Board, Oxford—which included the influential epidemiologist of smoking, Richard Doll—assessed the health impacts on UK veterans (Kendall 2004). They concluded that the overall mortality and cancer incidence rates among the veterans was similar to control groups, yet there was evidence of increased risk of leukaemia and multiple myeloma. A new study of UK veterans has suggested that further investigation of these findings is necessary, suggesting that even though there are higher observations of solid cancers, such as stomach, prostate, and bladder, and leukaemia among veterans compared to controls, there remain possibilities that the differences may be due to other factors. These may include lifestyle choices, including smoking and diet, exposure to asbestos, and other carcinogens in their day-to-day lives, separate from their participation in the UK's weapons tests (Kendall 2022).

Another recent survey, prompted by concerns among the families of veterans about their health and that of their descendants, conducted in collaboration between the London School of Hygiene and Tropical Medicine and Brunel University London examined 'Family Trios'—consisting of male veterans, their partners, and their children—for instances of increased germline mutations. Even though the survey reported that 20% of veterans self-reported instances of offspring with congenital malformations, compared with the 5% in control groups, the authors of the survey argue that there is no evidence of increased risks to the families of veterans from induced germline mutations caused by the weapons tests at Mone Bello (Moorhouse 2022; Rake 2022; Lawrence 2024). Yet, other surveys, particularly in relation to other international weapons tests, have found alternative data. Surveys conducted in collaboration between the University of Leicester, the Kazakh State National University, the University of Warwick, and the Institute of Health, Kyiv, found that radioactive fallout exposure had increased mutation rates among the populations of the nuclear weapons sites in Kazakhstan and Belarus, and also among post-Chernobyl families in Ukraine (Dubrova et al. 1997, 2002a, 2002b).

Historically, this chapter has highlighted the long-term nature of the debates regarding the relationship between radiation and genetic damage, and the arguments about ethical responsibilities and to whom those responsibilities are directed that emerged among post-war geneticists in the UK. The debates among these geneticists, particularly those such as Lionel Penrose and Charlotte Auerbach, underscored the potential long-term implications on populations genetics caused by radiation. Both Penrose and Auerbach's responses to the Monte Bello tests rested upon an argument that the ethical responsibilities prompted by the long-term nature of radiation exposure necessitated a comprehensive monitoring of radiation across generations and its effects. Throughout both their work on the MRC genetics effect panel and in their public outreach after the publication of the 1956 White Paper, they continually called for meticulous tracking of radiation levels, international regulations to control radiation sources, and emphasised that the health of future generations depended on scientific oversight and ethical governance.

Beyond the intertwined relationship between genetics research on mutations and the ethical responsibilities to safeguard the health of future generations, the debates on radiation exposure and genetic damage also reveal insights into the disciplinary changes within the field of human genetics in the UK, particularly in its attempts to overcome the legacy of eugenics. It is perhaps unsurprising that eugenic ideas were able to find a place within these debates, as research on human genetics moved outside of its disciplinary bounds and was coupled with speculations about the future in fields as diverse as ethics, public health policy, and demography. Yet, as Penrose's memos to the MRC panel reveal, there was a considered attempt to shift away from these eugenic ideas, such as ideas about the need to curtail the reproduction of certain families to affect qualitative changes among future generations, highlighting instead the natural occurrences of human diversity and the need for the discipline to reorient towards research on human health and the medical applications of genetics. The opposition to WMDs among post-war British geneticists was driven by research on the hereditary link between induced germ-cell mutations and concerns about their potential impact on population health. Their ideas about the cross-generational implications of such mutations were not limited to the immediate victims of these weapons but extended to their descendants and the generations belonging to an imagined irradiated future. Based upon all the knowledge that was available to them, it appears that their sincere hope was that if these responsibilities were taken seriously, they would never have to utter the words 'woe to you for being a grandchild.'

References

Alexander ML (1954) Mutations rates at specific autosomal Loci in the mature and immature germ cells of drosophila melanogaster. Genetics 39(3):409–428

Alexander ML, Wilson S (1955) Radiation in the developing germ cells of drosophila virilis. Proc Natl Acad Sci 41(12):1046–1057

Arnold L (2001) Britain and the H-bomb. Palgrave, Basingstoke

Arnold L, Smith M (2006) Britain, Australia and the bomb: the nuclear tests and their afterman. Palgrave, Basingstoke

Auerbach C (1945) The induction by mustard gas of chromosomal instabilities in drosophila melanogaster. In: Proceedings of the royal society of Edinburgh. Sect B: Biol Sci 62(3):307–320

Auerbach C (1956a) Biological hazards of nuclear and other radiations. Nature 178:453–454

Auerbach C (1957) Weh' dir, daß du ein Enkel bist. Kosmos, Stuttgart

Auerbach C, Robson JM (1944) Production of mutations by Allyl Isothiocyanate. Nature 81:154

Auerbach C, Robson JM (1946) Chemical production of mutations. Nature 157:302

Auerbach, C.: Genetics in the Atomic Age. Oliver and Boyd, Edinburgh and London (1956b).

Beale GH, Auerbach C (1995) Biogr Mem Fellows R Hist Soc 41:20–42

Burkett J (2013) The campaign for nuclear disarmament and changing attitudes towards the Earth in the nuclear age. Br J Hist Sci 45(4):625–639

Carlson EA (1981) Genes, radiation, and society: the life and work of H. J. Muller. Cornell University Press, Ithaca

Cooper MB et al (1983) The radiological status of the monte bello Islands; May 1983. Commonwealth Department of Health

Cooper MB et al (1990) Radiological hazard assessment at the Monte Bello Islands. Commonwealth Department of Health

Crew FAE (1969) Recollections of the early days of the genetical society. The Genetical Society-The First Fifty Years. Oliver and Boyd, Edinburgh, pp 9–15

Darlington CD (1948) The plasmagene theory of the origin of cancer. Br J Cancer 2(2):118–126

Darlington CD (1953) The facts of life. Allen and Urwin, London

Darlington CD (1955) The chromosome as a physico-chemical entity. Nature 176:1139–1144

Demarini DM (2020) The mutagenesis moonshot: the propitious beginnings of the environmental mutagenesis and genomics society. Environ Mol Mutagen 61(1):8–24

Demerec M (1947) Mutations in drosophila induced by a Carcinogen. Nature 159:604

Demerec M (1948) Mutations induced by carcinogens. Br J Cancer 2:114–117

Demerec M, Latarjet R (1946) Mutations in bacteria induced by radiations. Cold Spring Harb Symp Quant Biol 11:38–50

Dubrova YE et al (1997) Further evidence of elevated human Minisatellite mutation rate in Belarus eight years after the Chernobyl accident. Mutat Res 381(2):267–278

Dubrova YE et al (2002a) Nuclear weapons tests and human germline mutation rate. Am Assoc Adv Sci 295(5557):1037

Dubrova YE et al (2002b) Elevated Minisatellite mutation rate in post-Chernobyl families from Ukraine. Am J Hum Genet 71(4):801–809

Eiges NS (2013) The historical role of Iosif Abramovich Rapoport in genetics. Further studies using chemical Mutagenesis. Russ J Genet: Appl Res 3:316–324

Gray LH et al (1953) The concentration of oxygen dissolved in tissues at the time of irradiation as a factor in radiotherapy. Br J Radiol 26(312):638–648

Greene G (2017) The woman who knew too much: alice Stewart and the secrets of radiation. University of Michigan Press, Ann Arbor, pp 67–112

Haldane JBS (1935) The rate of Spontaneous Mutation of a human gene. J Genet 31:317–326

Biological Hazards of Radiation (1955) Nature 4460:695−697

Higuchi T (2020) Political fallout: nuclear weapons testing and the making of a global environmental crisis. Standford University Press, Redwood City

Huxley J (1942) Evolution: the modern synthesis. George Allen and Urwin, London

Kendall GM (2004) Epidemiological studies of UK test veterans: I. General description. J Radiol Prot 24(3):199–217

Kendall GM, Little MP (2022) The new study of UK nuclear test veterans. J Radiol Prot 42:020101

Kraft A (2018) Dissenting scientists in early cold war Britain: the "Fallout" controversy and the origins of Pugwash, 1954–1957. J Cold War Stud 20(1):58–100

Kraft A et al (2018) The Pugwash conferences and the global cold war: scientists, transnational networks, and the complexity of nuclear histories. J Cold War Stud 20(1):4–30

Latarjet R (1948) Intracellular growth of bacteriophage studied by roentgen irradiation. J Gen Physiol 31(6):529–546

Lawrence KJ (2024) M-FISH evaluation of chromosome aberrations to examine for historical exposure to ionising radiation due to participation at British nuclear test sites. J Radiol Prot 44:011501

Mayr E (1982) The growth of biological thought: diversity, evolution, and inheritance. Harvard University Press, Cambridge, pp 738–743

Mazumdar P (1991) Eugenics, human genetics and human failings: the eugenics society, its sources and its critics in Britain. Routledge, London

Medical Research Council (1956) The hazards to man of nuclear and allied radiations, vol 37. Her Majesty's Stationary Office, London

Moorhouse AJ et al (2022) No evidence of increased mutations in the germline of a group of British nuclear test veterans. Sci Rep 12(10830)

Moreno J (1999) Undue risk: state secret experiments on humans. W. H. Freeman and Company, New York

Morgan KZ (1953) Maximum permissible dose from ionizing radiation. Am Hyg Assoc Q 14(1):15–22

Muller HJ (1927) Artificial transmutation of the Gene. Science 66(1699):84–87

Muller HJ (1929) Parallel cytology and genetics of induced translocations and deletions in Drosophila. J Hered 20(6):287–298

Muller HJ (1930) The frequency of translocations produced by X-rays in Drosophila. Genetics 15(4):283–311

Muller HJ (1941) Induced mutations in Drosophila. In: Cold Spring Harbor Symposia on quantitative biology, vol 9. Cold Spring Harbor Laboratory Press, pp 151–167

Muller HJ (1950) Our load of mutations. Am J Hum Genet 2(2):111–176

Muller HJ (1954) The manner of dependence of the "permissible dose" of radiation on the amount of genetic damage. Acta Radiol 41(1):5–20

Muller HJ (1955) Radiation and human mutation. Sci Am 193(5):58–71

Murphy DP, Goldstein L (1929a) Etiology of the Ill-health of children born after maternal pelvic irradiation. Am J Roentgenol 22:207–219

Murphy DP, Goldstein L (1929b) The relation of maternal pelvic disease to deformities in the Newborn. Surv, Gynaecol, Obstet 49:804–805

National Committee on Radiation Protection (US) (1954) Permissible dose from external sources of ionizing radiation. US Government Printing Office

Neel JV, Schull WJ et al (1953) The effect of exposure to the atomic bombs on pregnancy termination in Hiroshima and Nagasaki. Science 118(3071):537–554

Nickerson S (2013) Taking a stand: exploring the role of the scientists prior to the first Pugwash conference on science and world affairs, 1957. Sci Can 36(2):63–87

Peace D (2021) From Galtonian Eugenics to biosocial science: the intellectual origins and policy implications of quantifying hereditary in interwar and post-war Britain (University of Kent, PhD, unpublished, 2021)

Peace D (2024) "A child of misfortune": Eugenics and children reception centres in post-war Britain. The Balkanistic Forum 33(3):55–85

Peace D (forthcoming, 2026) Britain's genetic empire: a century of eugenics. Bloomsbury, London

Penrose LS (1938) A clinical and genetic study of 1,280 cases of mental defect. Medical Research Council, HM Stationary Office, London

Penrose LS (1938b) Some genetical problems in mental deficiency. J Meteorol Soc Jpn 84(352):693–707

Penrose LS (1944) Mental defect. J Meteorol Soc Jpn 90(378):399–409

Penrose LS, Gunther M (1935) The genetics of Epiloia. J Genet 31:413–430

Penrose LS, Haldane JBS (1935) Mutation rates in man. Nature 135:907–908

Penrose LS (1958) Radiation, public health, and inspection for disarmament. Inspection for disarmament. In: Melma S (ed). Columbia University Press, New York, pp 100–108

Plummer G (1952) Anomalies occurring in children exposed in utero to the atomic bomb in Hiroshima. Pediatrics 10(6):687–693

Pontecorvo G (1968) Hermann Joseph Muller, 1890–1967. Biogr MemS Fellows R Soc 14:348–389

Rader KA (2006) Alexander Hollaender's postwar vision for biology: oak ridge and beyond. J Hist Biol 39(4):685–706

Rake C et al (2022) British nuclear test veteran family trios for the study of genetic risk. J Radiol Prot 42(2):021528

Richmond ML (2017) Women as public scientists in the atomic age: Rachel Carson, charlotte Auerbach, and genetics. Hist Stud Nat Sci 47(3):349–388

Roff SR, Holdstock D (1999) Mortality and morbidity of members of the British nuclear tests veterans association and the New Zealand Nuclear tests veterans association and their families. Med Confl Surviv 15(1):1–52

Sachse C (2018) The max plank society and Pugwash during the cold war: an uneasy relationship. J Cold War Stud 20(1):170–209

Schmidt U (2015) Secret science: a century of poison warfare and human experiments. Oxford University Press, Oxford and New York

Skin Lesions Produced by Mustard Gas (1918) Ann Surg 68(5):562−563

Stewart A et al (1958) A survey of childhood malignancies. BMJ 1(5086):1495–1508

Stone RS (1952) The concept of a maximum permissible exposure: carmen lecture. Radiobiology 58(5):639–661

Sturtevant AH (1955) The genetic effects of high-energy irradiation of human populations. Eng Sci 18:9–12

Sutow WW, West E (1955) Studies on Nagasaki (Japan) children exposed in utero to the atomic bomb; a roentgenographic survey of the skeletal system. Am J Roentgenol Radium Ther Nucl Med 74(3):493–499

Thoday JM (1954) Radiation-induced chromosome breakage, desoxyribose-nucleic acis synthesis and the mitotic cycle in root-meristem cells of Vicia Faba. New Phytol 53(3):511–516

Thoday JM, Read J (1947) Effect of oxygen on the frequency of chromosome aberrations produced by X-rays. Nature 160:608

Thoday JM et al (1946) Types of chromosome structural change induced by the irradiation of Tradiscantia microspores. J Genet 47:113–136

Thomlinson RH, Gray LH (1955) The histological structure of some human Lung Cancers and the possible implications for radiotherapy. Br J Cancer 9(4):539–549

Vucinic S et al (2015) Occupational and environmental mustard exposure, prevention and chemical weapons convention. In: Basic and clinical toxicology of mustard compounds, pp 359–387

Waddington CH (1938) Effect of X rays on hereditary mutation. BMJ 4028:642–643

Walker JS (2000) Permissible dose: a history of radiation protection in the twentieth century. University of California Press, Berkeley

Yamazaki JN et al (1954) Outcome of pregnancy in women exposed to the atomic bomb in Nagasaki. Am J Dis Child 87(4):448–463

Yanovskiy M, Socol Y (2023) Geneticist Iosif Rapoport: the Scientist versus the State. J Am Physicians Surg

David Peace is a Post-Doctoral Research Associate at the Centre for the Study of Health, Ethics, and Society at the University of Hamburg and is part of the ERC-funded "Taming the European Leviathan" project. His work explores the historical relationship between developments in quantitative analysis and the medical biosciences across the twentieth century, specialising in the influence of both eugenics and human genetics in population health policies. His work is embedded within various historiographical traditions, including the history of ideas, the history of science and medicine, and social and political history. He is co-editor of Leo Alexander at the Nuremberg Doctors' Trial (Franz Steiner, 2024) and Gender, Sex, and Medicine in Post-War Europe: Beyond Binaries (Bloomsbury, forthcoming 2026), and author of Britain's Genetic Empire: A Century of Eugenics (Bloomsbury, forthcoming 2026).

Part III
Participation of Civil Society in CWC Implementation

Chapter 10
Thirty Years of the Chemical Weapons Convention. How Far We've Come: The Colorado Experience

Irene Kornelly

Abstract This chapter reflects on Colorado's eight-decade involvement with chemical weapons and chronicles the state's pivotal role in implementing the Chemical Weapons Convention over the past thirty years. Focusing on the Pueblo Chemical Depot, it details the community-led resistance to incineration-based destruction methods and the eventual adoption of a neutralization and biotreatment approach under the Assembled Chemical Alternatives (ACWA) program. The chapter traces the complex, decades-long journey—from initial public distrust and environmental concerns to successful demilitarization of over 780,000 mustard agent-filled munitions. Key challenges included securing sustained funding, navigating environmental permitting, and adapting operations during the COVID-19 pandemic. The successful completion of the destruction in 2023 underscores how government transparency, public engagement, and innovative engineering can achieve both safety and environmental stewardship. Looking ahead, the chapter highlights the challenges and opportunities of depot cleanup and redevelopment, emphasizing a long-term vision to transform the site into an economic and community asset for future generations.

Keywords Pueblo Chemical Depot · Assembled Chemical Weapons Alternatives (ACWA) program · Depot cleanup · Redevelopment

For over eighty years, the State of Colorado has been involved with the chemical warfare program in the United States. In 1942 both the Rocky Mountain Arsenal and the Pueblo Ordnance Depot Activity constructed as a part of the government's efforts during World War II. The Arsenal began with the manufacture of mustard agent and later switched to the manufacture of Sarin agent. The Arsenal was an active facility into the early 1980s. The cleanup of the Arsenal was and still is a controversial issue. While officially closed in 1992, cleanup was not complete until 2010 and there are

I. Kornelly (✉)
Colorado Citizens' Advisory Commission for Chemical Weapons Demilitarisation U, South Denver, CO, USA
e-mail: ikornelly@pcisys.net
URL: https://cdphe.colorado.gov/hm/cocac

© The Author(s) 2026
B. Friedrich et al. (eds.), *Thirty Years of the Chemical Weapons Convention (CWC)*,
https://doi.org/10.1007/978-3-031-98854-7_10

still restrictions on water usage, residential and agricultural development. Most of the 20,000 acres located just east of Denver are managed now by U.S. Fish and Wildlife for wildlife refuge and minimal recreation.

With this legacy in mind, the Pueblo community was faced with the destruction of over 780,000 mustard agent-filled weapons in their backyard. The U.S. Army initially proposed to destroy the munitions with an incineration technology that was used at other stockpile sites in the country. This proposal was met with fierce opposition from citizens. Their concerns included pollution of air, water and soil with potentially toxic chemicals, loss of land value and the loss in value of agricultural crops and animal production. The community also had other requests that the chosen technology needed to apply. They wanted no incineration, a technology that could recycle the water used back into the plant and no off-site shipment of water or explosives. In other words, everything had to be completed on site.

Congressional legislation was passed in 1996 that established the Assembled Chemical Weapons Alternative (ACWA) program to look for alternatives to incineration for the Pueblo Chemical Depot in Colorado and Blue Grass Army Depot in Kentucky. The purpose of the ACWA program was to identify and demonstrate at least two technologies as alternatives to incineration. A national dialogue ensued which included representatives from all stockpile sites, environmental groups, NGOs, the Department of Defense and potential technology providers. Meetings were held quarterly with numerous small group meetings, site visits to review technologies and public presentations in Blue Grass and Pueblo. Many questions were asked, but the real question was, "Can chemical weapons be destroyed while preserving the environment and keeping workers and residents safe?" If not, what could be sacrificed in order to destroy the weapons. Six years later in 2002, the Department of Defense approved the technology of neutralization followed by biotreatment as the method to be used to destroy the stockpile at the Pueblo Chemical Depot.

The road to destroy the first weapon was a bumpy one. The decision to use the neutralization/biotreatment methodology was not universally accepted and there were some nay-sayers both in the Pentagon and the community. There was concern that the use of an untried technology could be more expensive and potentially less safe. The community had lost confidence in the Army and the representatives of the Department of Defense. Trust from both sides had to be established. Transparency was created so that everyone would be comfortable with the process. In the meanwhile, the Bechtel National Team (the chosen contractor) was designing a world class facility to destroy these munitions. Groundbreaking for the Pueblo Chemical-Agent Pilot Plant (PCAPP) was held in 2004 and in 2007 the final design was approved by the Department of Defense and construction began. However, the community had to give up their hope that the uncontaminated explosives would be destroyed on site. They were sent to the Anniston Army Depot for destruction in the Static Detonation Chamber (SDC).

Construction and environmental permitting continued in parallel. The workforce was hired and trained. When systemization of the facility began in 2010, construction was not finished and the complex environmental permitting was still on-going. The workforce began testing equipment and preparing the facility to destroy the munitions so that when construction and permitting were complete, operations could begin. The workers rigorously tested the plant over the next several years, fixing leaks in tanks, hoses and O-rings, repairing cracked concrete and making electrical and welding repairs. Like any new manufacturing plant, the design on paper did not always match reality.

Actual destruction of munitions began in 2015 with the use of the U.S. Army's Explosive Destruction System (EDS), which destroyed 951 items that were unsuitable for processing in the main PCAPP facility. The EDS completed this campaign in February 2016.

In September 2016, the PCAPP facility began 155 mm munitions destruction on a pilot test basis. The process used neutralization with hot water and sodium hydroxide, followed by biotreatment. This technology would eventually destroy the majority of the weapons. In July 2019, the facility was considered fully operational with all environmental permits approved. The 155 mm campaign ended in September 2020 and workers immediately began the changeover of the machinery to accommodate the 105 mm projectiles. That campaign began in December of 2020.

Soon after the 105 mm projectile campaign began, officials at PCAPP became concerned that the 4.2-inch mortars could not be safely destroyed in the main facility and if they could, it would be a slow process. The primary issue was draining the mortar shells. The agent would not sufficiently drain out of the shell. After a great deal of discussion with the public, the decision was made to use the Static Detonation Chamber (SDC) to destroy the 4.2-inch mortars. The public's main concern with this technology was that it was an incinerator and would have to be permitted as such. Other technologies were considered, but all were considered to have a slow destruction rate. The 2023 deadline was looming.

Three SDC units were constructed at PCAPP and destruction began in February 2022. Thus, the third campaign began at PCAPP even before the 105 mm projectile campaign was complete. As the 105 mm projectile campaign came to an end, a program was initiated to re-engineer the Cavity Access Machines to allow the mustard agent to be suctioned from the mortars, as opposed to drained. Thus 4.2-inch mortars also were destroyed in the main PCAPP facility.

With great fanfare, the last munition was destroyed in Pueblo on June 22, 2023. During the long process, injuries to workers were minimal, the environment was protected and the community was safe. Yes, chemical weapons could be successfully destroyed without destroying the environs!

There were numerous challenges in destroying the weapons. (1) Ensuring that sufficient funding for the project was available on an annual basis. This was particularly true in the beginning of the project when funding was often syphoned off to other destruction sites. (2) Permitting of the site for construction and operations was often long and tedious. Initial permitting was done through a Research, Development and Demonstration program which allowed for simultaneous construction, pilot testing and permitting. This was potentially risky for all involved, but it worked. (3) The global pandemic for COVID-19 added an additional challenge to the destruction program. Destruction of the munitions continued throughout the pandemic. While many of the office personnel worked from home or came into the office on a rotating basis, the workers involved in the destruction came in everyday, were tested for COVID-19 and went to work. Some workers did come down with COVID-19, but munition destruction continued. One negative occurrence was that the Center for Disease Control required that a substantial portion of the hydrolysate on site at the beginning of the pandemic be shipped off-site. This was the only major shipment of hydrolysate from PCAPP. All other hydrolysate was processed on site and water was recycled back into the facility.

The most important lessons learned at the site, which will hopefully be used at other controversial government sites, is that communities and the government can work together. The government can be transparent with the public without sacrificing national security. With education and personal interaction, the public can assist in the making of wise decisions. The workforce employed at PCAPP was a problem-solving team. They found a way to work through many challenges, make production more efficient and always keep safety in mind.

The future for the Pueblo Chemical Depot is bright. Closure of the plant will take 3–4 years and will encompass the disposal of all secondary wastes, decontamination and decommissioning of facilities and equipment and the demolition of buildings. Throughout the closure process, the safety of the workers, the public and the environment will continue to be a high priority.

The biggest challenge facing the Army, the Department of Defense and the community will be to cleanup all 23,000 acres of the depot to make the property available for development. Development will enhance the property tax base for Pueblo County and the State of Colorado, and create jobs within the community for many who will be losing their jobs as a result of the closure of the demilitarization facility. A recent poll of workers at the depot indicated that they would prefer to live in Pueblo if employment could be found.

Finding the money for the cleanup will also be a challenge. The cleanup must minimize or eliminate, if possible, the need for continuous maintenance. Cleanup should prevent the escape of hazardous wastes and constituents, leachate, and contaminated run-off into the ground water, surface water or the atmosphere. Cleanup has already been completed on many parts of the Depot and is underway in other parts. It is anticipated that the first parcel of land was turned over to the community in late 2024.

Community groups are already actively engaged in planning for the future. Pueblo-Plex, the organization established by the State of Colorado to receive and manage this property, has developed a plan for the area. A large railroad test facility, MxV, with a mission to test high speed rail for commercial and passenger use, is already constructing new facilities and remodeling old Army buildings for their use. Small manufacturing companies have moved to empty facilities for purposes of being closer to the Pueblo companies that they supply. A rail car transfer company has operated at the depot for many years, making use of the rail lines already at the depot. There are literally thousands for items of equipment that never were contaminated that could be turned over to the community for use. Utilities, water tanks, offices, warehouses, and a laboratory could also be reused. The potential is infinite.

The time needed to incorporate the land and buildings at the Depot into the development plan will extend into decades. Few if anyone currently working at the Depot will see this new completed activity. Just as the completion of the demilitarization was for future generations, so the redevelopment of the Depot will be for our children, grandchildren and great-grandchildren. Together we must all work today to build a better future for tomorrow.

Irene Kornelly has worked for the past twenty-five years towards chemical weapons destruction. She has participated in conferences in the United States and Europe emphasizing the importance of public involvement in the destruction and cleanup process to build trust and understanding between stakeholders. Irene is the chairman of the Colorado Citizens' Advisory Commission for Chemical Demilitarization.

Chapter 11
Civil Society Participation and Collaboration to Meet the Chemical Weapons Convention Goals: How Far We Have Come?

Deborah Klein Walker

Abstract Citizen participation and the involvement of non-governmental organizations (NGO) have been important in developing policy and implementing safe and effective elimination of chemical weapons stockpiles to meet the goals of the Chemical Weapons Convention (CWC). Because of opposition to incineration of stockpiles in the United States, citizen advisory boards were established in legislation in the 1990s to promote collaborative decision-making related to environmentally sound cleanup of chemical weapons and toxic chemicals. Although the chemical industry and scientists have been included in the work of the Technical Secretariat of the CWC since it began in 1997, participation of other members of civil society grew more slowly over time. After support from two Director Generals of the Organization of Prohibition of Chemical Weapons (OPCW), the CWC Coalition was established in 2009–10. Since the Third Review Conference in 2013, Coalition members have presented plenary statements and coordinated side events at each of the annual Conference of State Parties (CSP) and Five-Year Review Conferences. Financial and other support is required to sustain strong NGO participation from all parts of the world and ensure the participation of NGOs in important CWC issues during and between the CSPs. Recommendations are presented for future robust participation of NGOs and other members of civil society in OPCW deliberations and implementation of the CWC in National Authorities to sustain a world free of chemical weapons.

Keywords Citizen participation · Non-government organizations · Civil society · Outreach · Chemical weapons · OPCW · CWC

Never doubt that a small group of thoughtful citizens can change the world.
  Indeed, it's the only thing that ever has.
  —Margaret Mead

D. K. Walker (✉)
School of Public Health, Boston University, Boston, MA, USA
e-mail: deb.publichealth@gmail.com

American Public Health Association, Washington, DC, USA

© The Author(s) 2026
B. Friedrich et al. (eds.), *Thirty Years of the Chemical Weapons Convention (CWC)*,
https://doi.org/10.1007/978-3-031-98854-7_11

11.1   Introduction

The Chemical Weapons Convention (Convention on the Prohibition of the Development, Production, Stockpiling, and Use of Chemical Weapons and on Their Destruction) was open for signature in 1993 and entered into force in 1997. Its roots go back to the 1980s, including the 1989 Memorandum of Understanding between the government of the Union of Soviet Socialist Republics and the government of the United States of America regarding Bilateral Verification Experiment and Data Exchange Related to Prohibition of Chemical Weapons (Wyoming Papers) and the 1990 statements between the United States and the Union of Soviet Socialist Republics on Measures to Facilitate the Multilateral Convention on Banning Chemical Weapons, on Destruction and Non-Production of Chemical Weapons, and on Chemical Weapons Nonproliferation.[1]

Although the chemical industry and academic scientists were involved in the creation and implementation of the Convention, citizens and non-governmental organizations (NGOs) were not officially embraced until ten years after the Convention was brought into force. Citizen groups in the United States, however, were involved in the destruction of chemical weapons since the early 1990s. This paper will review the historical roots of citizen engagement in chemical weapons destruction in the United States and Soviet Union in the 1990s, and the involvement of NGOs in the deliberations of the Organization for Prohibition of Chemical Weapons (OPCW) annual meetings of States Parties and Five-Year Review Conferences in The Hague since 1997. Finally, we will summarize lessons learned and recommendations for the future for civil society participation in the OPCW and state party national efforts.

11.2   History of Citizen Participation in Chemical Weapons' Destruction

Citizen participation related to chemical weapons began in the United States before the CWC was open for signature in 1993. Activists in several states became alarmed around the issue of how to destroy the US stockpiles; citizens near the weapon stockpiles objected to eliminating the weapons by incineration, which was the preferred method of the US Army. Because of the intense opposition to incineration, the United States established Restoration Advisory Boards (RABs) to provide input into military base remediation, and Chemical Demilitarization Citizens' Advisory Commissions (CACs) were founded in 1992 to provide a forum for citizen and state concerns during the demilitarization of chemical weapons arsenals in the United States. Following the establishment of the citizen participation mechanisms, citizen advisory committees were set up in over 330 military and former military sites in the 1990s in the United States (Walker 1999). Similar committees also exist in European

---

[1] Path to the CWC https://www.cwccoalition.org/documents-on-disarmament-cwc/, last accessed 2024/7/4.

countries. The establishment of these committees can "empower local communities, overcome past mistrust and confrontational tactics by both citizens and the military, promote collaborative and cooperative decision-making, facilitate timely and cost-effective solutions, and allow safe, publicly accepted and environmentally sound cleanup of toxic pollution and chemicals, and in the longer run provide for sustainable conversion and redevelopment of valuable military lands" (Walker 1999, p. 3).

When the United States ratified the Chemical Weapons Convention (CWC) in 1997, it took on the legal obligation and the complex technical and engineering task of destroying its massive Cold War-era chemical weapons stockpile consisting of some 31,500 US tons (63,000,000 pounds) of deadly chemical agents, much of which were loaded in various forms of munitions. Because of the activism of citizens across the country, language about citizen participation in public decision-making was included in 1996 in the establishment of the Assembled Chemical Weapons Assessment (ACWA) program to establish research, development and demonstrate alternative, non-incineration technologies of the chemical weapons demilitarization process.[2]

Although a CAC was set up at each of the nine US stockpiles, two continue today at the last two sites of chemical weapon stockpiles: Pueblo, Colorado, and Blue Grass, Kentucky. The chairs of these two CACs. (Irene Kornelly, chair of the Colorado Citizens' Advisory Commission, and Craig Williams, co-chair of the Kentucky Citizens' Advisory Commission) have led their committees for over two decades to ensure the safe and effective clean-up using non-incineration methods. After years of delay, all U.S. declared chemical weapons stockpiles were irreversibly and verifiably destroyed in Pueblo in June 2023 and in Blue Grass in July 2023, two months ahead of the September 2023 deadline set by the Organization for the Prohibition of Chemical Weapons (OPCW), and six months ahead of the Congressional deadline of December 2023 (Kornelly 2023; Kornelly and Williams 2023; Phillipps and Ismay 2023). The event is a major milestone in disarmament: the United States was the last declared stockpile possessor state to complete its safe and permanent demilitarization of chemical weapons.

The Cooperative Threat Reduction (CTR) Act, known as the Nunn-Lugar Program, supported Russia in the destruction of their seven chemical weapon stockpiles (Walker 2010, 2016). None of the Russian stockpiles was destroyed using incineration as a primary destruction technology. As part of the effort to support Russia's destruction of chemical weapons, non-governmental organizations in Europe (Green Cross International), in the United States (Global Green USA) and Russia (Green Cross Russia) provided support for a citizen advisory committee at each Russian site, funded largely by the US CTR Program. These committees operated over more than a ten-year period (1996–2010) by providing advice and facilitating discussions

---

[2] History of PEO ACWA (Assembled Chemical Weapons Assessment): https://www.peoacwa.army.mil/about-peo-acwa/history-of-peo-acwa/, last accessed 2024/7/4.

between the government and citizens of the sites about the planning and implementation of the destruction process. The committees were disbanded by Russia shortly before the destruction process was completed in 2017.[3]

11.3 Non-Governmental Civil Society Participation in the OPCW

11.3.1 Early Years of the CWC

Representatives of NGOs from Europe and the United States attended the first meetings of the States Parties, but had no voice or role in the proceedings. Representatives sat in the back of the plenary hall with no access to written statements except through their country delegation. There were no side meetings of NGOs inside or outside of the meeting venue. Before the CWC Coalition was officially formed in 2010, Director General Rogelio Pfirter held luncheons of 15–20 NGO leaders to plan implementation of stronger NGO participation at the OPCW meetings. Based on Pfirter support, OPCW supported an NGO meeting in the Peace Forum in the Peace Palace in 2003. NGOs organized the agenda and made presentations in several main sessions related to CWC implementation, especially demilitarization efforts in the US and Russia. OPCW provided food and buses for States Parties; consequently, the event was successful and attended by over 250 individuals. The number of NGOs attending the meetings grew each year at the CSP meetings, using the "snowball technique" whereby a member would recommend and invite others around the Globe to participate. In subsequent years, OPCW held an Academic Forum and Industry Forum in 2007–2008.[4]

11.3.2 Establishment of the CWC Coalition

Discussions between Director General Pfirter and Paul Walker, one of the NGO leaders, about establishing a CWC Coalition came to fruition in 2009. Pfirter announced its launch in his introductory comments for the Opening Session of CSP-14 in November 2009[5]:

> Whilst addressing universality, I am extremely pleased to note that on the sidelines of this Conference, a group of more than 30 NGOs will hold a two-day meeting to discuss the establishment of a formal NGO coalition to support the Chemical Weapons Convention.

---

[3] Personal communication with Paul Walker, September 30, 2023 (2023).

[4] Personal communication with Paul Walker, September 30, 2023 (2023).

[5] Opening statement by the Director General to CSP-14, November 2009, pgs. 17–18 (2009). https://www.opcw.org/sites/default/files/documents/CSP/C-14/en/c14dg13_en.pdf.

This is the largest number of NGOs ever to attend an OPCW event and they include policy institutes, think-tanks, training centers, and academic institutions representing nearly every region of the world. The primary goal of the proposed coalition is to supplement the efforts of the OPCW with focused action by civil society to achieve universality of the Convention and complete chemical demilitarization, as well as to attain comprehensive national implementation of the Convention globally.

This initiative, which I applaud, is in accordance with the recommendations of the Second Review Conference, which underlined the importance of the involvement of all stakeholders and encouraged the development of such cooperation, with due regard to the role and responsibilities of States Parties and their National Authorities, and to do so on the broadest possible geographical basis. I therefore wish the NGOs every success in this very worthy undertaking and to reconfirm to the Conference that the Secretariat will provide its assistance to the NGO meeting.

The CWC Coalition was officially established in 2010 with over 40 NGOs attending the CSP-15. The Coalition evolved from an informal global network of NGOs working together to encourage and promote participation of civil society, industry, academia, and NGOs at the annual Conference of States Parties (CSP) to the Chemical Weapons Convention. The goals of the Coalition are to support safe and timely elimination of chemical weapons, prevent the misuse of chemicals for hostile purposes and promote peaceful chemical use.[6]

Paul Walker, who serves as the Coalition Coordinator, sought funds for operations and travel of Coalition members from States Parties, global organizations and foundations. Originally housed under Green Cross Switzerland in Zurich, Switzerland, the Coalition received funding from the Swiss Green Cross, the United States Ambassador to the CWC, and the Foreign Ministries of Norway and Germany. Although the amount was small, the funding provided for some travel to the CSPs, a CSP meeting recorder and limited administration time.

The Coalition organized one or more NGO Open Forums and/or side events with a planned theme during the week-long annual Conference of States Parties in The Hague. Since 2010, the Coalition has coordinated approximately 15 presentations at 3–4 side events, one Open Forum and 10 plenary statements each year among the NGOs attending the Conference of States Parties and Five-Year Review Conferences. These "side events" promote NGOs in presenting their work relevant to the CWC and foster interchanges with Conference delegates, the Technical Secretariat, and other stakeholders. The events also foster debate outside the normal dialogue among States Parties. These Forums are held in the OPCW headquarters, the World Forum Convention Center where the plenary sessions are held, or in other facilities in the vicinity of the OPCW headquarters (Ghionis 2023).[7]

The NGO Open Forum has held discussions on topics such as the challenges of chemical weapons destruction in the USA and Russia, the impact of chemical weapons on the health and well-being of survivors, regulation of riot control agents,

---

[6] https://www.cwccoalition.org/, last accessed 2024/7/4.

[7] Personal communication with Paul Walker, September 30, 2023.

destruction of buried and sea-dumped chemical weapons, and the mechanism of death in chlorine gas exposure. The United Nations 1540 Committee's Group of Experts in 2013 stated that *"The Open Forum [...] showed the range of civil society involvement and contribution to national CWC implementation or associated activities" and was "extremely rich in technical and substantive information [leading] to dynamic and meaningful discussions."*[8]

NGOs were allowed to speak for the first time in the Plenary Session beginning with the Third Five-Year Review Conference in 2013, especially thanks to the RC-3 Chair, Ambassador Krzysztof Paturej from Poland. Although only 10 statements were presented orally in the time allotted in the General Assembly plenary, all of the coalition member statements are posted on the OPCW and CWC Coalition websites every year. In 2021, for example, the Coalition organized statements across NGOs on ten key issues:

- Accountability for Old, Abandoned, and Sea-Dumped Chemical Weapons
- Regulation of Law Enforcement Use of Toxic Chemicals
- Chemical Weapons Victims' Medical Needs
- Organizational Governance and Evolution within the OPCW
- U.S. Chemical Weapons Stockpile Elimination
- Strengthening Education, Outreach, and Other Public Health Initiatives in All Member States
- The Chemical Weapons Convention and National Compliance
- CWC National Implementation
- Next Steps in Strengthening Controls Over Novichok Agents and Precursors
- Chemical Security and the Prevention of Chemical Terrorism

These are the themes mentioned across the almost 20 statements each year from 2014 to 2023.[9]

11.3.3 Recent CWC Coalition Activities

In 2021, the Coalition became housed in the Arms Control Association in Washington, DC under Paul Walker, Coordinator. A grant from the Global Affairs Canada Weapons Threat Reduction Program provided support for a part-time coordinator and full-time project assistant, as well as the establishment of regular newsletters, webinars and limited travel. In addition, the OPCW through its European Union grant provides funds for some travel of Coalition members to the annual meetings. The German Federal Foreign Office also supports some annual travel through the CWC Coalition. The funding has helped to significantly expand the activities and reach of the Coalition since its founding.

---

[8] https://www.un.org/en/sc/1540/about-1540-committee/group-of-experts.shtml.

[9] https://www.cwccoalition.org/past-ngo-statements/

To be a member of the Coalition, the organization or individual must engage in work relevant to the purpose of the CWC and must support the aims and mission of the Chemical Weapons Convention and the OPCW. The Coalition is working to continually increase gender and geographic diversity in our network. Currently, there are over 125 NGO members in the Coalition. Members include academic universities and centers, public health & other health organizations, environmental organizations, community citizen advisory boards, and global legal and human rights organizations. Although there are members from all continents and regions of the world, most are from Europe and North America. The CWC Coalition embraces the inclusion of NGOs representing victims of chemical weapons' attacks; the largest number of participating survivors are from Iran and Iraq, representing victims of Saddam Hussein's use of chemical weapons against Iran and the Kurds in the 1980s.[10]

The Coalition hosts a website (https://www.cwccoalition.org/), publishes a newsletter and holds webinars on relevant CW topics, organizes the drafting of plenary statements and Open Forum side events to promote networking opportunities among members of the Coalition, serves as a resource for NGOs interested in furthering the goals of the CWC, and encourages publications on relevant chemical weapon topics by Coalition members. Leadership of the Coalition works with OPCW leadership and staff between and during meetings.[11]

The Coalition was acknowledged for their work in 2022 at CSP 27 with the receipt of the OPCW-The Hague Award, given annually on the first day of the CWC CSP since 2013. The award was initiated in 2014 after the OPCW was awarded the Nobel Peace Prize in 2013 for verifying the abolition of all declared chemical weapon stockpiles; it seeks to recognize those individuals and organizations which have been active in promoting and implementing a world free of chemical weapons and thereby shares the 2013 Nobel Peace Prize. Dr. Paul Walker, Coordinator of the CWC Coalition, accepted the prize in The Hague on behalf of the CWC Coalition, and stated[12]:

> The CWC, the OPCW, and all delegations here play critical roles in helping us all build a world free of chemical weapons. But such far-reaching and historic multilateral treaties also require the global village to truly universalize and fully implement their mandates. We therefore will continue to underline the importance of civil society involvement at all levels—local, national, regional, and international—in order to build a much more safe and secure world, and to prevent any reemergence of such deadly weapons of mass destruction.

11.4   OPWC and Non-Governmental Civil Society Participation

---

[10] https://www.cwccoalition/org/, last accessed 2024/7/4.

[11] https://www.cwccoalition.org/, last accessed 2024/7/4.

[12] Paul Walker Comments, November 28, 2022: (2022) https://www.cwccoalition.org/opcw-the-hague-award/

11.4.1 NGO Participation

Since 2002 the three Director Generals (Ambassadors Rogelio Pfirter, Ahmet Uzumeu, and Fernando Arias) have been supportive of the participation of civil society organizations in the CWC. Since the Convention was opened for signature in 1993 to implement the Convention, the chemical industry and many scientists based in academia or NGOs have worked with the OPCW. As noted above, the inclusion of NGOs and individual citizens was modest at the beginning of the CWC, growing to many more over time with the formal creation of the CWC Coalition in 2010. Using funds from the European Union, the OPCW has supported travel for many NGOs to attend the meetings; these funds have been used primarily for participants from developing regions of the world.

NGOs must be approved by the General Committee, which consists of about 25 States Parties (out of the 193 members of the CWC). Once approved by the General Committee, three individuals of each NGO are allowed to attend the CSP or Review Conference. This process has rejected several NGOs each year with no explanation of why they were not accepted. Those not approved tend to be NGOs that have been active in the Middle East where there is much tension concerning the former use of chemical weapons.[13] The CWC Coalition issued a joint statement at both CSP-27 in 2022 and CSP-28 in 2023 about the rejection of NGOs, asking for more transparency in the process to understand why a NGO is rejected.[14] These statements point out that the NGO application process for CSP is not consistent with the general practices of other international WMD treaty conferences regarding weapons of mass destruction (WMD), and that the work of the rejected NGOs is important for the goal of achieving a world free of chemical weapons. In recent years, States Parties have sponsored side events outside of the official venues where these non-approved NGOs can present the important work they are doing related to chemical weapons.

The number of accredited (approved) NGOs has been collected by the Technical Secretariat each year from 1997 to 2023. The number has grown from an average of 10 NGOs during the first 12 CSPs to an average of 80 over CSP-23 to CSP-28; a steady increase in the number of accredited NGOs occurred from CSP 14 after the CWC Coalition was established. The number of NGOs attending the meetings is lower than the number accredited, largely because funding for travel is not available for all those accredited. Numbers of those NGOs and individuals are not available for each CSP. Similar to the CWC Coalition membership, the accredited NGOs are over-representative of Europe and North America. The themes of the focus of the NGOs in the data collected by the Technical Secretariat include chemical industry, chemical safety, development, environment, human rights, peace, public health, public policy,

---

[13] Personal communication with Paul Walker, September 30, 2023.

[14] CWC Coalition Joint NGO Statement at CSP-27—Regarding CSP rejections by the General Committee (November, 2022). The Hague, The Netherlands (2022); Joint NGO Statement at CSP-28—The Role of NGOs in advancing the CWC in the next quarter century (November, 2023) The Hague, The Netherlands (2023).

international affairs and law, regional cooperation, security, technology, victims, and youth (Ghionis 2023).[15]

Ghionis reports that the CWC Coalition and NGO Coordinator, also Chair of the Coalition, have "forged considerable improvements in the relationship between civil society and the OPCW" (Ghionis 2023, p.13):

> This includes both driving and supporting increased levels of CSO [civil society organizations] participation at sessions of the CSP, while also developing an active and vibrant interdisciplinary community attending to the CWC. The CWCC created a platform for intra-community engagement, networking, and collaboration. Often against strong political and financial headwinds, the leadership has generated a momentum over the last fifteen years which delivers a mandate to further strive to enhance civil society's relationship with the OPCW.

11.4.2 Partnerships with Chemical Industry and Disarmament Entities

The discussion of civil society engagement within the OPCW usually refers to industry, which is a part of a broader civil society, as a separate group (Ghionis 2023). The Technical Secretariat does not engage with other members of civil society (e.g., NGOs, academia) as it does the chemical industry. The major partnership with scientists and chemical industry is through the International Union for Pure and Applied Chemistry (IUPAC) and the Scientific Advisory Board (SAB). The Hague Ethical Guidelines were developed in 2015 by a Workshop on Ethical Guidelines for the Practice of Chemistry under the Norms of the CWC.[16] OPCW opened the new ChemTech Center in May, 2023, which will lead to sustained scientific accomplishments in the future.

In addition, since the beginning of the Convention, OPCW has worked closely with other arms control and disarmament entities, such as the Comprehensive Test Ban Treaty Organization (CTBTO), International Atomic Energy Association (IAEA), Biological Weapons Convention (BWC), United Nations Institute for Disarmament Research (UNIDIR), United Nationals Office for Disarmament Affairs (UNODA), and the World Customs Organization (WCO).

11.4.3 Advisory Board on Education and Outreach

The establishment of the Advisory Board on Education and Outreach (ABEO) in 2015 at CSP-20 evolved from recommendations of a working group of the Scientific

---

[15] https://www.cwccoalition.org/, last accessed 2024/7/4.

[16] The Hague Ethical Guidelines, November, 2015. The Hague, The Netherlands (2015) https://www.opcw.org/sites/default/files/documents/Science_Technology/Hague_Ethical_Guidelines_Brochure.pdf

Advisory Board. The Third-Year Review Conference (RC-3) in 2013 acknowledged that the role of education, outreach and awareness-raising is key to national implementation of the Convention. The ABEO, which consists of 15 individuals, is an entity within the Technical Secretariat to assist the OPCW to "prevent re-emergence of chemical weapons" in all activities and partnerships, especially after all declared stockpiles have been destroyed. The ABEO provides "advice on the development of education and outreach strategies, key messages and partnerships that support the implementation of the Convention, as well as identifying global education and outreach activities relevant to the Convention and those related to disarmament and non-proliferation." The specialized advice that the board provides aims to make education materials accessible to audiences in order to benefit the broadest range of target stakeholders. Furthermore, the role of the ABEO is to "provide advice on development of strategies and key messages for education and outreach, maintain an overview of global education and outreach activities relevant to the Convention, and provide advice to the DG on the development of maintenance of partnerships with relevant stakeholders and international organizations working in E&O area".[17] The AEBO was charged with giving advice to the DG on the development and maintenance of partnerships with relevant stakeholders, including the international organizations listed above.

The ABEO's progress through 2023 includes the articulation of goals for post-destruction phase of CWC that uphold norms and promote awareness of OPCW mission; partnerships established with individuals, civil society, scientists, industry, academia, and policymakers; provision of education & outreach activities that are effective, sustainable & cost effective to States Parties and National Authorities, and development of e-learning modules and content for OPCW's new learning management system. In addition, the ABEO is developing an international network of universities and think tanks and an inventory of international organizations.[18] Although the goals of the ABEO has potential to expand and engage NGOs and others in civil society, there is much more that the Technical Secretariat, States Parties, and National Authorities of the OPCW can do to fulfill their goals.

11.5 Summary and Lessons Learned

There has been progress in engaging NGOs, individuals and other members of civil society over the 30-year lifetime of the CWC. This is attributable to the leadership of the OPCW, individual States Parties and NGO Coalition members. Although there has been funding by a handful of States Parties for this work, the relative lack of

---

[17] Decision Statement—Establishment of an Advisory Board on Education and Outreach. CSP-20 (December 9, 2015). The Hague, The Netherlands (2015). https://www.opcw.org/sites/default/files/documents/CSP/C-20/en/c20dec09_e_.pdf.

[18] Report of the Fourteenth States of the Advisory Board on Education and Outreach. February 9, 2023. https://opcw.org/sites/default/files/documents/2023/04/abeo-14-01%28e%29.pdf

funding for a sustained effort at the conventions and between conventions and for travel to meetings has limited NGO engagement, which in turn limits their impact in implementing the Convention. Nevertheless, the CWC Coalition has been nimble and flexible in using the resources it has received to actively participate in meetings of the States Parties and some dialogue between meetings in The Hague. In addition to the lack of funding, the lack of universal support among States Parties for working with NGOs and civil society is an obstacle to future success of the CWC Coalition and the work of the Technical Secretariat and its various workgroups, committees and boards.

The CWC Coalition and other civil society members can help with National Authorities and government activities and support for the CWC mission in the future. Engaging civil society in Technical Secretariat activities and in the work of National Authorities in their countries needs to be strategically planned for the future to ensure "a world free of chemical weapons" now that all officially declared stockpiles have been eliminated. As Fernando Arias, Director-General of the OPCW stated in May, 2021[19]:

> The civil society community of non-governmental organizations, researchers, scientists, and other relevant stakeholders are essential partners in achieving the OPCW's mission and raising awareness about the risks posed by certain chemicals. The Chemical Weapons Convention Coalition has played a critical role in this regard by coordinating and supporting civil society engagement with the OPCW through the Conference of the State's Parties.

There have been many lessons learned from the work of citizen advisory committees and citizen engagement in destruction of chemical weapons in the United States and Russia. The chairs of the community advisory boards of the two final sites for destruction of chemical weapons in the United States state the following key insights (Kornelly and Williams 2023):

- Communication with citizens is critical; they can learn about the technical processes.
- Educating each other, while accepting that those running the program do not have all the answers, is important.
- Developing trust through dialogue is complicated.
- It is easy to say NO to proposals but hard to pick an alternative solution.
- Using accurate facts and information with the press and politicians is critical to success in dialog about solutions.

Finally, a research study on citizen advisory entities by Global Green USA in 1999 reported on what to do and not do in setting up and running a community advisory group (see appendix for list) (Walker 1999).

---

[19] Arias, F. Opening Remarks by the OPCW Director-General, at the Arms Control Association and Chemical Weapons Convention Coalition Webinar "Reinforcing the norm against chemical weapons: The April 20-22 Conference of States Parties to the Chemical Weapons Convention" (10 May 2021) The Hague, The Netherlands (2021). https://www.opcw.org/sites/default/files/docume nts/2021/05/20210510_DG%20Opening%20Remarks_ACA%20and%20CWCC%20Webinar_ WEB.pdf

11.6   Recommendations

The role for an engaged informed civil society, consisting of NGOs, individuals, academia, industry and other players, is critical for achieving full implementation of the Convention, especially in the era following the destruction of declared stockpiles. The OPCW and States Parties should support more collaboration with civil society national and international organizations to fully implement the next phase of the Convention focused on maintaining a world-free of chemical weapons.

11.6.1   Promote and Support Civil Society Participation Through the CWC Coalition

The OPCW, States Parties and others should support the future development of the CWC Coalition, currently housed in Washington, DC. Because the Coalition must be independent and able to present all options and ask all questions related to the implementation of the Convention, the Coalition should continue to be a free-standing entity, supported by but not part of the OPCW. Funds are needed, at a minimum, for travel to the OPCW CSPs and RCs, as well as meetings between the CSPs, and administrative functions such as maintenance of the website, newsletter and webinars. As more NGOs join the Coalition in the future, organization of the Coalition into subcommittees and functions, such as programs, fundraising, etc., would be possible with appropriate resources to sustain the Coalition structures (Ghionis 2023). The Coalition should continue to welcome all NGOs and individuals who are working on Chemical Weapons Convention issues, regardless of whether or not they are accepted by the OPCW as an official NGO to attend the CSPs or RCs. The Coalition should expand to include members of different genders, ethnic backgrounds, and geographic locations around the globe, and implement monitoring systems to track its membership and activities. Finally, inclusion of survivor groups and humanitarian issues has been vital to the energy of the Coalition and its work with the OPCW.

11.6.2   Continue to Develop Structures Within the OPCW to Embrace and Foster Civil Society Participation

All parts of the OPCW (e.g., Technical Secretariat, States Parties, Advisory Board on Education and Outreach, etc.) should develop a uniform strategy for working with all parts of civil society, including NGOs, academia, industry and committed individuals, to implement the Convention in all National Authorities of all members. The recent call by the German Ambassador for informal discussions about the role of

NGOs, including transparency about the selection process for CSP and RC participation, is welcome and may lead to improvements within the OPCW.[20] The following recommendations should be considered for improvements in working with civil society:

- The OPCW should develop a framework for monitoring the recruitment and participation of NGOs in OPCW activities so that future tracking of NGOs and other participants would include more information about the NGOs and individuals representing them.
- Much more support should be provided to include and coordinate activities of NGOs representing NGOs. Funds could be increased for use in the Victim's Trust Fund.
- The process for approving and accepting NGOs in the CSP and other meetings should be transparent; reasons for exclusion of an NGO, if any, should be provided.
- Collaboration with international organizations and entities within National Authorities that share the OPCW mission should be expanded to implement the Convention. The OPCW could maintain an updated directory of OPCW Memorandum of Understandings and collaboration activities with other organizations.
- After completing a needs assessment of what National Authorities need and request to implement the Convention, the OPCW should commit more resources and targeted support for collaboration activities so that capacity building, outreach and education "go to scale" in more regions and countries.
- The OPCW should expand the ABEO work on education e-learning modules and continue development of a global network of universities and think tanks

11.6.3   Engage with Public Health Associations and Government Entities

Members of the Chemical Weapons Convention Coalition (CWCC) support the strengthening of NGOs in every country to promote public outreach, education and public health functions. Well-funded public health systems can assist with the full implementation of the Convention in a transparent, science-based, and community-inclusive manner. Statements on the importance of public health associations and government entities committed to public health have been made at CSP-24 & RC-4 (November 2018), CSP-25 (November 2020), CSP-26 (November 2012), RC-5 (May 2023) and CSP-28 (November 2023).[21] These statements pronounced that the public health community is committed to:

---

[20] Statement by Ambassador Thomas Schieb, representative of Germany to the OPCW, at the 103rd Session of the Executive Council (July, 2023). The Hague, Netherlands (2023). https://www.opcw.org/sites/default/files/documents/2023/07/230710%20national%20Stat ement%20GERMANY-%20EC-103.pdf.

[21] Statement of Deborah Klein Walker, Past President, American Public Health Association at OPCW RV-4, November, 2018. The Hague, The Netherlands (2018).

- Quality, culturally-competent education and outreach
- Planning and training for emergency preparedness
- Comprehensive public health disease registries and surveillance systems to monitor disease and track chemical use
- Adequate health and other support for those injured by chemical weapons and toxic chemicals
- Elimination of chemical weapons and control of toxic chemicals in ways which do not disproportionally impact vulnerable populations
- Research on the impact of chemicals and safest methods of elimination
- Elimination of weapons and toxins in an environmentally safe manner.

All of these public health activities are embodied in the Global Charter for the Public's Health which was created by the World Federation of Public Health Associations in collaboration with the World Health Organization. Founded over fifty years ago, the World Federation of Public Health Associations (WFPHA) has over 130 member associations, which are mostly multidisciplinary national public health associations. Chemical weapons and other weapons of mass destruction have been discussed in plenary sessions at the most recent WFPHA Public Health Congresses held in Turkey, Ethiopia, India and Australia.

The Charter provides a framework for dealing with the public health challenges of addressing and eliminating chemical weapons, including guidance for protection, prevention, and promotion services using four core functions (governance, information, advocacy, and capacity) (Lomazzi 2016). The public health community acknowledges that environmental health is a major part of public health and that the elimination of chemical weapons is an important 'social determinate of health.' It is especially important that priority for support for these public health activities be given to low and middle income countries. Public health entities could work in collaboration with Responsible Care, a program of International Council of Chemical Associations in over 70 countries, to improve safe chemical management.[22]

The OPCW should therefore create a task force to foster a stronger connection to the WHO and its regional entities in order to support training and capacity for engaging public health to implement the Convention within National Authorities. The public health infrastructure within nations can play a significant role in meeting the OPCW mission "to implement the provisions of the Chemical Weapons Convention to achieve our vision of a world free of chemical weapons and the threat of their use, and in which chemistry is used for peace, progress, and prosperity."[23]

Lessons Learned—Selected Do's and Don'ts[24]

---

[22] Responsible Care, program of International Council of Chemical Associations. https://icca-chem. org/focus/responsible-care/

[23] OPCW Mission. https://www.opcw.org/about/mission.

[24] Walker, P. F. (1999). Citizen Advisory Boards and Public Involvement. Discussion Paper for Global Green/Green Cross Workshop, Moscow, March 23–24, 1999. https://www.cwccoalition. org/

Note: The following bullets underline key lessons learned in citizen advisory boards and public involvement processes. This is not meant as an all-inclusive list, but rather as an effort to stimulate thinking in developing additional models for public involvement in decision making.

DO involve citizens in important public decision making.

DON'T exclude any stakeholders from major decisions.

DO create citizen advisory boards to provide advice, support, criticism, and guidance.

DON'T ignore citizen input in decision making processes.

DO seek to involve citizens at an early stage.

DON'T make final decisions before engaging the public.

DO provide for a diversity of viewpoints on advisory committees.

DON'T exclude certain groups and/or individuals because of their opinions.

DO formalize advisory committees by state or federal legislation and/or decree.

DON'T expect ad hoc committees of citizens to have a legitimate voice.

DO have open meetings and seek full transparency of information.

DON'T limit access to information.

DO seek to build consensus in local, regional, and national communities.

DON'T expect full, 100% consensus.

DO provide some support for official advisory committees.

DON'T expect major financial or other support; advisory committees are traditional volunteer.

DO conduct regular meetings and make recorded minutes available to all.

DON'T meet irregularly and not record meeting discussions.

DO develop and maintain a list of interested citizens and groups for mailings.

DON'T ignore the need for regular outreach and advertising of meetings.

DO develop a set of ground rules for committee operations.

DON'T expect meetings to run smoothly without basic operating rules.

DO elect an energetic and committed chairperson to facilitate meetings.

DON'T expect to operate effectively with dedicated leadership.

DO make a commitment of time and energy in order to be effective.

DON'T think someone else will do your job on the committee.

DO carefully read and review all documents.

DON'T come to meetings unprepared.

DO expect to be an important and effective part of the public decision making process.

DON'T assume that a citizens' advisory committee will make final decisions.

References

CWC Coalition Joint NGO Statement at CSP-27—Regarding CSP rejections by the General Committee (November, 2022). The Hague, The Netherlands (2022); Joint NGO Statement at CSP-28—The Role of NGOs in advancing the CWC in the next quarter century (November, 2023) The Hague, The Netherlands (2023)

Ghionis A (2023) The OPCW and civil society: considerations on relevant themes and issues. Working Paper #10. (December, 2023) CBWNet, University of Hamburg, Germany. https://www.cbwnet.org

Kornelly I (2023) Thirty years of the chemical weapons convention: how far we've come—the Colorado experience. In: Presentation at 30 years of chemical weapons convention (CWC): histories, achievements, challenges, Narnack-Haus Berlin, October, 2023

Kornelly I, Williams C (2023) The long journey to eliminate the deadly U.S. chemical weapons stockpile. Webinar of the CWC Coalition, September 29, 2023. https://www.cwccoalition.org/2023/09/29/the-long-journey-to-eliminate-the-deadly-u-s-chemical-weapons-stockpile/, last accessed 2024/5/30

Lomazzi M (2016) A global charter for the public's health—the public health system: role, functions, competencies and education requirements. Eur J Pub Health 26(2):210–212

Phillipps D, Ismay J (2023) U.S. is destroying the last of its once-vast chemical weapons arsenal. New York Times, July 6, 2023. https://www.nytimes.com/2023/07/06/us/chemical-weapons-stockpile.html?searchResultPosition=3

Walker PF (2016) Cooperative threat reduction in the former soviet states: legislative history, implementation, and lessons learned. Nonproliferation Rev 23(1–2):115–129

Walker PF (2010) Abolishing chemical weapons: progress, challenges, and opportunities. Arms Control Today 40(9):22–30

Walker PF (1999) Citizen advisory boards and public involvement: a discussion of the role of citizens in public decision-making, Post-cold war demilitarization, and environmental clean-up. Discussion paper for Global Green/Green Cross Workshop, Moscow, Russia, March 23–24, 1999. https://www.cwccoalition.org/

Statement of Deborah Klein Walker (2018) Past President, American Public Health Association at OPCW RV-4, November 2018. The Netherlands, The Hague

Webography

Arias F (2021) Opening remarks by the OPCW director-general, at the arms control association and chemical weapons convention coalition webinar "Reinforcing the norm against chemical weapons: the April 20–22 conference of states parties to the chemical weapons convention" (10 May 2021). The Hague, The Netherlands

https://www.opcw.org/sites/default/files/documents/2021/05/20210510_DG%20Opening%20Remarks_ACA%20and%20CWCC%20Webinar_WEB.pdf

Decision Statement—Establishment of an Advisory Board on Education and Outreach. CSP-20 (December 9, 2015). The Hague, The Netherlands (2015). https://opcw.org/sites/default/files/documents/CSP/C-20/en/c20dec09_e_.pdf

The Hague Ethical Guidelines, November, 2015. The Hague, The Netherlands (2015). https://www.opcw.org/sites/default/files/documents/Science_Technology/Hague_Ethical_Guidelines_Brochure.pdf

History of PEO ACWA (Assembled Chemical Weapons Assessment): https://www.peoacwa.army.mil/about-peo-acwa/history-of-peo-acwa/

OPCW Mission. https://www.opcw.org/about/mission

Opening statement by the Director General to CSP-14, November 2009, pgs. 17–18 (2009). https://www.opcw.org/sites/default/files/documents/CSP/C-14/en/c14dg13_en.pdf

Path to the CWC https://www.cwccoalition.org/documents-on-disarmament-cwc/

Paul Walker comments, November 28, 2022: (2022) https://www.cwccoalition.org/opcw-the-hague-award/

Report of the Fourteenth Session of the Advisory Board on Education and Outreach. February 9, 2023.

https://www.opcw.org/sites/default/files/documents/2023/04/abeo-14-01%28e%29.pdf

Responsible Care, program of International Council of Chemical Associations. https://icca-chem. org/focus/responsible-care/

Statement by Ambassador Thomas Schieb, representative of Germany to the OPCW, at the 103rd Session of the Executive Council (July, 2023). The Hague, The Netherlands. (2023). https://www.opcw.org/sites/default/files/documents/2023/07/230710%20nati onal%20Statement%20GERMANY-%20EC-103.pdf

https://www.cwccoalition.org/

https://www.cwccoalition.org/past-ngo-statements/

https://www.un.org/en/sc/1540/about-1540-committee/group-of-experts.shtml

Deborah Klein Walker is a behavioral sciences researcher and public health practitioner with more than 50 years of experience in academia, state and federal government, and consulting. She is currently an Adjunct Professor at the Boston University School of Public Health and at the Tufts University School of Medicine. She is a former Associate Commissioner in the Massachusetts Department of Public Health and a Vice President for Public Health and Epidemiology at Abt Global. She is a former president of the Global Alliance for Behavioral Health and Social Justice and the American Public Health Association, a member of the CWC Coalition. She is the author of three books and over 100 peer-reviewed articles. She is a graduate of Mount Holyoke College (BA) and the Harvard Graduate School of Education (EdM, EdD).

Chapter 12
Voices from the Margins: The Unseen Impact of Chemical Weapons on Civilian Populations

Homeyra Karimivahed

Abstract This chapter explores the long-term and largely overlooked impact of chemical weapons on civilian populations, focusing on the 1987 mustard gas attack on Sardasht, Iran. As one of the first chemical assaults targeting civilians, the Sardasht bombing left enduring physical, psychological, and socio-economic consequences for thousands of survivors. Drawing from historical records, health studies, and personal narratives, this study examines the chronic physical ailments—particularly respiratory, dermatological, and ocular conditions—as well as the psychological trauma experienced by survivors and their children. The chapter highlights the compounded social and economic burdens, especially for women, and the persistent stigma that deepens isolation and suffering. It critiques the muted international response to Iraq's use of chemical weapons during the Iran-Iraq war and calls for stronger global action to support victims. Through advocacy, medical intervention, and legal reform, the chapter urges the international community to recognize Sardasht's suffering and take meaningful steps toward justice, prevention, and comprehensive survivor care.

Keywords Iran-Iraq war · Mustard gas attack on Sardasht · Iran · Chronic physical ailments · Psychological trauma

12.1 Introduction

On June 28, 1987, my beloved hometown of Sardasht was devastated by a brutal chemical weapons attack. That day claimed countless innocent lives and left a permanent scar on our hearts. For 37 years, we have carried the heavy burden of this tragedy, enduring relentless physical suffering from chronic respiratory issues, unbearable skin conditions, and other severe health problems. The mental and emotional wounds

H. Karimivahed (✉)
The Rotary Peace Centre, University of Bradford, Bradford, West Yorkshire, UK
e-mail: karimi_vahed@yahoo.com
URL: https://www.bradford.ac.uk/mlss/rotary/

© The Author(s) 2026
B. Friedrich et al. (eds.), *Thirty Years of the Chemical Weapons Convention (CWC)*,
https://doi.org/10.1007/978-3-031-98854-7_12

are just as profound, as many of us are haunted by trauma, grappling with anxiety, depression, and Post Traumatic Stress Disorder (PTSD). Our community's social fabric remains frayed, with families continuing to mourn their lost loved ones and struggling with the aftermath.

This article aims to shed light on the unseen impact of chemical weapons on civilian populations, focusing on the long-term consequences observed in Sardasht, Iran. By examining the historical context, health implications, and the urgent need for international support, this study underscores the necessity for comprehensive assistance to chemical weapons survivors and the imperative to prevent future atrocities.

12.2 Historical Context

Following World War II, the most extensive chemical attacks on human beings ever occurred in violation of the 1952 Geneva Protocol, during the eight-year Iraq-Iran war. A UN fact-finding team confirmed the use of mustard gas and nerve agents against Iranian troops (Ahmadi et al. 2010). During the eight-year conflict, Iraq employed a variety of chemical agents, including mustard gas, Tabun, and Sarin, against Iranian forces and civilian populations. The deployment of approximately 1,800 tonnes of mustard gas, 140 tonnes of Tabun, and 600 tonnes of Sarin had catastrophic effects, with over one million Iranians exposed to these lethal substances. Immediate deaths from these attacks were estimated at around 5,500, but the long-term health complications affected tens of thousands more (Nasiri et al. 2021).

Sardasht, a civilian Kurdish town (Fig. 12.1) located in the mountainous region of northwestern Iran, near the border with Iraq, became a tragic focal point in the history of chemical warfare. On June 28, 1987, four bombs filled with mustard gas were dropped on the town. The attack was carried out in two waves, targeting both the densely populated residential areas and the central bazaar, which were crowded at the time. The chemical bombing of Sardasht was one of the most catastrophic uses of chemical weapons against civilians (Khateri and Soroush 2005).

250 kg Mustard warheads that burst in the center of town, exposing about 4,500 civilians to the toxic gas. According to unofficial estimates, there were more than 8000 victims, and more than 120 people died as a result of the violence (Hashemian et al. 2006). This marked the first time in history that a civilian population had been subjected to a chemical weapons attack. The consequences of the attack were felt immediately and have continued to affect the community for decades. Mustard gas causes severe blistering of the skin and mucous membranes upon contact. Victims suffered from painful burns, respiratory issues, and long-term health problems, including increased risks of cancer and chronic respiratory diseases. The psychological impact was equally severe, with many survivors experiencing long-term mental health issues such as PTSD, anxiety, and depression. Image by Google Maps.

The international response to Iraq's use of chemical weapons was largely muted during the war and despite clear evidence and numerous reports, including findings

Fig. 12.1 Sardasht geography position

from United Nations inspection teams, the global community failed to take decisive action against Iraq. This lack of accountability emboldened Iraq to continue its chemical attacks with impunity.

Following that, the March 1988 genocide in Halabja, which killed 4,000–7,000 Kurdish women, children, and men, was one of the most heinous gas attacks on a civilian population in modern history (M-Hasan 2011). Cornish says that chemical weapons were most recently employed by Iraqi Baath Military Forces in their war with Iran and against the Kurdish community of Halabja in 1988, killing approximately 5,000 people (Cornish 2007).

12.3   Health Implications

The victims of the Sardasht attack suffer from a myriad of health issues that have persisted for decades. The eyes, skin, and lungs are the most commonly affected organs. Survivors experience chronic respiratory issues, skin disorders, and eye problems. Mustard gas, the chemical agent used, is known to cause severe blistering, respiratory distress, and long-term carcinogenic effects.

12.3.1   Chronic Physical Conditions

Respiratory complications are among the most debilitating consequences faced by survivors. Many suffer from chronic obstructive pulmonary disease (COPD), asthma, and other severe lung conditions that require constant medical attention. Skin lesions and severe scarring are also prevalent, often leading to secondary infections and long-term dermatological issues. Ocular complications, including chronic conjunctivitis and corneal opacities, further impair the quality of life for many victims.

12.3.2   Psychological Disorders

Beyond physical ailments, psychological disorders are prevalent. Anxiety, depression, and PTSD are widespread among survivors, exacerbating their already challenging living conditions. The psychological impact is profound, as survivors recount the horrors of the attack, the loss of loved ones, and the ongoing struggle with their health. Studies indicate that the children of survivors also suffer from psychological issues, displaying higher levels of PTSD compared to their peers. A study by Ahmadi et al. (2010) revealed that 93% of children of chemical warfare survivors showed moderate to severe levels of PTSD, highlighting the transgenerational impact of such atrocities.

12.4   Social and Economic Impact

The attack's impact extends beyond health, deeply affecting the social and economic fabric of Sardasht. The traditional role of women in this community has made them particularly vulnerable. Expected to manage household duties and care for their families, many women lack the strength to fulfil these roles due to the debilitating effects of chemical exposure. The stigma associated with their visible injuries further isolates them from social gatherings, leaving them feeling alienated and alone.

12.4.1   Economic Challenges

Economically, the attack has left Sardasht struggling. The lack of a specialized medical center forces victims to travel long distances for care, adding a financial strain to their already difficult lives. Many families have lost their primary breadwinners, leading to increased poverty and dependency on inadequate government support. This economic strain hinders the town's overall development and perpetuates the cycle of hardship.

12.4.2   Social Stigma and Isolation

The social stigma attached to the visible symptoms of chemical exposure exacerbates the isolation of survivors. Constant coughing, skin discoloration, and other symptoms make it difficult for victims to participate in community life. They are often treated as outcasts, not only by society but sometimes by their own families. This lack of social support compounds their suffering, making it harder for them to lead normal lives.

12.5   International Response and the Need for Support

The international community's response to the Sardasht attack was minimal at the time, with little condemnation or assistance provided. This lack of action highlights the need for robust international mechanisms to support victims of chemical warfare. Organizations such as the World Health Organization (WHO) and the Organization for the Prohibition of Chemical Weapons (OPCW) must prioritize financial and medical support for survivors. Research into long-term health effects and innovative treatments is crucial.

12.6   Medical and Psychological Support

Providing adequate medical care to the survivors of Sardasht is of paramount importance. This includes not only treating physical ailments but also addressing psychological trauma. Mental health services need to be integrated into the overall care plan for these victims. International organizations can play a pivotal role by offering financial support and expertise to establish specialized medical centers in Sardasht.

12.7   Advocacy and Awareness

Raising awareness about the plight of Sardasht's survivors is essential to garnering international support. Advocacy efforts should focus on educating the global community about the long-term consequences of chemical warfare and the urgent need for ongoing assistance. This can be achieved through various means, including academic publications, media campaigns, and collaboration with international human rights organizations.

12.8 Policy and Legal Measures

International policy and legal measures are necessary to prevent the recurrence of chemical warfare. Strengthening international treaties and ensuring strict enforcement can deter the use of chemical weapons in the future. Moreover, providing a legal framework for the prosecution of those responsible for such attacks can bring a sense of justice to the victims.

12.9 Conclusion

The tragedy of Sardasht serves as a grim reminder of the devastating effects of chemical weapons. The survivors' enduring physical and psychological suffering, coupled with significant social and economic challenges, underscores the need for comprehensive international support. By recognizing and addressing the plight of chemical weapons victims, we can honor their resilience and work towards a future where such atrocities are never repeated.

Our survivors, resilient and courageous, desperately need ongoing medical care, mental health support, social assistance, and justice. We urgently call upon the international community to recognize our plight, provide essential aid, and ensure such horrors are never repeated.

This article calls upon the international community to acknowledge the suffering of Sardasht's survivors, provide necessary aid, and work collectively to prevent the use of chemical weapons. Through continued advocacy, research, and support, we can ensure that the memory of Sardasht lives on and contributes to a world free from the horrors of chemical warfare.

References

Ahmadi K, Reshadatjoo M, Sepehrvand N, Ahmadi P, Yaribeygi H (2010) Evaluation of vicarious PTSD among children of Sardasht chemical warfare survivors 20 years after Iran-Iraq war. J Appl Sci 10(23):3111–3116

Cornish P (2007) The CBRN system. Assessing the threat of terrorist use of chemical, biological, radiological and nuclear weapons in the United Kingdom. Chatham House, London, pp 1–28

Hashemian F, Khoshnood K, Desai MM, Falahati F, Kasl S, Southwick S (2006) Anxiety, depression, and posttraumatic stress in Iranian survivors of chemical warfare. JAMA 296(5):560–566

Khateri S, Soroush M (2005) Assessment of psychosocial impacts of a chemical weapons attack on civilian population of Sardasht, Iran. Prehosp Disas Med 20(1):94–95

M-Hasan F (2011) A quantitative analysis about the prevalence of PTSD after the chemical attack in Halabja. Europ Psychiatry 26(2):164–164

Nasiri L, Vaez-Mahdavi M, Hassanpour H, Kaboudanian Ardestani S, Askari N (2021) Sulfur mustard and biological ageing: a multisystem biological health score approach as an extension of the allostatic load in Sardasht chemical veterans. Int Immunopharmacol 101(Pt B):108375

Homeyra Karimivahed, the International Secretary of the Organization for Defending Sardasht Victims of Chemical Weapons (ODVCW), is a dedicated advocate with over 15 years of voluntarily empowering underserved cultures and experience representing victims of chemical weapons from diverse backgrounds. She is the second generation of Sardasht chemical weapons attack (the first civil human community) and the daughter of a family who lost 11 members in Sardasht's 1987 chemical attack (8,000 victims). She has also spoken out strongly in favor of more medical understanding and support for victims of chemical warfare.

Chapter 13
Voices from the Centre: A Mission to Destroy Chemical Weapon Stockpiles Worldwide

Paul Walker and Ulf Schmidt

Ulf Schmidt:

Why don't we make a start then, if that's all right?

I wanted to go back, Paul, to the time when you worked in intelligence. I'd like you to tell me how you became interested in this field. You once told me that you worked in a particular Senate committee. Perhaps you could tell me about this. You were in Germany at the time; I'd love to hear a bit more about that.

Paul Walker:

Well, that's a long story too. But, to make a long story short.

I was at Holy Cross College, which is a Jesuit College in Massachusetts, well known, and I was thinking about graduation, and what I would do after graduation, and someone mentioned that the draft was still on, and this was the height of Vietnam, 1967–68.

I had been before in Germany as an exchange student in 1963, so I was relatively fluent in German, and I had also studied my junior abroad at the University of Vienna, in Austria, 1966–67. And, so, my major, when I went back to Holy Cross was German, but I didn't know what to do with German. So, I finally decided I would go on in International Affairs and Foreign Policy Studies. I was supposed to go to Johns Hopkins School for Advanced International Studies in Washington. But,

---

The interview between Paul Walker and Ulf Schmidt has been slightly edited to enhance readability.

P. Walker
Chemical Weapons Convention Coalition and Arms Control Association, Washington, DC, USA

U. Schmidt (✉)
Centre for the Study of Health, Ethics, and Society, University of Hamburg, Hamburg, Germany
e-mail: ulf.schmidt@unihamburg.de

I realized suddenly, when I went back to Holy Cross in 67, that I faced the military draft.

So, I looked around for options, but when I graduated, everything was closed, except the US Army, and everyone knew the US Army was a direct ticket to Vietnam. So, I looked around and nothing seemed to work. So, when I graduated in 1968, with a Bachelor of Arts in German, I was sort of handed my degree in one hand, and my A1 draft card in the left hand. And I said, well, I'll appeal this.

So, I appealed the draft to the local draft board first, and then to the state board, and I went on to start study at Johns Hopkins. I got my draft notice in the middle of the semester, saying report next week for duty. My roommate was in the same position. We wrote the local board, and said "Could we please at least postpone it a couple of months until we finish the semester, so that we don't lose our tuition and the like?", and she said "Yes". But, when January came, and we told them the schedule of the final exams, she said, "Well, I'll send you your draft notice again, and you'll have to report very soon after that."

So, we finished our exams on Wednesday afternoon, and we departed. This was in Washington DC, and we were to report for duty in Boston the next morning at 5. So, we packed up. We had to be pre-packed, you know. As soon as we finished the exam, we started to drive home. It is about an 8 to 9-h drive. We drove home, slept for three hours, and then, reported for duty.

So, I went. I think that's part of serving two- and a-half/three years in the military. It was really an important part of my evolution. And when I got out of the service, I really was intent on studying Military Foreign Policy, International Affairs, International Security Studies.

In the military, because I was fortunate enough to be good in languages, they sent me to Russian language school. By the skin of my teeth, I stayed out of Vietnam. We flunked out the lowest guy in Russian language class in Monterey every month and sent him to Vietnam. So, there was a lot of incentive to study well and do well. In Russian, I was sent to Germany of all places, for about a year, and I was in Russian intelligence, that is, American intelligence focused on Warsaw Pact forces in East Germany,[1] for a year. Basically, we called it human electronic and human intelligence, but it was listening to Warsaw Pact troop movements and if anything came up, if it was a surprise, you had to immediately translate and send it off to Washington, you know, to the National Security Agency.

---

[1] The Warsaw Pact was a political and military alliance treaty signed in Warsaw in May 1955 by the countries of Central and Eastern Europe and the Soviet Union. In addition to the Soviet Union, the original signatories of the alliance comprised Bulgaria, Czechoslovakia, East Germany, Hungary, Poland, and Romania. See Haigh (1968). Albania withdrew from the Treaty in 1968. See Lalaj (1968). Signatories of the Pact pledged to defend each other if one or more of the other signatories were attacked. The Pact meant that these countries were obliged to follow the Soviet Union's strategic policies and came to represent what was known as the Eastern or Soviet Bloc. It entailed the inauguration of a military Joint Command headed by the Soviet general Ivan Konev, which legalized the presence of Soviet troops within the Soviet Bloc states. The Warsaw Pact stood in contrast to NATO, the military alliance developed in 1949 by the United States, Canada, and other Western European nations to establish a joint defense against the Soviet Union. See Haigh (1968). The Warsaw Pact was disbanded in February 1991 after the collapse of the governments of the Soviet Bloc countries. See https://history.state.gov/milestones/1953-1960/warsaw-treaty, last accessed 10/04/25.

I got out, finished. I was given an "early out," because President Nixon was cutting back on Vietnam troops tremendously by then, and the services were anxious to get rid of troops as quickly as they could. So, they offered anybody who wanted to go into public service, go back to school, do teaching, police, fire, or things like that, even hospital work, they'd give up to six months early out. So, I was fortunate in that I got a five-month early out.

Ulf Schmidt:

Did chemical and nuclear warfare ever play a role in these intelligence gatherings?

Paul Walker:

No, not really. Not really.

Ulf Schmidt:

It was basically human intelligence gathering, old school monitoring of troop movement, is that what it was?

Paul Walker:

Yeah, troop movements, it was war games, things like that. Yeah.

To make a long story short, I finished John Hopkins. I also did two internships at The U.S. Arms Control and Disarmament Agency (ACDA).[2] We called it ACDA back then, which was founded by John F Kennedy and negotiated all the arms control up until then. During that time, we were in the middle of SALT I and SALT II negotiations,[3] with Gerard Smith, who was the chief negotiator and head of the Arms Control and Disarmament Agency.

---

[2] The aim of the United States Arms Control and Disarmament Agency (ACDA) has been to establish and put into practice arms control and disarmament policies through research, advice, and international negotiations. See https://history.state.gov/departmenthistory/people/principaloff icers/director-us-arms-control-disarmament-agency, last accessed 28/03/25. It was inaugurated as an independent agency by the Arms Control and Disarmament Act in 1961, and abolished in 1999. See Arms Control and Disarmament Agency (1999).

[3] SALT (Strategic Arms Limitation Talks) I and II were two sets of strategic arms limitations discussions that took place during the 1970s between the United States and the Soviet Union that aimed to result in the limitation of the manufacture of strategic missiles that could carry nuclear weapons. See Cameron (2022); https://history.state.gov/milestones/1969-1976/salt, last accessed 08/07/25. The outcome of SALT I was the signing of two agreements in May 1972 in Moscow: the Anti-Ballistic Missile (ABM) Treaty and the Interim Agreement on Strategic Offensive Forces. See Cameron (2022). The ABM Treaty aimed to reduce ABM (a surface-to-air missile created to eliminate in-flight ballistic missiles) deployment to strategically unimportant levels. See https:// www.armscontrol.org/factsheets/anti-ballistic-missile-abm-treaty-glance. The aim of SALT II was to produce a long-term, detailed treaty setting out wide limits on strategic attack weapons systems. The main aims of the SALT II negotiations were to establish provisions for the same amount of strategic nuclear delivery vehicles for the United States and the Soviet Union, to start reducing these delivery vehicles, and to establish limitations on qualitative developments that could pose a threat to future stability. See https://2009-2017.state.gov/t/isn/5195.htm, last accessed 08/07/25. SALT II treaty discussions were postponed after the Soviet Union invaded Afghanistan in 1979, and the treaty was never put into place nor renounced. See Cameron (2022).

So, all that gave me interest, a lot of interest, in arms control and disarmament.

When I finished Hopkins a year and a half later, 1973, I applied to MIT[4] and was accepted. MIT was one of the most prominent, still is, I think, in arms control and disarmament, and foreign policy, and technical issues regarding, peace and disarmament, and the like.

And I was at MIT for five years, completed the doctorate in 1978, and went on to a post-doc at Harvard.

Ulf Schmidt:

In Sociology?

Paul Walker:

Political Science, and MIT called my focus "Security Studies." Yeah, we nicknamed it "Bombs and Bullets." I went on to Harvard for a post-doc, and wrote a book called The Price of Defense,[5] which argued that we could cut the defense budget by 1/3 to 1/2 and have better security than we had back then. So that was it.

After I finished at Harvard and the postdoc, I was hired by a new group called the Union of Concerned Scientists, UCS,[6] funded by a Nobel laureate from MIT, a physicist called Henry Kendall. Henry had just set up UCS in Harvard Square. We had a funky little office, over a fortune teller, in the Square. UCS was set up in 1979 to address Three Mile Island,[7] because he worked in nuclear power and nuclear engineering and was a nuclear physicist, and was very concerned over the vulnerability of nuclear power to accident. So, UCS was doing work in that field, but he also wanted to get into nuclear weapons. He was an abolitionist and believed in

---

[4] MIT refers to the Massachusetts Institute of Technology. See https://www.mit.edu/about/, last accessed 08/07/25.

[5] The *Price of Defense* was published in 1979 with the Boston Study Group by Times Books. The book assessed the United States' military capabilities and defence budget necessities. See The Boston Study Group (1978); Goss (1980).

[6] The Union of Concerned Scientists was established in 1969 by scientists and students at the Massachusetts Institute of Technology. The founding vision of the non-profit organisation was for scientific research to focus on providing solutions for urgent environmental and social issues rather than on military technologies. See https://www.ucs.org/about/history, last accessed 08/07/25. A document produced by the Massachusetts Institute of Technology in December 1968, signed by 50 faculty members, led to the founding of the Union of Concerned Scientists. The document stated that 'misuse of scientific and technical knowledge presents a major threat to the existence of mankind' and called for a 'critical and continuing examination of governmental policy in areas where science and technology are of actual or potential significance'. It also expressed 'determined opposition to ill-advised and hazardous projects such as the ABM system, the enlargement of our nuclear arsenal, and the development of chemical and biological weapons'. See https://www.ucs.org/about/history/founding-document-1968-mit-faculty-statement, last accessed 08/07/25.

[7] Three Mile Island is a reference to the Three Mile Island Unit 2 nuclear reactor, located near Middletown in Pennsylvania, the United States, which melted down in part on 28 March 1979. The accident was the most serious one in the history of commercial nuclear power plant operation in the United States. The radioactive releases from the accident had no traceable health implications on workers at the plant or members of the public. Significant regulatory changes including emergency response planning and radiation protection were introduced after the accident by the United States Nuclear Regulatory Commission. See https://www.nrc.gov/reading-rm/doc-collections/fact-sheets/3mile-isle.html#impact, last accessed 28/03/25.

nuclear arms control and disarmament, and wanted us to do something along those lines. So, I started there working. Debbie,[8] my wife, was an Assistant Professor at Harvard at the time, School of Public Health. And the first thing we did was launch a campaign against the MX missile.[9]

The MX at that point was a proposal that a couple of administrations had put together. I think Reagan was certainly in favor of it when I came on board, but most of all Jimmy Carter was. The original plan for the MX was to deploy 100 new missiles, each with 10 warheads, and then to develop 1000 launch points in Utah and Nevada, and shuttle the missile between all the launch points. So, you'd have all these missiles roaming around on trucks. Originally, the idea was on a train, but trucks came later, all over the desert in Utah, Nevada; it was going to take up about 1/3 of both States just to deploy missiles and the garages. We were going to take up so much concrete that we couldn't build anything else in the country for 10 years; while it was being built. It would monopolize concrete production and it was the most harebrained scheme I've ever seen. And we organized a UCS road trip through Utah and Nevada for about two weeks and spoke in every little town and village. I still recall Tonopah, Utah, as one stop. Most people were against the MX system at the time. You know, these are conservative Republicans, and most people were really against the land taking, and the use of water out there, and certainly the use of cement and everything else.

We wound up the two-week speaking tour in Las Vegas and organized a nationally televised TV dialogue, chaired by Bill Moyers, who was a well known political commentator back, then. It was a wonderful discussion and we had millions watching on television. The thing I remember most, is one fellow got up and he said "There are a lot of modes in all this nuclear stuff we're talking about". He said, "There's the air-based mode, there's the sea-based mode, there's the land-based mode". He said, "We're talking about a land-based missile here." And, he said, "But I have a fourth mode and that's the commode, you know, the commode," and everyone got a big laugh out of that. And actually, six months after that, the mobile mode was cancelled, but the Air Force still went forward with the missile. It replaced some of the Minuteman ICBMs[10] we had. But we were very successful in helping cancel the MX.

And I moved on from there to the next thing we did at the UCS, was building the first national teach-in on nuclear weapons and war. It was 1981, November 11th, Veteran's Day. And we organized the teach-in, as we had done during Vietnam years

---

[8] Deborah Klein Walker is a behavioural sciences researcher and public health practitioner. She is an Adjunct Professor at the Boston University School of Public Health and at the Tufts University School of Medicine. She formerly served as Associate Commissioner in the Massachusetts Department of Public Health and a Vice President and Senior Fellow at Abt. Associates. See https://www.arborresearch.org/2023/08/04/deborah-klein-walker/, last accessed 08/07/25.

[9] Plans were announced for the development of the MX system in 1979. See https://historytogo.utah.gov/mx-missile-project/, last accessed 28/03/25.

[10] ICBMs refers to intercontinental ballistic missiles. These weapons have ranges between 6,000 to 9,300 miles and could strike any target around the globe. See https://www.nps.gov/articles/series.htm?id=09AF828C-1DD8-B71B-0B7EF976E874E182, last accessed 08/07/25.

earlier. And we produced over 100 hundred symposia, at 100 universities, with 1000 speakers, all the same weekend nationwide.

Ulf Schmidt:

Interesting.

Paul Walker:

And it really, really was well done. We got public discussion going on nuclear weapons and nuclear war. This was the time of the nuclear freeze campaign. I was part of drafting the nuclear freeze proposal, with Randall Forsberg,[11] George Sommaripa,[12] Philip[13] and Phylis Morrison,[14] and Martin Moore-Ede.[15] So, I went on from there.

Ulf Schmidt:

So, would you have seen yourself as part of a broader sort of disarmament movement at the time, maybe linking up also with Pugwash[16] and things going on in the UK, you know the campaign against nuclear warfare?

---

[11] Randall Caroline Forsberg (1943–2007) was founder and director of the Institute for Defense and Disarmament Studies in Massachusetts, the United States. Between 1968 and 1974, Forsberg carried out a comparative study of military research and development programmes around the globe at the Stockholm International Peace Research Institute. From 1973 to 1983, she worked as a consultant on the preparation of United States and Soviet Union strategic weapons estimates. See Ms Randall Forsberg (1989). In 1979, she co-authored *The Price of Defense* with the other members of the Boston Study Group. See Goss (1980).

[12] George Sommaripa was a peace activist. See Biography of George Sommaripa (1980). He was a member of the Boston Study Group. See Goss (1980).

[13] Philip Morrison was a theoretical astrophysicist. From 1973, he was Institute Professor at the Massachusetts Institute of Technology. He published multiple books that aimed to communicate science to the public, including *The Ring of Truth* (1987), written in collaboration with his wife, Phylis Morrison. See https://news.mit.edu/2005/morrison, last accessed 04/04/25. He was a member of the Boston Study Group. See Goss (1980).

[14] Phylis Morrison was a teacher and graphics artist. She was also a member of the Boston Study Group. See Goss (1980).

[15] Martin Moore-Ede was a Harvard Medical School professor and member of the Boston Study Group. See Goss (1980). He is a renowned expert on circadian clocks and their control by light. He has extensively studied the health problems such as sleep disorders, fatigue, diabetes and cancer generated by life at night. See https://thelightdoctor.com/about/, last accessed 10/04/25.

[16] Pugwash refers to a movement that has the fundamental aim of striving for the elimination of all nuclear weapons and other weapons of mass destruction by using science and reason to understand the catastrophic impact on humans these weapons can have. The name 'Pugwash' originates from the name of a fishing village in Nova Scotia, Canada, where the first meeting of scientists concerned with eliminating nuclear weapons was held. See https://pugwash.org/about-pugwash/, last accessed 08/07/25; Rotblat (1985). Since this first meeting in 1957—that was attended by scientists from across the ideological divide of the Cold War—Pugwash has organised over 400 meetings. See https://pugwash.org/history/resources/, last accessed 08/07/25.

Paul Walker:

Yeah, in the 80s, I came to Germany probably ten times speaking on the INF treaty and nuclear weapons Pershing and Ground Launched Cruise Missiles (GLCMs)[17] in Germany. I was in Britain a bunch of times. So, it was really an international movement, particularly in Europe, but also in the United States, against nuclear weapons and war, and people were fearful of nuclear weapons.

Ulf Schmidt:

I remembered it personally, in particular, it was very sensitive in the Federal Republic of Germany, and people went onto the streets.

Paul Walker:

Yeah, we had marches. We marched. I think it was 1981. We must have had a million or more people in New York marching in favor of a nuclear freeze, at least a nuclear freeze.

But anyway, by the early 1990s, I also, at that point, was very active in writing speeches in all for Ron Dellums who was a sort of left wing, very liberal, democrat, congressman, from Oakland, California. He was in the class of 1979, originally, when elected to Congress. So, he was young at that point. The class of 1979 was really an anti-Vietnam War class. We elected Dellums. We elected Pat Schroeder[18] from Colorado, and a bunch of others. And I wrote some of his speeches and presentations. I also worked on the Jesse Jackson campaign[19] and became a stand-in for Ron Dellums, a tall African American, ex-Marine by the way, and a social worker from Oakland. He was Jesse Jackson's foreign policy advisor. And Ron asked me if I would help out with Jesse and, be his substitute, basically. So, I followed Jesse Jackson in his first campaign and advised him. That's a whole story too unto itself. But eventually,

---

[17] The Intermediate-Range Nuclear Forces (INF) Treaty, passed in 1987, stipulated that the United States and the Soviet Union must destroy and reject all their nuclear and ground-launched ballistic and cruise missiles that had ranges between 500 to 5,500 kilometres. The treaty was the first agreement between the two countries to decrease their nuclear arsenals, destroy a category of nuclear weapons, and introduce wide-ranging on-site visits for verification. See https://www.armscontrol. org/factsheets/intermediate-range-nuclear-forces-inf-treaty-glance, last accessed 04/04/25.

[18] Patricia Schroeder was a congresswoman elected in 1972. She was the first woman to serve on the Armed Services Committee for the subsequent 24 years she was in Congress. Schroeder was a supporter of women's rights; during her time in Congress, she advocated for many issues affecting women, such as pay equality, employment safety for family and medical leave, and women's health.
   See https://history.house.gov/People/Listing/S/SCHROEDER,-Patricia-Scott-(S000142)/, last accessed 04/04/25.

[19] In 1984, Jesse Jackson tried to secure the Democratic Party nomination for the presidency, an attempt that was unsuccessful. He tried again in 1988, and was again unsuccessful. Jackson was also a civil rights activist, a minister, and a global hostage negotiator. See https://www.nps.gov/fea tures/malu/feat0002/wof/Jesse_Jackson.htm.
   Last accessed 04/04/25.

Ron Dellums became chairman of the Armed Services Committee.[20] He and Pat were put on the Armed Services Committee in 1979 when they joined the House of Representatives, but the problem was they didn't have enough chairs. It apparently enlarged the Committee by 1. So, they offered Ron and Pat to share a chair.

Ulf Schmidt:

A joint one?

Paul Walker:

A joint one. So, they would both sit together in the Committee room, in kind of protest. They would both sit on the chair, with one cheek each on the chair. The first black on the Committee and the first woman on the Committee. And they were really good to do that, and extraordinarily good on the Committee. But Ron, after 12 years, made it to Chairman, very different from his experience in 1979.

Ulf Schmidt:

Interesting.

Paul Walker:

He worked way up in very hierarchical fashion; he worked his way up to the chairman-ship amongst 55 or 56 members of the Committee. There were about 75 committee staff. And he asked me if I would come down and help serving on the Committee, to which I said "Great, wonderful opportunity to do that."

I became Policy Director of the Committee in the early 1990s. And eventually, I had over 100 issues on my plate. I was "Mr. Arms Control." You can spend about 10 min on any one issue. One of the issues I had to deal with was chemical weapons.

Ulf Schmidt:

Were you getting on people's nerves?

Paul Walker:

Yes, very much so. Very, very much so. But I was there under the Democrats, you know, and Democrats had the majority. So, it was, it was kind of fun, 'cause we had control of everything and we could put together a budget. You know, Dellums was interested in cutting the budget. The budget at that point we thought was extraordinarily high, was $225 billion, a year.

Ulf Schmidt:

Billion or million?

---

[20] Ronald Dellums was a congressman who aimed to reduce the military budget. He served on the Armed Services Committee, a standing committee of the United States House of Representatives, from 1973–1999. See https://history.house.gov/People/Detail/12109, last accessed 04/04/25.

Paul Walker:

Billion. Billion, yeah. We don't talk in millions in defense policy, it's only billions, you know. But compared to today, we limited a few things. We made a few cuts here and there, weapon systems, and the like, ABM.[21] We discussed a lot of ballistic missile defense.[22] It was kind of in its early heyday. But, today, the defense budget is over 900 billion. So, it's grown by a factor of four… or more.

But one of the things I had to cover was chemical weapons, and chemical weapons were… hum… had come up, because nobody, basically, the Soviets, and ourselves, didn't want them. We had built sizable arsenals. We had 31,500 US tons. The Russians had 40,000 metric tons, but nobody knew what to do with them. We didn't have launch systems anymore. We stockpiled them in nine stockpiles in the United States, seven major stockpiles in the Soviet Union. And, we had met, in the late 80s, bilaterally, in Wyoming, of all places, in Jackson Hole, Wyoming, to discuss unilateral and reciprocal reductions in chemical weapon stockpiles and their eventual elimination.

Ulf Schmidt:

With the Soviets?

Paul Walker:

With the Soviets.

Ulf Schmidt:

They came over?

Paul Walker:

Yes, Soviets came and they agreed. They agreed in 1988. This is just before the, you know, a year or two before the Soviet Union collapsed and several years before the Chemical Weapons Convention was opened for signature.

Ulf Schmidt:

Under Gorbachev?

Paul Walker:

With Gorbachev. This is all Gorbachev's doing. We agreed to get rid of the weapons in the late 1980s.

Ulf Schmidt:

And the United States would help fund the destruction? Fund some of them too?

---

[21] An anti-ballistic missile.

[22] Ballistic missile defence systems aim to defend a specific area from attack by establishing the location of an incoming ballistic missile, tracking it, and then setting in motion an interceptor to eliminate the missile before it reaches the area of destination. See https://armscontrolcenter.org/fact-sheet-u-s-ballistic-missile-defense/, last accessed 08/07/25.

Paul Walker:

And the US and West would help fund destruction too.

Ulf Schmidt:

Interesting.

Paul Walker:

And, of course, there had been negotiations for, almost, you know, 80 years at that point in a way, because the 1925 Geneva Convention, and the Protocol, which banned use of chemical weapons, didn't ban anything else, but prohibited their use in warfare. And that's one of the reasons I think we got away with very limited use, you know, over 60 years or, so from World War One to the 1980s, and the use of by Saddam Hussein, and later Syria.

But so, we had agreed to get rid of them, but nobody knew how to do it. You know these were primarily old chemical stockpiles; some of them, maybe ten percent, were 10 or 15 years old, but for the most part, they were old World War II vintage weapons: mustard, lewisite, sarin, soman, and nobody knew really the right way to do it.

So, in 1990, just before I had joined the Committee, the US started to destroy its chemical weapons stockpile, far ahead of the Russians. And, we built the world's largest incinerator, on Johnston Atoll, in the Pacific Ocean, 750 miles West of Hawaii.[23] That was a prototype. We started experimenting with burning of live mustard agent. And, that proved to be a mixed experience; we scrubbed the incinerator with enormous wet and dry scrubbers, but it still didn't screen emissions fully, and the incinerator, from time to time, would burp out live agent.

And we also found things in the chemical weapons that we examined we went through them, which surprised us. For example, when you open the weapon, for example mustard artillery shells, they would sort of champagne, they would spurt out, and splash all over the disassembly line, opening them. And we also found, which made things very dangerous of course, mercury in the mustard weapons. So, we realized we had to investigate all the weapons and what was actually inside. No one ever dreamed there would be mercury inside the weapons, and that was because, when they were made you had to gauge the temperature, mustard freezes at a very high temperature, higher than freezing point. The US apparently used thermometers. and after we filled the weapon, we just dropped the thermometer, with a mercury thermometer, into the weapon, and just left it there. These were going to be used on the battlefield, we didn't care about any mercury, you know, that was the least of our worries.

---

[23] Johnston Atoll was where the United States army destroyed over 400,000 redundant chemical weapons procured from Okinawa and additional United States military bases located in the Pacific Basin and West Germany between 1971 and 1991. The Johnston Atoll Chemical Agent Disposal System (JACADS) was the first of its kind in the world. The chemical weapons were eliminated at the molecular level. By 2003, the Army has finished demolishing the incinerator buildings.

See https://archive.epa.gov/region9/features/jacads/web/html/index.html, last accessed 08/07/25.

So, the Army, when I got on the House Armed Services Committee said to me, and we're still working on the Chemical Weapons Convention[24] too at the same time: "We're going to transport all the weapons to three centralized facilities, it will be Tooele, Utah], the biggest arsenal in the US, just West of Salt Lake City; it will be Jacads, on Johnston Atoll, in the Pacific; and it'll be either Alabama or Arkansas, that had two sizable stockpiles too." I told the Army back then that this was really not a realistic plan, I said "Have you done all your studies on movement? Is it by railroad, or ship, or plane, or truck, or what?". And they said, "well most of it would be moved by ship, and railroad and truck." So, I said, then, of course "You're going to evacuate every community along the way?". They said "Well we didn't think of that, we thought, you know, this would be secret, dead of night". And I said "That just is not going to fly, I'm sorry that's not to play these days, this has to be public, there has to be public dialogue."

Ulf Schmidt:

It shows the importance of transparency because the general public will eventually find out.

Paul Walker:

People will find out you know, "what are these secret transports going through the middle of town?".

Ulf Schmidt:

And people start rumors.

Paul Walker:

And, you know, inevitably, there'd be an accident somewhere, or there might be a natural catastrophe, who knows. I said, if any of this failed, and there was an accident, and a release of VX...[25]

Ulf Schmidt:

You could never explain that, exactly.

---

[24] The Chemical Weapons Convention (CWC) of 1992 is a disarmament and arms control treaty that came into effect in 1997. See Macleod (2017). The main goal of the CWC is the international banning of chemical weapons and the complete elimination of all such weapons that still exist. See https://www.bafa.de/EN/Foreign_Trade/Chemical_Weapons_Convention/chemical_wea pons_convention_node.html, last accessed 07/04/25. The CWC was the result of twenty years of negotiations in the Geneva Conference on Disarmament and an additional four years of work of the Preparatory Commission in the Hague. The CWC sparked the start of a global chemical weapons disarmament regime led by the Organisation for the Prohibition of Chemical Weapons (OPCW). See Trapp (2017).

[25] VX, one of a series of V agents, is a variant of a toxic nerve gas. It is one of the most toxic chemical warfare agents of the 20th century. See Schmidt (2015).

Paul Walker:

And I said, we'll potentially kill thousands of people, you know. So, I said, you've got to operate on site. And the Army, very reluctantly, finally went back to onsite disposal. But of course, they stuck with incineration. And I explained to them that burning is not a good idea. It works very well, but it's like a big red flag to the public health and environmental communities. And, I think, I said, if we're going to burn, and this may be the most efficient way to do it, but it's going to be very controversial. It's wrong. I'm not so worried about, you know the physical danger, but I think communities, local communities, are going to be worried about what's coming out of the smokestack.

Ulf Schmidt:

Yes.

Paul Walker:

So, the Army was going through turmoil, saying that they were in a hurry with the Chemical Weapons Convention breathing down their necks with specific destruction deadlines. The original plan was: four years of destruction, finished by 1994, and a two-billion-dollar cost you know, totally off base.

But in the meantime, we're talking to the Russians and planning for the first onsite inspection of a Russian site. This was one of the biggest Russian sites, the easternmost, east of the Ural Mountain range. The other six stockpiles were west of the Urals. This one was in Siberia and called Shchuch'ye and held 5,400 metric tons of nerve agent.

Ulf Schmidt:

And before you went, did you have any idea what you would encounter?

Paul Walker:

We did. We were briefed.

I was one of the staff people on the inspection team. We had an Assistant Secretary of Defense; we had General Orton, Head of US chem weapons in the US Army; we had two congressmen, and their wives actually go with us too. We were shown satellite pictures of Shchuch'ye and they told us, in general, what it would be like, a lot of dangerous shells and agents.

But when we went, you know, we took a flight from Moscow to Chelyabinsk, and then a two-hour drive to Shchuch'ye. It's a very rural farming area, a few villages of a few thousand people each, spread out around the area. So about 14,000 people, within a 50-mile range.

We entered the stockpile in the morning, and got dressed in Russian green military uniforms. I don't know for what reason. They had us wear long green underwear as well, very hot in July.

Ulf Schmidt:

Really? So, you had to wear Russian military uniform?

Paul Walker:

Yeah, for some reason. We were fitted with gas masks as well. The funniest thing was testing the gas mask. We all stuck our heads, with gas masks on, in a tent filled with tear gas and saw if the gas masks work. But looking at each other in these bug-eyed masks, you know, here we are with General Orton, and the Assistant Secretary of Defense, and the others around the mask, we all said "What are we're feeling like?", "What are we doing here?", you know, "This is the weirdest."

Then they put us on these old decrepit school buses, and they drove us around this stockpile, which looked like it hadn't been, you know, opened in 50 years. It had 3 feet of grass, which is a real fire danger. It had a 16-year-old guard at the front gate that, when we talked, told he hadn't been paid in 6 months and he was glad we're here because now they might make up a couple of months' salaries for him.

The place was very insecure. As far as I could see, warehouses with stockpiles inside, all corrugated metal above ground, warehouses locked with bicycle padlocks, and swinging front doors, and broken windows, and holes in the roof. The Russians had mowed a path to two or three of the warehouses so we could go in. You could see these five calibers of artillery shells preloaded with VX nerve agent. You'd see them as far as the eye would go. They were just enormous numbers of shells. They also had two calibers of missile warheads for FROG[26] and Scud short-range missiles,[27] and these were on railroad dollies; they had about 1,000 of those.

Ulf Schmidt:

What happened when you entered this room with your group, did anyone say something?

Paul Walker:

No, we had to wait outside, you know, until they said, the warehouses were safe. You know, they had a little bird cage with them, and they sent a birdie in first, to see whether the bird would come back, you know. The parakeet tested the atmosphere, and then flew back in the cage, so the man said it was OK, and we went in. Everyone was just astounded at the site, so large, so insecure, so close to the Kazakhstan border, and so very deadly.

Ulf Schmidt:

You once told me that you've never felt so close to God.

Paul Walker:

In some ways... [Silence]. In some ways that's right. [Silence]. I mean, we couldn't imagine...

---

[26] FROG stands for Free Rocket Over Ground. It is the NATO reporting name for a Soviet short-range unguided battlefield missile that was first seen in 1957. See Martin (1974).

[27] The Scud is a surface-to-surface tactile missile (to be launched at ground or sea) that was developed by the Soviets and first came into service in the 1960s. It was developed as a heavy artillery rocket. See Zaloga (2006). Scud is the NATO reporting name. See Geiger (1967).

Ulf Schmidt:

Was that the realization that this was a possibility—that humankind could completely destroy itself?

Paul Walker:

We all realized this at that point... Excuse me, Ulf.

Ulf Schmidt:

No, but it's so moving, when you told me the story, I couldn't believe it.

Paul Walker:

Yeah, in some ways, we realized... You know... Any one of these million weapons could wind up tomorrow, in, you know, Tel Aviv, or Berlin, or, you know, London, or New York, or Washington, anywhere else, if it wasn't taken care of. And there was no security. I mean, the important thing is there was no security on the site, at all.

Ulf Schmidt:

And it's, as you said, it's one thing to do it abstract, as a political scientist, and another, actually realizing, here this is the stockpile that could potentially decimate thousands and thousands of people, is that right?

Paul Walker:

Yeah, there were millions of artillery shells.

Ulf Schmidt:

It must have been unbelievable.

Paul Walker:

It was unbelievable. And we all returned from this inspection, the next couple days, and wrote a classified memo to the Secretary of Defense immediately, and said "We've got to help get security on these sites." We met afterward, privately, with some of the leaders of the local villages and asked them if there'd be any chance we could, we said, hypothetically, obtain a weapon from the stockpile, and they said, "Sure, just name your price, you know." This was 1994, Russia was in terrible shape, and this was far away from Moscow in Siberia; people were desperate for food, and they were picking up scrap metal in the fields trying to figure out how to make fifty cents a day. They were excited about possibly selling a weapon to the West if they could, and they said, you know, it didn't matter, "These weapons weren't any good anymore, we're going to destroy them anyway."

So, they were half serious when they said, "Sell us a weapon," you know. There was chain link fence around the stockpile site. There were no guard towers, no night lights, nothing at all. So, we went back and the first thing we implemented was a security program for two of the sites, both holding portable artillery shells. The one we were visiting was Shchuch'ye and, the second site was Kizner in the Udmurt

Republic. But it took us from 1994 until 2002 to implement those security measures. By then, we had trained the troops, put a higher fence around, barbed wire and the like, including guard towers, and we had installed video and electronic surveillance, all paid for by the West—some \$20 million. In the meantime, the Russians were figuring out how to destroy the site. The Assistant Secretary of Defense, during this inspection, had offered the Russians an incinerator, and he said, "I'll build you the same thing we're building in the United States, incinerators". And the Russians, immediately, both at Shchuch'ye, but also in Moscow meetings with the Russian military and civilian hierarchy. They all said: "No, we don't like incineration. It's expensive, it's controversial, has some environmental impact, public health impact, high maintenance, high cost." So, we started a program called the "JREP," the Joint Research and Evaluation Program, to figure out what technology would be best for the Russians, and that's what they eventually wound up with—neutralization rather than incineration.

So, we started plans a few years later for an industrial site in Russia to destroy Shchuch'ye, and planning started too for the other 6 sites in Russia. The Russians wanted a site far away from the villages, for safety purposes, so no citizens would get hurt. But the construction went on for about 10–12 years, where we had to clear and dewater a site about twenty miles from the stockpile; this included cutting down 20–25,000 trees, a whole Siberian forest, and building a dedicated railroad between the stockpile and the site. And eventually the industrial site was built, the stockpile was neutralized, and, of course, the Russians finished destruction at all seven sites in 2017.

Ulf Schmidt:

And was it all paid for by the West?

Paul Walker:

No, the West paid probably about twenty percent of Russia's demilitarization effort. Western contributors included the British, the Germans, the Italians, and others. The Americans were the major contributor with around two to three billion dollars, most of it through the Cooperative Threat Reduction (CTR) program. Both the UK and Germany were also major contributors. The Russian program itself eventually cost, I would estimate, ten billion dollars. So the West paid about a quarter to a third of the Russian program.

Ulf Schmidt:

But they also contributed to some of it?

Paul Walker:

Yeah.

Ulf Schmidt:

The Russians themselves, they also could, as best as they could, at the time?

Paul Walker:

The Russians contributed as well. So, the Russians paid, you know, probably, in the end, I've never seen numbers, but I would estimate probably six or seven billion dollars. And we paid, we, and the British, and the Germans, and the Canadians, and many others, paid 3 billion dollars, and that all came from cooperative threat reduction program.

I'm just looking at the time.

Ulf Schmidt:

We can continue this conversation at another time, but I think we covered some of the key aspects already. I think it is an absolutely incredible story, Paul. I'd love to talk to you further on it.

Paul Walker:

We can talk further. I realized, after we spoke last, I wrote up the onsite inspection in Shchuch'ye, about 20 years ago in Topic Magazine. It was a local Harvard student publication.[28] I tried to locate that last night, but I couldn't locate it. I have it somewhere in hard copy.

Ulf Schmidt:

I'd love to see the article.

Paul Walker:

I'll look for it, if I don't have it electronically.

Ulf Schmidt:

I mean it's a conversation we can continue at some point.

But I think it's absolutely incredible what you started there. And, I mean you, you're one of the people who have been in this area for 50 years, really. It's absolutely extraordinary what you've achieved in that period. And I think, what we're going to see now, are some of the successful results. I mean, yeah, as you said, imagine if any of this would have got into the wrong hands, ever.

Paul Walker:

Yeah, it was. It's an extraordinary accomplishment, I think. But chemical weapons destruction took much longer than we thought. You know, 33 years in the United States, because we started in 1990 and finished in 2023. Thirty-three years is a long time. But it got done and nobody was hurt. Nothing was lost and the Russians finished in 2017. They actually started demilitarization in 2002 in Gorny, a destruction facility built primarily by the Germans.

---

[28] Walker (2002).

Ulf Schmidt:

I find it shows that international cooperation is possible when political and military leaders put their differences aside, and that things can be achieved, if there is a common goal to save the world from disaster. But it needs people like yourself. I mean, if you think about the commitment, and the enormous amount of time you have invested. I can imagine that the Russians were not easy negotiators.

Paul Walker:

No, no. I remember sitting across during lunch in one of our visits to Russia with an ex-military General and, he said, Doctor Walker, I realize what you're all up to. He said, "You're up to undermining Soviet military strength, and you're not going to destroy your stockpile, you're just waiting for us to destroy our stockpile." He represented the Soviet right wing with strong skepticism of the West. The Russian generals, however, and the whole military hierarchy, were all in favor of getting rid of their chemical weapons. It was a pain in the neck for the Russians and very expensive to maintain, I'm sure.

Although when I was at Shchuch'ye, it was apparent that nothing was being spent on security. So, we were fortunate to get security on those sites and to see the stockpiles safely destroyed. But fortunately, they finished in 15 years. We did a lot of public outreach, too, under Green Cross International: we organized public outreach offices; we had public dialogues; we had annual national hearings in Moscow. I'm sure the Russians might say today that none of that was necessary, but I'm sure today they wouldn't be finished without any of that happening.

Ulf Schmidt:

Absolutely.

Paul Walker:

And they got a lot of money from the West. As I say, it was the West, you know, who funded about a third of the destruction of the stockpiles. So, it was, overall, I think, a very successful program.

References

Arms Control and Disarmament Agency (1999) Federal Register 64(62):15686

Biography of George Sommaripa (1980) In: Hearing before the subcommittee on arms control, oceans, international operations, and environment of the committee on foreign relations united states senate ninety-sixth congress. First Session on The arms control implications of current national defense programs. Washington: United States Government Printing Office

Boston Study Group (1978) The price of defense: a new strategy for military spending. New York, New York Times Press

Cameron J (2022) Soviet-American strategic arms limitation and the limits of co-operative competition. Diplomacy Statecraft 33(1):111–132

Geiger GJ (1967) Russia's SCUD Missile. Ordnance 51(282):610–611

Goss C (1980) Review: the price of defense by the boston study group. Pres Stud Q 10(1):133–135

Haigh P (1968) Reflections on the Warsaw Pact. The World Today 24(4):166–172

Lalaj A (1968) The Prague Spring and the Albanian 'Castle'. In: McDermott K, Stibbe M (eds) Eastern Europe in 1968. Responses to the Prague Spring and the Warsaw Pact Invasion, pp 235–255. Cham, Palgrave Macmillan

Macleod R (2017) The Genie and the Bottle: reflects on the fate of the Geneva Protocol in the United States, 1918–1923. In: Friedrich B et al (eds) One Hundred Years of Chemical Warfare: Research, Deployment, Consequences. Springer, Cham, pp 189–213

Martin JM (1974) "Soviet Nuclear Tactics" from the Ordnance Magazine, May-June 1970. In: Hearings before the subcommittee on U.S. security agreements and commitments abroad and the subcommittee on arms control, international law and organization of the committee of foreign relations United States Senate, Ninety-Third Congress. Second Session on U.S. Nuclear Weapons in Europe and US-USSR Strategic Doctrines and Policies. 213–217. Washington: United States Government Printing Office

Ms. Randall Forsberg (1989) In: The proceedings of the thirty-first air force academy assembly, 14–17 March 1989, 30. Colorado Springs, U.S. Air Force Academy

Rotblat J (1985) The Pugwash conferences on science and world affairs. Med War 1(1):51–54

Schmidt U (2015) Secret science: a century of poison warfare and human experiments. Oxford University Press, Oxford

Trapp R (2017) The use of chemical weapons in Syria: implications and consequences. In: Friedrich B et al (eds) One Hundred Years of Chemical Warfare: Research, Deployment, Consequences. Springer, Cham, pp 363–275

Walker P (2002) The road to Shuch'ye (Shchuch'ye). Topic Magazine, Harvard Student Magazine, summer

Zaloga S (2006) Scud ballistic missile and launch systems 1955–2005. Bloomsbury, London

Paul Walker coordinates the Chemical Weapons Convention (CWC) Coalition, is Vice Chair of the Arms Control Association, and is a member of the US Department of State International Security Advisory Board. He has worked on chemical weapons demilitarization since undertaking the first US on-site inspection of a Russian CW stockpile in 1994 when he was a Professional Staff Member of the US House of Representatives Armed Services Committee. He led the Green Cross International Program on Environmental Security and Sustainability for 20 years and was awarded the Right Livelihood Award in 2013 "for working tirelessly to rid the world of chemical weapons." Walker holds an MA from Johns Hopkins SAIS and a Ph.D. in international security from MIT; he is also a US Army Vietnam-era veteran.

Ulf Schmidt is Senior Professor of Modern History at the University of Hamburg, founding-director of the Centre for the Study of Health, Ethics, and Society, and a Fellow of the Royal Historical Society. His interests are in the history of modern medical ethics, warfare, and policy in twentieth-century Europe and the United States. He is especially interested in the history of authoritarian regimes and modern dictatorships. He is the author, among others, of *Secret Science. A Century of Poison Warfare and Human Experiments* (OUP, 2015) and is the co-editor of *Propaganda and Conflict: War, Media and the Shaping of the Twentieth Century* (Bloomsbury, 2019). He is Principal Investigator of a six-year ERC Synergy Grant on "Taming the European Leviathan: The Legacy of Post-War Medicine and the Common Good".

Part IV
CWC—A Model for the Abolition of Weapons of Mass Destruction (WMD)

Chapter 14
From Disarmament to Abolition: Learning to Maintain the CWC Model

Alexander Ghionis

Abstract The Chemical Weapons Convention (CWC) has been successful in overseeing the destruction of declared chemical weapon stockpiles, but its role as a model for the permanent abolition of all chemical weapons requires careful consideration. Drawing on the speaking notes for the author's presentation to the Conference that this volume records, this paper reflects on thoughts and dynamics that may aid efforts to maintain and develop the CWC as a sustainable model of chemical weapons abolition. It briefly unpacks the expectations and assumptions surrounding the CWC's potential as a model for WMD abolition, suggesting that approaches to reviewing relevant science and technology, outreach and engagement, and efforts to leverage expertise and networks can be informed by the experiences of other international regimes in pursuit of maintaining a permanent abolition of chemical weapons beyond and including those classed as WMD. The paper suggests that learning from other regimes and adapting to new realities can strengthen the CWC's role as a living, evolving model for abolition. However, realising this potential will require a shared vision and commitment from all stakeholders, including states parties and civil society, as well as a willingness to invest in the necessary capabilities and partnerships.

Keywords Chemical Weapons Convention · OPCW · Science and technology

14.1 Reframing the Question

Is the Chemical Weapons Convention (CWC) a model for the abolition of weapons of mass destruction (WMD)? This was the question the author was asked to consider for the conference that this present volume records.[1] While the CWC may indeed serve as a model for the disarmament of a certain category of declared weapons

A. Ghionis (✉)
The Harvard Sussex Program, The Science Policy Research Unit (SPRU), University of Sussex, Falmer, UK
e-mail: A.Ghionis@sussex.ac.uk

[1] https://www.un.org/pga/wp-content/uploads/sites/108/2024/01/SOTF-Co-Facilitators-Zero-Draft_Pact-for-the-Future-circulation.pdf, last accessed 2024/05/04.

© The Author(s) 2026
B. Friedrich et al. (eds.), *Thirty Years of the Chemical Weapons Convention (CWC)*,
https://doi.org/10.1007/978-3-031-98854-7_14

of mass destruction, it is crucial to recognise that the CWC's role extends beyond this initial achievement to the much wider task of preventing the (re)acquisition and re-emergence of chemical weapons in general.

The CWC's success in achieving the destruction of declared chemical weapon stockpiles is undoubtedly a significant accomplishment. However, this represents only one aspect of the CWC's broader mission. The CWC aims to "completely exclude the possibility of the use of chemical weapons", as stated in its sixth preambular paragraph. This requires not only the irreversible disarmament of existing, declared, stockpiles but also the prevention of their re-emergence by constraining opportunities and suppressing actors' motivations and intentions to (re)acquire them. The first function, to essentially disarm and destroy, has a quantitative and time-bound character to it. The goal is to identify, remove, and destroy existing stockpiles, and subsequently verify their absence; the task has a start point, and in practical terms, an endpoint: the job can be completed. This function must remain on stand-by to deal with chemical weapons stockpiles if they (re)emerge. The second function is different, insofar as it is less quantitative (the risk may be hard to grasp), less easily identifiable (there may be no 'declaration' of their existence), and constant (assuming that there may be pressures today and in the future that may incite actors to seek to acquire chemical weapons, there can never be any lapse in vigilance or effort). This dual-function focus on disarmament and prevention is essential for creating a world free from the threat of chemical weapons.

Furthermore, it is essential to recognise that not all chemical weapons fall under the category of WMD, and conversely, of course, not all WMD are chemical weapons. The CWC's scope extends beyond the abolition of chemical weapons that may be considered WMD, as it seeks to prevent the use of any toxic chemical used to inflict death or harm through its action on life processes, regardless of scale or type. This nuance is crucial in understanding the comprehensive nature of the CWC's prohibition, and accordingly the challenges it faces in maintaining a world free of chemical weapons. Chemical weapons are not composed of only a select group of chemicals—rather, any chemical can be a chemical weapon if an actor intends to use it for purposes other than those purposes not prohibited under the CWC. Therefore, while the CWC may serve as a valuable model for the disarmament of declared WMD stockpiles, its role as a model for the *permanent* abolition of WMD (or, more accurately, a whole class of prohibited weapons) is less easily stated.

As such, this paper briefly unpacks some of the expectations and assumptions that make the initial question tricky to answer, and then reflects on some dynamics that can aid efforts to maintain and develop the implementation of the CWC toward a sustainable model of chemical weapons abolition. These include reviewing relevant science and technology, outreach, engagement, and efforts to leverage expertise and networks. In exploring these aspects, the paper explicitly recognises the value of learning from the experiences of other treaty regimes. Rather than solely focusing on what the CWC can offer as a model, it is equally important to consider what lessons can be drawn from other organisations and regimes. By engaging with these experiences and adapting to new realities, the CWC can strengthen its role as a living, evolving model for the abolition of chemical weapons. The emergence of what we may call a model treaty in this context relies on its ability to remain relevant and effective in the face of emerging challenges.

As this paper recalls a short presentation, its purpose is not to provide a comprehensive examination but rather to provide a partial, conversational overview that may inspire more discussion and research in this direction. The main purpose is to suggest that there may be value in critically examining how the CWC's role both as a model for the abolition of a subset of WMD and the wider chemical weapons category may be informed by perspectives and lessons from other regimes. In doing so, the underlying principle may be identified as being one that recognises that, rather than passively expecting others to learn from the CWC experience, it may be fruitful instead to consider what can be learned from the experiences of others, as implementation of the CWC begins to unfold over a much broader landscape than was otherwise visible when timely destruction was the guiding goal.

14.2 Unpacking 'a Model for WMD Abolition' in the CWC Context

Taking a historical perspective, there is ample evidence to suggest that the CWC has been—and may remain—an effective model for abolishing stockpiled chemical weapons, many of which have been considered militarily significant WMD.[2] The connotation may then be that other efforts towards abolishing certain categories of WMD may learn something from the CWC's approach, potentially reproducing aspects identified as leading to positive outcomes.

Yet, maintaining disarmament through preventing (re)acquisition, and therefore reemergence, is a continuous process. If we ask whether the CWC can *remain* a model for WMD abolition, and what that might entail in the future, we find it not so easy to answer. This, in part, may be because we need to imagine how the abolition can be *sustained* in a way that the CWC remains the model approach. To begin to consider the areas that might contribute to this sustainability, reflecting on whether we mean 'maintaining the abolition of a specific category of WMD', those being chemical weapons, or whether we mean 'maintaining the abolition of all chemical weapons', including those deemed as WMD, is important. Not all chemical weapons are WMD, and of course not all WMD are chemical weapons. Indeed, the CWC strives to abolish and prevent all chemical weapons, not just those with an order of effect high enough to be deemed as WMD in its general meaning.[3] Thus, by seeking a sustainable abolition of chemical weapons, we may assume that that must also include those chemical weapons considered WMD; however, focusing only on those chemical weapons that may be considered WMD leaves the CWC open to failure by

---

[2] See, for example: Black (2016), Manly (2007), Robinson (2008).

[3] Carus (2012), Robinson (2002).

conditioning a view in which chemical weapons are defined by notions of lethality—a term, it is worth recalling, that does not appear in the CWC as a qualifier of the definition of a chemical weapon.[4]

14.2.1 The CWC as a Model for Abolishing a Specific Category of WMD

To view the CWC as a model for abolishing a specific category of WMD is not a challenging task—nor should it be—as the considerable efforts put into its negotiation and implementation have borne impressive fruits. For example, the conference and presentation to which this paper builds from occurred just five months after the Organisation for the Prohibition of Chemical Weapons (OPCW) announced that the last declared stockpiled chemical weapon had been irreversibly destroyed in the United States.[5] Moreover, the CWC has proven to be flexible and robust enough to facilitate the non-routine destruction of Syria's declared chemical weapons stockpiles, within politically and physically difficult and dangerous conditions. In terms of the abolition of existing chemical weapons stockpiles in the general sense, we can find evidence right from 1997 through to the present day to argue that the OPCW has navigated delays, the expiration of destruction deadlines, and the significant scientific and technological hurdles demanded of irreversible disarmament through creative, cooperative, and goal-oriented diplomatic and technical work.[6] All of this happened within the legal frameworks provided or supported by the CWC.

The evidence suggests that the CWC can indeed accommodate the seriously difficult task of eliminating a whole class of existing, declared, stockpiled, chemical weapons. A cursory examination of the chemical weapons that have been, in the main, destroyed through the provisions of the CWC reveal that they are certainly of a type (and often quantity) that may fairly be termed as being potential WMD. Without digressing too far into the specifics of the CWC, the chemicals, types, and quantities that are included for verification purposes within the CWC's Annex on Chemicals can largely be attributed to purposes and utilities that reflect relatively indiscriminate, wide-area use, and which rely on their highly toxic properties to inflict serious damage, if not death, upon humans. In particular, we see this with the nerve agents and sulphur mustard. For the purposes of this discussion we can encapsulate these varied agents and their utilities as broadly being representative of

---

[4] The term does, however, appear as part of a set of guideline criteria used to consider whether a chemical may be placed on the Schedules of Chemicals and thus subjected to declarations under the CWC's verification regime.

[5] 'OPCW confirms: All declared chemical weapons stockpiles verified as irreversibly destroyed' News 7 July 2023 https://www.opcw.org/media-centre/news/2023/07/opcw-con-firms-all-declared-chemical-weapons-stockpiles-verified, last accessed 2024/05/08.

[6] Asada (2017).

a Cold War strategic stockpiling and employment of chemical weapons which have the potential to cause mass human destruction.[7]

14.2.2 The Need for a Dynamic View of Abolition

Our view of abolition, however, cannot be meaningfully defined only in a static, historical sense. It is true, and will remain so, that various actors—both state and non-state—may find utilities in chemical weapons, and will seek to (re)acquire them. If this happens, and if our prevention and response is less effective than we might argue has been the effective function of the CWC for its first 30 years, then the treaty model provided may be questioned. Do we hold ourselves hostage to fortune if we say the CWC is a model, and then we find that challenges (re)emerge and they are unsuppressed and uncontained? This is surely one reason why asking ourselves how we maintain the model may be our safest bet, especially so in contemporary conditions that have seen the use of highly toxic nerve agents, such as sarin, VX, and Novichok, as well as sulphar mustard and chlorine, and allegations of the use of riot control agents in war.

14.2.3 Disaggregating Chemical Weapons and WMD

Indeed, this contemporary resurgence of chemical weapons renders it appropriate to disaggregate our often-synonymous understanding of chemical weapons as being WMD, or as noted above, of 'the Cold War era' concept of mass employment. A great many chemical weapons do indeed have the potential to be WMD through the types, quantities, and purposes to which they are employed, and it is to these that the verification regime of the CWC is designed to control and prevent. The Annex on Chemicals remains open, and new chemicals can be added to the Schedules if states parties feel they pose a significant threat to the letter and spirit of the CWC.[8]

Yet, not all chemical weapons are employed on a large scale, nor are they employed because of their aggressively lethal toxic properties, nor are they composed of scheduled chemicals. In light of an actor's objectives, chemicals may be used for their assumed utility in enabling death, incapacitation, area denial and area clearance, punishment, psychological and physical terrorisation, assassination, provocation, framing, harassing, and so forth. Other utilities may be offered. In essence: the use of chemical weapons does not need to equate to metrics of mass destruction, nor do they need to be of the types and quantities outlined within the CWC's annexes, those in particular being chemicals that have been negotiated to be subject to industrial verification. As such, actors may use chemicals that do not appear in the scheduled

---

[7] Trapp (2018).
[8] Kelle (2022).

chemicals lists for violent utilities (i.e. not for those purposes not prohibited by the CWC). These still constitute chemical weapons, as articulated through the CWC's general purpose criterion, but may not be viewed as WMD. The use of chlorine in a residential neighbourhood, or chloropicrin in close combat, might be relevant examples to reflect on.

14.2.4   Looking Forward: Preventing (Re)acquisition and Re-Emergence

Taking these aspects in mind, we find that we can say, at the least, two things. First, the CWC has been successful so far in providing for the international verification of the destruction of stockpiled chemical weapons and the concomitant international verification of the non-production of chemical weapons that are derived from scheduled chemicals listed in the CWC Annex on Chemicals. The second is that as we celebrate these achievements, we must also consider what the CWC offers in terms of preventing the (re)acquisition and re-emergence of chemical weapons in a sense broader than the activities supporting the international verification measures applied to scheduled chemicals. This broader sense not only speaks to the whole range of potential chemicals that can be used as chemical weapons, but to the efforts to prevent and constrain actors' motivations to perceive their utilities and employ them.

These aspects enable our perspective to begin to shift from seeing the implementation of the CWC not as a historical effort but rather a continuous, evolving one. Moreover, it enables us to see the CWC less as a treaty directed to relatively narrow perspectives of WMD, and instead to a wider framing of controlling the purposes to which chemistry is put. The distinction is both conceptual and practical: language is important for defining the scope, and the scope sets the boundaries for what we believe to be the proper meaning of chemical weapons under CWC, and what we should do to implement it. As such, chemical weapons, both those thought of as WMD and those which are not, should be treated as one to prevent a duality of meaning emerging that enables demarcations to be drawn and a creeping legitimisation of the use of certain chemical weapons but not others.

With this view, it becomes difficult and perhaps unwise to claim the CWC as a model if its fundamental purpose is not to abolish those chemical weapons that are WMD, but rather all chemical weapons, irrespective of their type or quantity. And here, then, we might wish to reconsider the question as to whether the CWC provides a model for the abolition of WMD and, instead, consider how the CWC can extend its success in removing the presence of chemical weapons to maintaining their absence.

This point does not invalidate the question as to what the CWC can offer others, but it does create space to think about how efforts in other international regimes, treaties, and mechanisms might provide useful insights that contribute to the CWC's ongoing work. By drawing lessons from the successes and challenges of other treaties and their attendant regimes, can we identify approaches that may be valuable in strengthening the CWC's role in abolishing a subset of WMD specifically, and in a whole class of weapons generally? Given that the sheer scale of risk that the CWC must be implemented to guard against goes well beyond the immediate knowledge, practice, and experiences of our efforts so far, looking outside at how others work in similar spaces might have value. To start that conversation, two thematic areas that appear relevant in view of learning from other regimes are now sketched below, in section three and four.

14.3   Reviewing Relevant Science and Technology

As implementation of the CWC (and thus the work of the OPCW) becomes more explicitly balanced toward preventing (re)acquisition and re-emergence, assurances that emerging threats and challenges are adequately reviewed and considered are required. How the CWC accommodates technological change has been, and will remain, an important enabler of success. The need for scientific and technological advice is clear, insofar as we might understand the CWC to be a treaty concerned with controlling particular technologies, and as such Article VIII paragraphs 21(h) and 45 of the CWC require the establishment of a Scientific Advisory Board (SAB). The SAB plays a central and continuous role in the operation of the CWC. It is worth considering, then, how the OPCW's relationship with, and operation and expectations of, the SAB can remain effective moving forward, especially in light of the evolving contexts of abolition. What may be required to ensure the SAB can continue to deliver advice relevant for states parties as they grapple with the challenges of preventing the (re)acquisition and reemergence of chemical weapons?

14.3.1   Re-Examining Fundamental Questions of Science Advice: The BWC

The contemporary discussions around a science advisory mechanism for the Biological Weapons Convention (BWC) provide a good starting point for reflecting on the fundamental questions that may be posed in relation to reviewing relevant science and technology (S&T). Efforts by state parties to the BWC highlight the difficulties in establishing an effective S&T review mechanism without formal treaty-based provisions or strong institutional frameworks. The complex array of questions being asked, and the challenging conceptual and practical analysis required, demonstrate the contextual nuances and regime topographies that a mechanism must sit within. That the OPCW has a treaty-based mechanism with a proven track record in its implementation is certainly cause for celebration.

However, while these nuances may be significant arbiters of what is deemed possible and appropriate for S&T advice mechanisms, a more fundamental point emerges when comparing the world to which both the SAB and a potential BWC mechanism must frame, review, and reflect on. As the OPCW grapples with preventing the (re)acquisition and re-emergence of chemical weapons, it may find itself navigating a much broader landscape of potential scientific and technological possibilities, where the scope of what is potentially relevant is continually growing. This expanding scientific (and, thus, potential risk) landscape shares some characteristics with the discussions in the BWC to establish an S&T mechanism, particularly around scope, focus, and identification of relevant topics. Both mechanisms are looking out to an increasingly complex, converging, and capacious socio-technical world, both trying to make sense of an ever-expanding spectrum of potential and emerging risks.

If we are to accept that there may indeed be some alignment in their viewpoints, could the ongoing efforts by states parties to the BWC to develop a science advice mechanism offer insights that may be useful for the OPCW? How might the OPCW's S&T advice requirements evolve in contemporary and future contexts, and will those requirements demand an approach that captures an expanding range of scientific and technological developments? Might a future form and function of a BWC S&T mechanism provide useful templates for addressing emerging S&T challenges from the perspective of the OPCW? There may be no clear or binary answer to this, yet there may well be aspects that bear relevance to the SAB that emerge over time, either in terms of providing new options or ideas, or reinforcing the value of what the OPCW already has.

14.3.2 Enhancing Scientific and Technological Stakeholder Engagement

These are broad brushstrokes and are simply suggestive of whether there is a value in thinking a little more deeply about the direction of travel. If we take these seriously enough to follow them through to their practical implications, then we may begin reflecting on how the OPCW can ensure the SAB can have its potential maximised and its capacity to maximise deepened. Established and functioning science advice mechanisms in other regimes may provide some thoughts as to how that maximising of potential might be approached or what it might look like.

One avenue of thinking about this can centre on ways to complement the work of the SAB through additional opportunities for multidisciplinary engagement. For example, the Comprehensive Nuclear-Test-Ban Treaty Organization's (CTBTO) science conferences offer a potential model for broader engagement, bringing together diverse experts to discuss technological advancements and their implications for the treaty. By fostering a multidisciplinary dialogue, the CTBTO science conferences are intended to support bridging between scientific knowledge and policy decision-making with the goal of ensuring that the treaty remains responsive to evolving challenges. Such an approach could be particularly valuable as a wider network of experts and stakeholders become visible as the OPCW's work to implement the CWC broadens. The OPCW does have a history of hosting workshops and conferences, both through the SAB and on other strategic matters such as chemical terrorism. What may be interesting to explore is how a programme of scientific conferences that bring together policy-makers, scientific experts, and other communities from civil society, could be deployed during the inter-sessional periods between review conferences.

To further strengthen its scientific advisory capabilities, the OPCW could explore the establishment of standing thematic bodies to address specific scientific and technological issues. These specialised, longer-standing bodies could bring together experts from different disciplines to tackle targeted challenges over longer periods than may currently be foreseen in the SAB's temporary working group format. In this sense, such a standing body would sit conceptually between the SAB's temporary working groups and the Advisory Board on Education and Outreach, although the body would report to the SAB. The purpose would be to embed a structured and stable approach to expanding the SAB's expertise in, for example, artificial intelligence, that may help ensure its ability to provide comprehensive advice on issues that will have enduring impacts on the OPCW's operational environment and work. In a sense, the International Atomic Energy Agency's 'International Nuclear Safety Advisory Group' may provide one such template in terms of its scope and mandate.

In all cases, state parties would be required to see value in these additions, and to support them both politically and financially. Maintaining the primacy of the SAB, and ensuring such additional activities fall within its mandate without changing the implication of such advice, may be one way to anchor support.

14.3.3 Strengthening Capacity to Enhance Science Advice

The COVID-19 pandemic has highlighted the potential of digital platforms to facilitate remote participation and collaboration, offering new opportunities for the OPCW to engage with a broader range of experts and stakeholders. By leveraging virtual meeting tools, online collaboration platforms, and other digital technologies, the OPCW could expand its reach and tap into a global pool of scientific expertise. For the ideas briefly sketched in the preceding subsection and later in this paper, developing virtual capacity will increase the ability to deliver without significant increases in financial resources. That said, the OPCW's SAB provides invaluable science advice for its states parties, and increasing the financial resource to allow the SAB's practical implementation to be expanded to allow additional standing bodies, events, and activities to take place both in person and digitally may be deemed a worthwhile investment, especially in light of the evolving challenges and opportunities that new, potentially transformative technological advances may bring.

To complement any such effort in terms of the SAB's activities, deepening and broadening science policy expertise within the Technical Secretariat would also bring benefit. Strengthening the organisation's in-house science policy capacity could help support the SAB's work—especially if efforts to expand its footprint were to be pursued. This may involve adjusting staffing and divisional priorities to emphasise scientific capacity, recruiting experts from diverse fields, and providing ongoing training and professional development opportunities for staff. Restructuring to support scientific capacity can also help. For example, creating science officer roles in each major division, who work both vertically in their division, and laterally with other science officers under the guidance of the Science Policy Adviser in the Office of Strategy and Policy may enable both a cross-cutting approach to organisational work, as well as creating increased capacities to support an expanded SAB footprint.

14.3.4 Maintaining Core Strengths While Adopting to New Realities

While it is important to consider potential enhancements to the SAB, we must also be mindful of the adage 'don't try to fix something that isn't broken.' The SAB's proven track record suggests that any changes should be carefully considered and balanced against the need to maintain its core strengths. So as not to undermine the SAB's existing strengths, effectiveness, and international credibility, the reflections here should be understood within the vein of building continuity of relevance and impact rather than seeking change for change's sake.

Ultimately, the key to maintaining and strengthening the OPCW's scientific advisory capabilities may lie in striking a balance between continuity and change, between preserving what works and adapting to new realities. By considering how the experiences of other international organisations may or may not be useful, leveraging the potential of emerging technologies and collaborative platforms, and building on the SAB's proven strengths, the OPCW can position itself to provide the high-quality, responsive, and impactful scientific advice needed to support its mission in the years to come.

14.4 Outreach, Engagement, and Leveraging Expertise Relevant

As the OPCW more squarely sits on a prevention footing, reflecting on how and why outreach and engagement may need to evolve could be useful. As the scope of what is considered relevant science and technology expands, so too will the range of experts, stakeholders, and interlocutors that the OPCW needs to engage with increase, revealing deep socio-technical networks that prevention efforts must be embedded within. While the specific issue of accreditation of civil society to the Conference of State Parties has been a focus in recent years, there are deeper questions to consider about what engagement really means in this new context, what its purpose is, and how it can be most effectively conducted.[9] These questions are particularly salient in light of the evolving implementation landscape outlined in Sect. 14.2 and the need for adaptive and inclusive approaches to ensuring the CWC's continued relevance and effectiveness.

14.4.1 Drawing Lessons from Other Regimes

Looking to other regimes, we find a spectrum of approaches to outreach and engagement that may offer valuable insights. The BWC, for example, has seen attendance by civil society expand in recent years, with a significant number of civil society organisations attending the Ninth Review Conference in 2022. These attendees cover and represent a wide range of scientific and social issues. As the OPCW moves into the non-acquisition era, the profile of civil society seeking to engage (or needing to be engaged) may begin to more closely resemble that of the BWC, especially as the work of the OPCW potentially interfaces with a wider swath of potentially relevant science and technology. Of course, the institutional and organisational structures that facilitate civil society engagement with the CWC and the BWC are considerably different—yet, if the BWC's recent experience shows us anything, as the scientific, technological, and risk landscape opens up for the OPCW (and as the primacy of destruction recedes), more communities than before may believe they have something to contribute, or become important to reach out to. This may well relate to those of the scientific community, but it may also relate to those who also play active roles in shaping the socio-technical landscape in which emerging technologies play out, for example developers, entrepreneurs, data scientists, and others: there will be an expanding range of key stakeholders and enablers that can work with the OPCW. Engaging with these diverse communities will be crucial for ensuring the CWC's continued relevance and effectiveness in the face of evolving challenges, although the strategic approach and practical requirements are not necessarily self-generating or obvious.

Indeed, exploring how other international organisations conceptualise, strategise, and implement such efforts may help stimulate ideas as well as provide new key connections. It can also help better inform understandings of what is possible or desirable. For example, the CTBTO might provide some ideas for structured and targeted engagement. Initiatives like the CTBTO Youth Group and the Group of Eminent Persons demonstrate a sense of purpose and messaging, aimed at building long-term support for the treaty and its goals. As the OPCW looks to engage younger generations and ensure the continued relevance of the CWC beyond its disarmament framing, could similar approaches be adapted to its context? What role might youth networks or ambassador programs play in the OPCW's outreach strategy?

---

[9] Ghionis (2023).

14.4.2 Developing a Multidimensional Engagement Strategy

More broadly, as the OPCW's focus shifts from the relatively narrow, contained issues of destruction and verification to the multi-faceted challenges of prevention, adaptive and responsive outreach strategies can be beneficial. As part of this, engagement should not be viewed as a homogenous activity, but rather a variegated one that engages with different groups for different purposes. For example, scientific experts, industry representatives, public health officials and law enforcement agencies all have important roles to play in preventing, and responding to, re-emergence and (re)acquisition challenges. So too do academics, educators, social and spiritual leaders, although the approaches and practical implications may be very different.

To effectively support this, the OPCW may need to develop a multidimensional engagement strategy that aligns with its evolving priorities and leverages the strengths of different stakeholder groups. This could involve, for example, creating dedicated forums or platforms for engaging with particular non-governmental communities, deepening partnerships with industry to promote best practices in chemical safety and security, collaborating with public health and emergency response agencies to strengthen preparedness and resilience against chemical threats, or finding ways to collaborate on academic research. The key will be to develop an engagement approach that is both comprehensive and strategic, guided by a clear sense of purpose over time. Examining how other international organisations parse and conduct different engagement strategies for various non-governmental communities could be illuminating, for example looking at the International Monetary Fund's Civil Society Policy Forum or its Youth Fellowship; the UN's Office on Drugs and Crime's Civil Society Unit; or by considering how the Food and Agriculture Organization has sought to strengthen its engagement and partnership with civil society within its wider strategic frameworks and goals.

In several cases, the OPCW already pursues these activities and outcomes, although its direct engagement with different non-governmental communities can be fragmented. Ensuring that an underlying, flexible, strategy and purpose unites efforts across all functional or topic areas can bring about enhanced and sustainable outcomes.

Of course, expanding and diversifying engagement also brings challenges. How can the OPCW ensure that it is engaging the right stakeholders, in the right ways, at the right times? How can it manage the logistical and resource implications of a broader engagement portfolio? Here again, lessons from other regimes may prove instructive, especially those that have implemented a strategic effort to enhance engagement within narrow resource envelopes. Further research and consultation in this area may prove fruitful.

14.4.3 Leveraging Technology, Networks, and Expertise

If preventing the (re)acquisition and re-emergence of chemical weapons is to be the defining challenge of the OPCW's future, then leveraging technology, networks, and expertise will be essential to meeting that challenge. Just as science and technology continue to advance at a rapid pace, offering both risks and opportunities, so too must the OPCW's capabilities and partnerships keep pace.

Take, for example, the use of open-source information and analysis for verification and investigation purposes. Other organisations, such as the IAEA and CTBTO, have been pioneering the use of satellite imagery, environmental sampling, trade data, and other publicly available information to complement and enhance traditional verification methods. As the OPCW looks to strengthen its ability to detect and deter violations of the CWC, particularly in view of chemical weapons composed of non-scheduled chemicals, could similar approaches be adapted? What role might open-source tools and emerging technologies like machine learning play in future verification and investigation efforts?

The OPCW could also look to other regimes for lessons on how to effectively leverage and expand its networks of expertise. The OPCW's designated laboratories, for instance, are a key asset in its verification regime, providing critical analytical capabilities. But as the types of samples and analysis required to support non-acquisition efforts evolve, how can this network adapt and grow? Could the experience of the CTBTO in building and maintaining its International Monitoring System and associated radionuclide labs offer valuable insights? What lessons might be drawn from the IAEA's efforts to develop its Network of Analytical Laboratories (NWAL) and expand its capabilities in areas like environmental sampling and nuclear forensics? Considering these questions are, at the least, ways to expand our view as to what may be desirable to ensure the OPCW has the technical capabilities and expert networks needed to respond to evolving challenges and maintain the CWC's effectiveness.

Beyond formal laboratory networks, the OPCW may also need to think creatively about how to harness scientific and technical expertise more broadly. The OPCW's SAB already plays a key role in providing technical advice and assessment, but could its role be complemented by other mechanisms, such as shared scientific working groups or expert rosters designed between regimes? What role might partnerships with scientific professional societies, academies, or research institutions continue to play in expanding the OPCW's access to knowledge and innovation?

These sort of questions focus attention on partnerships and synergies with other international organisations that have science and technology review bodies, or which rely on input through different means, such as those already mentioned or those within, for example, the Food and Agriculture Organisation (FAO), the International Criminal Police Organization (INTERPOL), the UN Environment Programme (UNEP), the World Health Organization (WHO), the World Organisation for Animal Health (WOAH). These organisations often grapple with similar challenges at the intersection of science, technology, society, and policy, and may have valuable insights and best practices to share. Indeed, the OPCW already interfaces with these and other international and regional organisations on a number of other issues, for example through inter-agency coordination on chemical and industrial accidents, safety and security, or terrorist attacks. This international institutional 'quilting' of different organisational competencies can bring different communities together with the aim of sharing, learning, and aligning.[10] By forging collaborative relationships with these bodies from a science policy perspective, the OPCW can tap into a wealth of cross-disciplinary expertise, stay abreast of the latest developments in relevant fields, and ensure that its scientific advice is informed by a comprehensive under-standing of the global landscape.

Indeed, we might take some inspiration from the UN's zero draft Pact for the Future, ahead of September 2024's Summit of the Future.[11] Paragraph 92 calls for, inter alia,

> cross- and trans-disciplinary collaboration and a strong science-policy-society interface in order to build trust in science. We encourage the United Nations system to take an active role in forging closer links with national and multilateral science advisory bodies to optimally leverage science, technology and innovation […]

14.4.4 Investing in Human Capital and Institutional Capabilities

Of course, leveraging technology and expertise also requires investment in the OPCW's own human capital and organisational capabilities. As the demands on the Technical Secretariat evolve and expand in the non-acquisition era, how can it attract and retain the right mix of skills and knowledge? What lessons from other international organisations on issues like staff development, knowledge management, and succession planning may prove valuable for the OPCW? In particular, how do other international organisations integrate and train staff for roles which there may have been little precedent for in the past? And, crucially, who may have useful experiences about enhancing institutional memory and stand-by knowledge, skills and capacity, in the OPCW's case for future destruction-related activities?

Ultimately, preventing the re-emergence of chemical weapons will require not just political will and legal norms, but also practical tools and capabilities. By learning from the experiences of other regimes and investing in the continuous strengthening of its technological edge, expert networks, and human capital, the OPCW can position itself to meet the challenges of the future. This will require both vision and resources, as well as a willingness to adapt and innovate in the face of rapid change. But if the history of the CWC has taught us anything, it is that the cost of failure is too high to ignore.

---

[10] Ghionis (2022).

[11] Pact for the Future' Zero draft of 24 January, Summit of the Future, September 2024. Available at https://www.un.org/sites/un2.un.org/files/sotf-co-facilitators-zero-draft_pact-for-the-futurepdf, last accessed 2024/05/04.

14.5   Concluding Remarks

As we reflect both on the impressive achievements of the OPCW in implementing the CWC over the past quarter-century and on its role in preventing (re)acquisition and re-emergence, it is clear that while the CWC has served as a useful template for the destruction of a category of WMD, its ability to permanently abolish all forms of chemical weapons remains to be seen. Our experience so far suggests that the CWC itself has the potential to be such a model, but ensuring its continued relevance and effectiveness will require a proactive and adaptive approach to its international implementation—to say nothing of the significant challenge posed by national implementation, which falls outside the scope of the present paper, but deserves focused attention. Aside from this, what is obvious is that the real challenge we face is how to ensure the CWC is sustainably and effectively implemented. The discussion in this paper has sought to underscore the value of looking beyond the CWC itself for guidance on how to strengthen the OPCW's ability to implement it.

Strengthening the OPCW's science and technology review and advice mechanisms, including through an expanded role for the Scientific Advisory Board and increased scientific capacity within the Technical Secretariat, could support efforts to keep pace with technological change. This may involve exploring new approaches to stakeholder engagement, such as multidisciplinary conferences and thematic standing bodies, as well as investing in digital platforms and collaborations to expand the OPCW's reach and capabilities.

Adapting the OPCW's outreach and engagement strategies to the evolving nature of CWC implementation, including through the development of a multidimensional approach aligned with prevention priorities, can help ensure the continued relevance and effectiveness of these efforts. By learning from the experiences of other regimes, the OPCW may be able to develop more targeted and impactful strategies for building support and expertise.

Leveraging lessons from other regimes on the use of advanced technologies, the optimisation of expert networks, and the retention of critical expertise can enhance the OPCW's capabilities and better position it to address future challenges. This may involve exploring the potential of open-source information and analysis in a more expansive view of verification, augmenting and adapting the OPCW's laboratory network, and investing in staff development and knowledge management.

However, realising this potential will require more than just technical fixes. It will require a shared vision and commitment from all stakeholders, particularly the states parties to the CWC. The success of the CWC ultimately depends on the political will and financial support of its states parties. They must believe in the mission of the OPCW, not just as a historical achievement, but as an ongoing necessity in a world where the threat of chemical weapons persists.

States parties must recognise that investing in the OPCW's scientific and technological capabilities, its outreach and engagement efforts, and its human capital is a strategic imperative. Science, technology, and people are the keys to unlocking successful prevention. The OPCW and its member states must be willing to invest in a long-term vision, and provide the necessary resources and support to realise it.

Civil society also has a crucial role to play in shaping the future implementation of the CWC. Non-governmental organisations, academic institutions, industry associations, and other stakeholders must continue to enable, visualise, analyse, and consider these futures, both to inform states and to inspire action. They can provide valuable expertise, perspectives, and networks that complement and enhance the OPCW's efforts. They can also serve as a bridge between the OPCW and the wider public, helping to build understanding and support for its mission.

Ultimately, the success of the CWC will depend on the ability of all stakeholders to work together towards a common goal. The OPCW can no longer afford to see itself as working in isolation, insulated from the broader challenges and opportunities of the international system. Instead, it must embrace a networked approach, forging partnerships and collaborations with other regimes, organisations, and actors. It must be prepared to learn from others, to adapt to new realities, and to innovate in the face of complex challenges.

By embracing the lessons of others—be they things to do or things to avoid—and investing in the continuous improvement of its capabilities and practices, the OPCW can not only strengthen its own effectiveness but also serve as a powerful model for how international institutions can adapt to new challenges. This effort will require further research and reflection to understand how specific approaches can be adapted to the unique context of the CWC and the OPCW, as well as the establishment of dedicated resources and mechanisms to support ongoing learning and evolution.

The CWC's role as a model for WMD abolition is not a static achievement nor a stand-alone goal. It is, rather, a dynamic and ongoing process. Moreover, such a model is achieved and maintained only as part of the success of the wider effort to achieve the complete and lasting abolition of chemical weapons.

References

Asada M (2017) 'The OPCW's arrangements for missed destruction deadlines under the Chemical Weapons Convention: an informal noncompliance procedure. Amer J Int Law, 108

Black R (2016) Development, historical use and properties of chemical warfare agents. In: Worek F, Jenner J, Thiermann H (eds) Chemical warfare toxicology: Volume 1: Fundamental Aspects (The Royal Society of Chemistry; Cambridge) p 12

Carus WS (2012) Defining "Weapons of Mass Destruction"' Occasional Paper, No. 8 (National Defense University Press; Washington, D.C.)

Ghionis A (2022) Change and continuity in the organisation for the prohibition of chemical weapons. Doctoral Thesis (University of Sussex; Falmer) p 265ff

Ghionis A (2023) The OPCW and civil society: considerations on relevant themes and issues. Working Paper 10 (CBWNet; Berlin)

Kelle A (2022) Adding Novichok nerve agents to the CWC Annex on Chemicals: a technical fix and implications for the chemical weapons prohibition regime' WMD Compliance Enforcement Series, Paper 13 (UNIDIR; Geneva)

Manly R (2007) Verification under the chemical weapons convention. In: Bellany, I. terrorism and weapons of mass destruction responding to the challenge (Routledge; Oxon)

Robinson JP (2002) What should be the scope of the CWC? A workshop report. CBW Conv Bull, 55

Robinson JP (2008) Is the OPCW implementing the OPCW definition of chemical weapons?. The CBW Conv Bull, 78

Trapp R (2018) The future chemical weapons convention—outdated model or CWC 2.0. In: Crowley M, Dando M, Shang L (eds) Preventing chemical weapons: arms control and disarmament as the sciences converge (Royal Society of Chemistry; London)

Alexander Ghionis is a Research Fellow in Chemical and Biological Security at the Harvard Sussex Program, based in the Science Policy Research Unit (SPRU) at the University of Sussex. His work explores the evolving landscape of chemical and biological weapons (CBW) control, with a current focus on the implications of artificial intelligence for the CBW Conventions. Recent analysis also includes an examination of OPCW-civil society dynamics and of the 2023 Fifth Review Conference of the Chemical Weapons Convention. Beyond his primary role, he serves as a non-resident Fellow for UNIDIR's Weapons of Mass Destruction Program, an associate researcher on the German CBWNet project, and a member of the Chemical Weapons Convention Coalition.

Chapter 15
The Chemical Weapons Convention as a Model for Verification of the Biological Weapons Convention

Alexander Kelle

Abstract The prohibitions of chemical and biological weapons have been codified in two international treaties, the Chemical Weapons Convention (CWC) and the Biological Weapons Convention (BWC). Against the background of renewed discussions among BWC states parties about compliance and verification measures for the BWC, this paper asks to what extent the CWC with its complex verification system might serve as a model for verification of the BWC. To this end the first section of the paper will outline similarities and key differences between the two treaties. This will be followed in section two by a brief discussion of the CWC verification system with both routine and non-routine verification elements. Section three will analyze the organizational aspects of verification under the CWC and section four will highlight the impact of science and technology on the evolution of verification under the CWC. The final section will summarize the argument and provide an assessment of the suitability of the CWC as a model for verification under the BWC.

Keywords Verification · Chemical weapons convention · Biological weapons convention

15.1 Similarities and Key Differences Between the CWC and BWC

A. Kelle (✉)
Institute for Peace Research and Security Policy (IFSH), University of Hamburg, Berlin Office, Germany
e-mail: kelle@ifsh.de

© The Author(s) 2026
B. Friedrich et al. (eds.), *Thirty Years of the Chemical Weapons Convention (CWC)*,
https://doi.org/10.1007/978-3-031-98854-7_15

173

15.1.1 Core Obligations Under the Two Treaties

There is significant overlap with respect to the core obligations under the CWC and BWC.[1] In the CWC, which entered into force in 1997, these are contained in Article I of the treaty and encompass several norms, understood here as "standards of behavior in terms of rights and obligations" (Krasner 1982, p. 186) to guide state party behavior. More specifically, according to Article I of the Convention states possessing chemical weapons (CW) undertake to destroy their stockpiles and all states parties equally agree to refrain from the development, production, acquisition, stockpiling, transfer and use of chemical weapons. In order to put the CW destruction norm into effect, CWC states parties agree in Article III on detailed declarations regarding chemical weapons, including old and abandoned CW, and CW-related facilities. States parties have to submit these declarations to the Organization for the Prohibition of Chemical Weapons (OPCW), where they are processed by its Technical Secretariat (TS). Articles IV and V contain the standards of behavior for CW possessor states concerning chemical weapons and CW production facilities, including their declaration and inspection obligations. Articles IX and—to some extent—Article X of the Convention express the inspection norm, for whose realization challenge inspections and investigations of alleged use of CW are the key instruments foreseen in the treaty. Taken together the verification provisions of the CWC define both the behavioral standards in terms of norms, rules and procedures for the conduct of verification measures to be undertaken, and the verification goals to be achieved.

The Biological and Toxin Weapons Convention contains similar normative guidance in relation to the basic prohibitions related to biological and toxin weapons. These concern the non-acquisition, non-retention, non-transfer, and non-use norms contained in Article I of the BWC. Although there is a rudimentary destruction norm contained in Article II of the treaty, this is expressed in the form of an expectation that possessor states will have destroyed their BW stockpiles at the latest nine months after the convention has entered into force for them. However, the BWC does not contain any normative guidance for states parties to submit declarations or accept inspections. Furthermore, there is no international organization established under the BWC. These gaps in the structure of the BW prohibition regime have their origins in the time and context in which the treaty was negotiated in the late 1960s/early 1970s, a process dominated by the then superpowers, the United States of America and the Soviet Union (Kelle and Siegmann 2023).

In sum, similarities exist between CWC and BWC as far as the broad scope of the treaties is concerned. Both CWC and BWC prohibit a category of weapons of mass destruction (WMD). They require that possessor states destroy existing CW and BW stockpiles and that all states parties refrain from acquiring, producing etc. new

---

[1] The text of the Chemical Weapons Convention can be accessed on the OPCW website at https://www.opcw.org/chemical-weapons-convention/articles, the text of the Biological and Toxin Weapons Convention at https://front.un-arm.org/wp-content/uploads/2020/12/BWC-text-English-1.pdf, last accessed 2024/02/23.

ones. However, this is as far as the similarities in the prohibitions formulated extend. The two treaty-based regimes differ significantly with respect to the organizational support that is available to support treaty implementation, including a science advice mechanism. The CW and BW prohibition regimes differ also regarding agreement on a verification principle to inform the norm-based rights and obligations for regime participants, i.e. CWC and BWC states parties, and concerning the object of prohibition under each treaty. The following sub-section will address the latter two of these four areas of differentiation.

15.1.2 Key Differences Between CWC and BWC

The object of prohibition. Although often treated together as CBW, for the most part, the respective prohibitions under the CWC and BWC have a significantly different focus. While both prohibitions are comprehensive in nature and not limited by a list, for example the Schedule of Chemicals annexed to the CWC or some other control list for verification or export control purposes, they target different objects. According to Article II of the CWC, these are primarily "toxic chemicals and their precursors, except where intended for purposes not prohibited under this Convention, as long as the types and quantities are consistent with such purposes". The purposes not prohibited include *inter alia* "industrial, agricultural, research, medical, pharmaceutical or other peaceful purposes", as well as protective purposes and military purposes "not connected with the use of chemical weapons and not dependent on the use of the toxic properties of chemicals". All these definitions encompass both manmade toxic chemicals and those of natural origin, i.e. toxins produced by bacteria (*chlostridium botulinum*), plants (ricin) or marine algae (saxitoxin). The latter two have also been included in the CWC Schedules.

The BWC also prohibits development, stockpiling, production, or transfer of biological agents and toxins of "types and quantities" that have no justification for protective or peaceful use. However, the BWC does not explicitly prohibit the use of biological weapons, an omission that was later addressed by subsequent review conferences. Given the lack of a verification system and the resulting absence of the need to operationalize this with a categorization of biological agents of concern, the treaty is also silent on which biological agents it covers. Usually, biological agents of concern are divided into bacteria, rickettsiae, viruses and toxins, the first three categories of which are replicating agents and would therefore pose different challenges to a verification system than those faced by one addressing toxic chemical compounds. Yet, any verification system would require a prior understanding among states concerned that verification of treaty compliance is useful and possible in ascertaining treaty compliant behavior, i.e. a verification principle.

(Dis-)Agreement on verification principle. International institutions, such as the CBW prohibition regimes, are based on a set of principles about what is right and wrong and how to govern a particular issue area, or, in the words of Stephen D.

Krasner, "principles are beliefs of fact, causation and rectitude" (Krasner 1982, p. 186). The CBW prohibition regimes share three such principles (Kelle 2014), but crucially differ on the verification principle, which is present in the CW prohibition regime but absent in the issue area of BW prohibition. According to this principle the treaty-compliant behavior of regime participants, that is states parties to the CWC, has to be verified for the whole system to work. This principle then informs the declaration, inspection and investigation norms of the CW prohibition regime, which are spelled out in CWC Articles III, IV, V, VI, IX and—to a lesser extent—X. Without this basic agreement about the necessity of verification, and the three additional principles that (1) the use of CW constitutes an abhorrent act of warfare, (2) the defense against CW is permitted, and (3) the civilian uses of chemistry are permitted, none of the regime norms that provide guidance for state action under the CWC could have been agreed.

As indicated above, the USA and Soviet Union strongly influenced negotiations for the BWC at the time. When the United Kingdom presented the first draft treaty text for what would become the BWC, this contained verification provisions. However, neither US president Richard Nixon nor the Soviet leadership were interested in retaining such a provision in the treaty. While the former had a rather specific set of interests in a quick negotiating success (Tucker 2002), the Soviet Union would not accept any verification measures involving on-site inspection activities until after the mid-1980s (Leitenberg 1996). BWC states parties undertook the first attempt to seek agreement on verification measures, such as declarations, visits, inspections and investigations, which would have required a verification principle in the second half of the 1990s in the context of the so-called Ad-hoc Group negotiations (see Sect. 15.2.2).

15.2 Verification Measures Under the CWC

15.2.1 Routine and Non-Routine Verification Measures

Verification measures under the CWC are guided by the declaration, inspection and investigation norms. In practical terms, states parties have to submit a variety of declarations to the OPCW Technical Secretariat and have to accept OPCW inspections on a subset of the facilities declared. In particular, states parties have to declare the CW they possess as well as related facilities for their production, storage and destruction (Article IV and V). States parties also have to declare certain activities not prohibited (Article VI). Both sets of declarations then form the basis of OPCW inspections of relevant facilities under the corresponding treaty provisions. These two broad elements form the CWC's routine verification system. In a nutshell, one deals with the verification of destruction of declared CW stockpiles, related facilities, and old and abandoned CW, and the other verifies the compliance of states parties with non-prohibited activities.

Since the CWC entered into force in 1997, verification activities have taken up the largest part of the OPCW's resources. Among these, verification of CW destruction through systematic verification consumed the majority of funds. This changed only after 2017 when the Russian Federation completed its destruction program. It is also noteworthy that none of the declared CW possessor states met the destruction deadline ten years after the treaty had entered into force, as specified in the Convention, and a few also missed the extended deadline another five years later in April 2012 (Kelle 2014). Verification activities under Article VI, covering mostly industry facilities, were started in a staggered manner with higher-risk Schedule 1 facilities being inspected immediately after CWC entry-into-force and so-called other chemical production facilities only in the year 2000, i.e. three years later. More recently, changes in the global distribution of chemical industry and in the make-up of industry facilities, the latter of which have also been caused by scientific and technological advances, has triggered discussions among CWC states parties on how to adapt the industry verification system to changing circumstances, so that it can continue to provide continued confidence in compliance (Ghionis and Kelle 2024).

Non-routine verification activities have to be requested by CWC states parties in case they suspect non-compliance by another state. Such a request can take the form of either a challenge inspection (CI) under Article IX or an investigation of alleged use of CW (IAU) under Articles IX and X of the treaty. However, in the more than quarter century since the Convention entered into force, no state party has made such a request, notwithstanding the occasional accusation of treaty violations at CWC review conferences.

Instead of relying on the verification tools foreseen in the CWC, its states parties have supported alternative, and more flexible, verification tools implemented by the OPCW Technical Secretariat. These were created in order to clarify the declaration submitted by the Syrian Arab Republic after it had acceded to the convention in October 2013, and to establish the facts surrounding repeated reports of CW use in that country. The so-called Declaration Assessment Team (DAT) and Fact-Finding Mission (FFM) were established by the then Director-General Ahmet Üzümcü in 2014. The initial agreement on the DAT and FFM was concluded between OPCW secretariat and the Syrian government, with subsequent endorsement by the OPCW Executive Council. Over the past decade, the DAT has conducted 24 rounds of consultations in country and two more limited deployments. As a result, the DAT has identified "a total of 24 outstanding issues related to the gaps, inconsistencies, and discrepancies noted in the initial and subsequent declarations by the Syrian Arab Republic." Over the same period, the FFM has investigated over 70 cases of alleged CW use in Syria and confirmed the use of chlorine in 14 instances, the use of sulphur mustard in three cases, and the use of sarin in another three cases.[2]

While it was the FFM's task to determine whether CW had been used, the OPCW-UN Joint Investigative Mechanism (JIM) was set up in August 2015 by UN Security Council Resolution 2235 (2015) to identify the perpetrators of such use. The JIM's

---

[2] OPCW, Note by the Technical Secretariat. Accession of the Syrian Arab Republic to the Chemical Weapons Convention: Ten Years On, document S/2213/2023, The Hague (2023), quote on p. 3.

mandate was extended once for 12 months, but blocked by Russia in the fall of 2017. During its short-lived existence, the JIM issued several reports and identified both the Syrian government and the so-called ISIL/Da'esh as responsible for CW use. After Russia had blocked the continuation of the JIM's work, several CWC states parties pursued the internalization of its attribution function into the work of the OPCW. The resulting Investigation and Identification Team (IIT) within the technical secretariat was set up in 2018 and issued three reports in 2020, 2021, and 2023, respectively. In all CW use cases investigated, the IIT identified the Syrian armed forces as responsible for the attacks (Kelle 2023).

In sum, the CWC provides the foundation for an elaborate verification system composed of routine and non-routine verification measures. Routine verification between 1997 and 2023 was undertaken with the twin goals of verifying the destruction of declared CW stockpiles in possessor states and providing continued confidence in compliance of all states parties with the provisions of the treaty, in particular its Article VI. With the completion of CW destruction activities in the US in July 2023, the industry verification system and its adaptation to changing realities will increase in relative importance. In the realm of non-routine verification measures, it is noteworthy that the tools foreseen in the CWC have never been used. Instead, the OPCW developed verification instruments to deal with particular use cases. In the case of the IIT, this has also shown a significant shift in the focus of the verification measures, going well beyond the confirmation of CW use to include the identification of perpetrators. As a result, CWC states parties have upheld the declaration, inspection and investigation norms, but continue to adapt their implementation to changing needs as perceived by CWC states parties.

15.2.2   Implications for the Verification of the BWC

BWC states parties agree to "never in any circumstances to develop, produce, stockpile, or otherwise acquire or retain [...] microbial or other biological agents, or toxins whatever their origin or method of production of types and in quantities that have no justification for prophylactic, protective, or other peaceful purposes."[3] While this text is a forerunner of the general obligations contained in the CWC, the lack of specificity and absence of definitions that are contained in the CWC, creates some gray areas at the margins of the scope of the treaty. The BWC does not, for example, contain a prohibition on biological weapons' research. It also does not provide any guidance on where the dividing line between research and development is situated. Similarly, the BWC allows for prophylactic and protective purposes, but does not specify what would constitute types and quantities of any BW agent for which a justification would exist. These uncertainties concerning the exact nature of prohibited and permitted activities creates potential problems for treaty verification. However, given the political realities of the time, the BWC does not contain any provisions for the verification

---

[3] See fn. 1.

of states parties' compliance with the treaty. In case a state party suspects a treaty violation, it has two options available under the BWC. According to Article V, it can launch a consultation process and seek to clarify its compliance concerns. The treaty provision was invoked only twice: by Cuba in 1997 (Zilinskas 1999) and by Russia in 2022 (Zanders 2023). In neither case did the consultation process result in a consensual resolution of the compliance concern. BWC states parties can also lodge a complaint with the UN Security Council under BWC Article VI, which is what Russia did following the lack of a clear outcome of the Article V process it had pursued. However, the Security Council rejected this further Russian attempt to discredit the United States and Ukraine over alleged offensive BW activities, with only China supporting the Russian draft resolution.[4]

Given the lack of a verification system under the BWC, beginning at the Third BWC Review Conference in 1991 its states parties engaged in a systematic effort to remedy this gap in the normative treaty structure. An ad hoc group of governmental experts to identify and examine potential verification measures (VEREX) met in 1992 and 1993 and submitted a report on over twenty verification measures to strengthen the BWC. On this basis a Special Conference of BWC states parties in 1994 agreed to set up another ad hoc group (AHG) for negotiations on a legally binding protocol to strengthen the effectiveness and improve implementation of the BWC (Sims 2001).

The mandate for such a protocol contained a complex set of four areas to be considered: compliance measures, confidence-building measures (CBM), definitions and objective criteria, and Article X of the BWC, the latter of which expresses the cooperation norm of the BW prohibition regime. As one participant in the negotiations later noted, "Instead of launching a focused negotiation, the compromise mandate rolled Trojan horses within the walls of the Biological Weapons Convention" (Ward 2004, p. 186). Negotiations began in 1995, and in 1997 moved to a rolling text containing elements of potential protocol language submitted by states parties, and, in the later stages of the negotiations, also drafted by the AHG chairperson. By July 2001, when the US administration of George W. Bush formally ended the protocol negotiations, several issues in each of the four areas of the AHG mandate were still controversial. Here the focus will be on compliance measures, i.e. those related to the establishment of potential declaration, inspection and investigation norms.[5]

Declarations under a BWC protocol. BWC states parties found it easy to agree on the inclusion of declarations into the compliance protocol for two reasons. Such reporting was commonplace under the CWC, as well as in the context of peaceful nuclear activities reported to the International Atomic Energy Agency. Secondly, most BWC states parties saw the CBM agreed upon in 1986 and expanded in 1991 as a positive step in enhancing confidence in compliance (Chevrier and Hunger 2000).

---

[4] United Nations, Security Council Rejects Text to Investigate Complaint Concerning Non-Compliance of Biological Weapons Convention by Ukraine, United States. Meetings Cover-age Security Council SC/15095, 9180th meeting (2 November 2022), https://press.un.org/en/2022/15095.doc.htm, last accessed 2024/02/29.

[5] The following three subsections are largely based on Littlewood, J., *The Biological Weapons Convention: A Failed Revolution*, Ashgate, Aldershot (2005).

This however, is not to suggest that agreement on the scope of declarations could be achieved easily. As Littlewood has noted, a minimalist approach would have sought to put the politically binding CBMs on a firmer legal basis through inclusion in the BWC Protocol. He lists as a baseline "mandatory declaration requirements" for half a dozen BW-related activities or facilities, including maximum biological containment facilities, biodefense research and development programs, and outbreaks of infectious disease. Although Littlewood concludes that the set of declaration requirement contained in the final version of the draft protocol text goes beyond such a minimalist approach, he also notes the absence of transfer declarations related to certain listed agents, toxins or equipment—as requested by the Non-Aligned Movement during AHG negotiations. He also highlights that the scope of declarations on biodefence, on high biological containment and on production facilities remained contested throughout the protocol negotiations. The weak biodefence declaration requirements proposed by the US would in fact have made it more difficult to shed light on the lingering concerns about a continuing clandestine Russian BW program.

Inspections (by another name) under a BWC protocol. The equivalent to routine on-site inspections under the CWC were discussed under the label of "visits" in the context of the AHG negotiations. This was mostly due to the US maintaining that the BWC is not verifiable, unlike other arms control treaties, and hence a different terminology was applied. Although many BWC states parties regarded on-site routine activities by an implementing organization as an important element of the BWC protocol being negotiated, in terms of substance there were numerous variations on the theme of visits and the exact purpose they would have. In general terms, visits were conceptualized as a follow-up activity to declarations submitted by states parties, but BWC states parties already on the question whether there should be a conditionality established between the declarations and routine visits, or whether one should put a greater emphasis on randomly-selected visits. As noted by Littlewood, diverging interests led to a situation in early 2001 where the debate on visits was still complicated, with several key issues still not settled. These included *inter alia* the purpose of visits, their mandatory, or alternatively, voluntary nature, the types of facilities to be visited, and the types of on-site activities that would be allowed. Based on the final version of the draft protocol text, Littlewood summarizes that its provisions for visits "was weaker than the comparable function under the CWC and much weaker than the original vision proposed by reformists." Over the course of the AHG negotiations BWC states parties promoting a minimalist approach to visits, such as the US and others, had prevailed.

Investigations under a BWC protocol. Of the three norms relevant for ensuring compliance with the BWC, the investigation norm is the one most clearly spelled out in the treaty text itself. However, the investigation provisions contained in BWC Article VI suffer from the shortcoming of having to involve the UN Security Council. Thus, any such investigation would depend on none of its five permanent members exercising its veto. Due to this discriminatory element and dependency on the Security Council, the efforts to strengthen the BWC with a compliance protocol featured investigations as a central element. Not surprisingly, diverging views among states parties could be observed in this area, too. According to Littlewood, Russia took the minimalist position in the AHG negotiations by asserting that the available procedure under Article VI only needed to be developed further. This should be done in a way to not diminish the role of the UN Security Council. In contrast, the majority view held that a substantial investigation process by an independent organization should be included in the BWC protocol. The latter view prevailed, and as a result both field investigations and facility investigations found their way in to the AHG chairperson's composite draft protocol text. While not identical to the challenge inspections and investigations of alleged use under the CWC, certain similarities do exist. Lengthy AHG debates revolved around the initiation of the different types of investigations and the scope of an investigation, i.e. the degree to which a receiving state party could "manage" the access of an inspection team. On balance, Littlewood concludes, those pressing for more far-reaching reforms of the BWC prevailed and the protocol would have been likely to contain a useful investigation mechanism.

A renewed mandate for talks about verification and compliance. When the US administration of George W. Bush in July of 2001 withdrew its support for the BWC Protocol negotiations, the topics of verification and compliance "had been off the official BWC agenda and at the centre of intense political disputes in the BWC regime" (Jakob 2023). Only at the 9th BWC Review Conference in late 2022 did its states parties agree to establish a working group that would look inter alia at compliance and verification measures as part of the intersessional process for 2023 to 2026. This became possible after the US administration of Joseph Biden announced its willingness in late 2021 to engage in such talks. These are not formal negotiations leading to a legally binding instrument, as they were pursued by the AHG 25 years ago. Rather, the US is willing to "examine possible measures to … enhance assurance of compliance".[6] However, this is just one of the issue areas that the BWC working group has to address in the four-year period up to the 10th review conference in 2027. Given the time-limited mandate and competing issues such as assistance and protection, and a science and technology review mechanism, it remains to be seen how much progress can be achieved on concrete verification measures for the BWC.

15.3   Organizational Aspects of Verification Under the CWC

15.3.1   Collaboration Between CWC States Parties and OPCW Technical Secretariat

Article VIII of the CWC creates the Organization for the Prohibition of Chemical Weapons.[7] The OPCW is composed of three main organs: the Conference of the States Parties as "the principal organ of the Organization", the Executive Council as

---

[6] United States of America, *Statement by Under Secretary of State Bonnie Jenkins to the 2021 Biological Weapons Convention Meeting of States Parties*, Geneva, November 22, 2021, https://doc uments.unoda.org/wp-content/uploads/2021/11/US-FINAL-updated.pdf, last accessed 2024/02/29.

[7] See fn. 1.

its executive organ, and the Technical Secretariat, which supports the Conference and the Council in performing their tasks and "shall carry out the verification measures provided for in the Convention."[8] All states parties to the CWC automatically become member states of the OPCW and undertake to cover the costs of the organization according to a specific distribution key. The states parties are also subject to the verification mechanisms of the organization, and agree to certain chemical industry facilities being inspected by the TS.

Achieving the verification goals under the CWC therefore requires the collaboration between CWC states parties and the OPCW Technical Secretariat. In this division of labor, states parties have to submit declarations and receive inspections. The OPCW TS, on the other hand, receives initial and annual declarations (and checks their accuracy), selects facilities and plant sites for inspection (according to CWC provisions, political guidelines by SPs, and a selection algorithm created on this basis), conducts inspections (both industry and CW-related), reports on inspections (both individually and in summary fashion in the annual Verification Implementation Report (VIR) and maintains readiness to conduct CI and IAU.

15.3.2 Implications for the Verification of the BWC

The BWC does neither contain verification measures, nor does it create an international organization. As a result, it remains silent on the organizational aspects of potential verification activities. This absence of organizational support for the implementation of the BWC was increasingly noted after the first CBMs were introduced at the 1986 BWC review conference. The gap was subsequently a topic of discussion both at the third BWC review conference and during the VEREX process. Hence, when the negotiations of the AHG started, the gap in regime structure plus the increasing availability of organizational support for other arms control treaties, such as the CWC, or the Comprehensive Test Ban Treaty under negotiation, led to a widespread expectation that an organization to implement the BWC protocol would be established. However, not all states parties shared this view: as noted by Littlewood, Iran proposed to entrust the conduct of inspections under the protocol to the World Health Organisation and Russia played down the need for a new professional organization, as it sought to preserve the primacy of the UN Security Council in dealing with breaches of the BWC.

During AHG negotiations it became clear quickly that the size and composition of a new organization for the prohibition of biological weapons would depend on the functions the organization would expect to perform. However, given that there was no expectation of BW stockpiles to be declared and inspections of a BW destruction process to be conducted, many expected a "lean and mean" organization to emerge. Littlewood highlights two organizational issues in particular as controversial during AHG negotiations. One was the composition of the executive council of the future

---

[8] Ibid.

organization. The other one was more directly related to the question of verification activities and concerned questions of immunity and liability of the organization and its staff. In the end, more draconian measures that would have entailed waiving the immunity of the organization or its Director General in case of a breach of confidentiality, which would compromise a state party's national security or confidential business information, were not included in the final draft version of the protocol text.

Following the collapse of the AHG process, successive review conferences from 2002 onward have managed to agree on an inter-sessional work program with topics for annual meetings of experts and of states parties. In the course of this process, states agreed in 2006 that consistent organizational support was required for these regular meetings. This led to the creation of the Implementation Support Unit (ISU), whose main functions include administrative support and assistance, including the preparation of background documentation for annual meetings as well as BWC review conferences, national implementation support and assistance, and support and assistance in the processing of the CBMs. Over time, the ISU has grown to a total staff of four, a far cry from the OPCW, which still has more than 400 posts in its technical secretariat.[9]

15.4 Science, Technology and Treaty Verification

15.4.1 The Science and Technology Ecosystem Under the CWC

The relevance of science and technology (S&T) for the CWC and its implementation, including its verification system, is widely acknowledged, including in the treaty text itself. Article VIII[10] of the CWC contains the requirement for CWC states parties to adapt to scientific and technological (S&T) developments of relevance to the treaty when they convene for CWC Review Conferences every five years. It states that beginning with the first such conference treaty "reviews shall take into account any relevant scientific and technological developments. At intervals of five years thereafter, unless otherwise decided upon, further sessions of the Conference shall be convened with the same objective".[11]

In addition to this procedural element, the CWC in Article VIII (21) (h) also establishes an organizational structure in the form of the Scientific Advisory Board (SAB), to enable the OPCW Director General to provide S&T advice to the policymaking organs and member states of the organization. Originally composed of 20 scientific

---

[9] United Nations Office for Disarmament Affairs, Implementation Support Unit Homepage, https://disarmament.unoda.org/biological-weapons/implementation-support-unit/, last ac-cessed 2024/02/29.

[10] OPCW, Decision: Addition of Five Seats to the Scientific Advisory Board, document C-9/DEC.13, The Hague (2004).

[11] See fn. 1.

experts, the Conference of States Parties decided in 2004 to add an additional five seats to the SAB, thereby increasing its membership to 25. Since its establishment, the SAB has provided advice on many S&T-related verification topics. This has happened as part of the reporting on its regular board meetings (taking place twice a year), in the form of Temporary Working Groups (TWG) of the SAB, utilizing external expertise going beyond the SAB membership, and in the context of the SAB reports for CWC Review Conferences. One TWG in 2015, for example, dealt explicitly with verification issues, addressing *inter alia* verification technologies and methodologies in other arms control contexts that could benefit verification under the CWC, and new and emerging technologies that may enhance existing OPCW verification capabilities.[12]

The comprehensive report submitted by the SAB on developments in science and the response to this report by the Director-General are key inputs for this. The most recent SAB report to the 5th CWC Review Conference in May 2023 again contained a section on science and technology relevant to verification, including 17 detailed recommendations.[13] The Director-General's response provides his interpretation of the SAB report and recommendations and highlights several points for states parties' consideration.[14]

In sum, the OPCW science and technology review and advice mechanism has over time evolved into a well-developed eco-system that monitors S&T advances relevant to the CWC through the SAB's regular work, addresses specific issues including those related to verification through the in-depth work of TWGs and regularly, at least every five years on the occasion of a review conference, issues recommendations for the consideration of CWC states parties.

15.4.2   Implications for the Verification of the BWC

Mirroring the gaps in regime structure and treaty-based guidance in relation to verification in general and in the area of organizational governance, science advice in the BWC context is much less developed than in the CWC. As the BWC is a much less operationally defined treaty, the space devoted to the need to review science and technology is minimal to non-existent. Because the BWC does not have an organizational backbone like the OPCW, the treaty text does not provide any role

---

[12] OPCW, *Verification Report of the Scientific Advisory Board's Temporary Working Group*, The Hague (2015).

[13] OPCW, *Report by the Director-General: Report of the Scientific Advisory Board on Developments in Science and Technology to the Fifth Special Session of the Conference of the States Parties to Review the Operation of the Chemical Weapons Convention*, document RC-5/DG.1, The Hague (2023).

[14] OPCW, *Report by the Director-General: Response to the Report of the Scientific Advisory Board on Developments in Science and Technology to the Fifth Special Session of the Conference of the States Parties to Review the Operation of the Chemical Weapons Convention*, document RC-5/DG.2, The Hague (2023).

assignments or processes for S&T review, let alone advice, in support of treaty implementation, except for the stipulation that the quinquennial review conferences "shall take into account any new scientific and technological developments relevant to the Convention."[15]

During negotiations by the AHG for the BWC protocol, S&T issues were discussed also in organizational terms. However, only some vague statements found their way into the final version of the draft protocol text. Those few references to S&T stated that "in undertaking its activities the Organisation shall consider measures to make use of advance in science and technology." In addition, the draft protocol foresaw that the Director-General may be directed to establish a Scientific Advisory Board. The extent to which the board members "expertise and experience in the particular scientific fields relevant to the implementation of this Protocol" would also include verification related areas was not specified.[16]

Over more than 50 years of BWC implementation, the way in which review conferences have addressed S&T issues has developed from generic exhortations that Article I of the BWC covers new S&T developments of relevance to the treaty to more detailed annual discussions among states parties—from 2012 onward—in the intersessional periods between successive BWC review conferences (Kelle 2014, pp. 90–97).

The organizational support available to BWC states parties and their meetings by the ISU has led to regular compilation for review conferences of a background document on S&T of relevance to the treaty. The most recent one of these compiles information provided by four BWC states parties: Cuba, Switzerland, the United Kingdom and the United States of America. Not surprisingly, none of these broad-based S&T reviews from different national perspectives directly addresses S&T developments related to the verification of the BWC. Yet, most of the contributions offer some relevant insights by highlighting developments in inter alia disease surveillance and monitoring, and vaccine development.[17]

In parallel to the periodic review of relevant S&T developments, BWC states parties over the past decade have also made progress towards the establishment of an S&T review mechanism under the convention. The 9th BWC review conference in late 2022 came very close to establishing a scientific advisory board. Under the proposed plan, the board would have a hybrid structure with the larger advisory group open to experts nominated by BWC states parties and a "limited-size Scientific Reporting Committee."[18] With the basic structure thus formalized, the

---

[15] See fn. 1.

[16] United Nations, *Protocol to the Convention of the Prohibition of the Development, Production, and Stockpiling of Bacteriological (Biological) and Toxin Weapons and on Their Destruction*, document BWC/AD HOC GROUP/CRP.8, (2001).

[17] Ninth Review Conference of the States Parties to the BWC, *New scientific and technological developments relevant to the Convention. Background information document submitted by the Implementation Support Unit*, document BWC/CONF.IX/7, Geneva (2022), https://documents.un.org/doc/undoc/gen/g22/611/90/pdf/g2261190.pdf, last accessed 2024/02/29.

[18] North Macedonia, Draft Terms of Reference and Rules of Procedure toward the development of a BWC Science and Technology Advisory Process, document BWC/CONF.IX/WP.65,

newly-established working group would only have to determine details related to the board's mandate, composition, etc. However, due to the negotiation dynamics at the conference, the decision to establish the scientific advisory board was not included in its final document.

15.5   Conclusions: The CWC as Model with Limited Utility for BWC Verification

This chapter set out to assess the degree to which the verification system under CWC could serve as a model for the future verification of the BWC. There are clear similarities between the CWC and BWC in terms of the scope of their prohibitions, but key differences exist that are highly relevant for the topic under discussion. While the CWC reflects agreement among its negotiators that treaty compliance of its states parties with the provisions of the treaty has to be verified, no such agreement existed at the time of negotiating the BWC. Efforts to negotiate a compliance protocol to the BWC ended unsuccessfully in 2001.

In the CWC, this verification principle has led to the adoption of declaration, inspection and investigation norms, which form the basis of the routine and non-routine verification system under the treaty. Routine verification measures under the CWC deal with both the verification of destruction of CW stockpiles declared by possessor states and so-called industry verification measures under Article VI of the treaty. Non-routine verification measures foreseen in the treaty have never been used by states parties. Instead, additional, more flexible measures were established to deal with the gaps in the Syrian declarations to the OPCW and the continued use of CW in that country. As the complex negotiations for a BWC protocol in the Ad-hoc Group of BWC states parties during the second half of the 1990s have demonstrated, simply transferring the CWC model to the BWC context is very difficult, and most likely not the best option to address compliance concerns under the BWC. Verification of destruction of BW arsenals should be a moot point, given the relevant BWC provisions, and intrusive routine verification measures in the biotechnology sector or civilian laboratories will be highly contested in light of the growing interests of several states in developing their bio-economy. This leaves non-routine verification measures as the most plausible element of the CWC verification system to be applied in future BWC verification/compliance agreements. As the experience of CWC implementation shows, such challenge inspections or investigations of alleged use will benefit from in-built flexibility to adapt to a variety of future investigation scenarios under the BWC.

The organizational dimension of a future verification system under the BWC will be determined by the scope of such a system and the range and number of activities foreseen. Given that elements of the verification system foreseen under the CWC

Ge-neva (2022), https://documents.un.org/doc/undoc/gen/g22/614/16/pdf/g2261416.pdf, last accessed 2024/02/29.

are not applicable in the BWC context, a correspondingly smaller organizational footprint can be expected in the latter context. In addition, past discussions among BWC states parties about adding even very small numbers of additional staff to the BWC Implementation Support Unit do not suggest a big appetite for establishing a large secretariat in analogy to the OPCW TS. Thus, BWC states parties are also unlikely to replicate the CWC model in the BWC context as far as the organizational support for verification activities is concerned.

The S&T advice ecosystem that has developed since the CWC entered into force has provided useful pointers for treaty implementation, including implementation and adaptation of the verification system. The scientific advisory board currently under discussion in the BWC working group could significantly close the gap between the two treaties in this area. The proposed hybrid structure would represent a compromise solution between those advocating for an open-ended nature of the advisory body and the need to have a workable instrument available that would be in a position to report on S&T developments of relevance to the BWC and provide advice to its states parties. In the absence of an implementing organization for the treaty, acting on any S&T related recommendations would be left exclusively to BWC states parties.

References

Chevrier MI, Hunger I (2000) Confidence building measures for the BTWC: performance and potential. Nonproliferation Rev 7(3):24–42

Ghionis A, Kelle A (2024) The chemical weapons convention after its fifth review conference: key issues for the European Union, EU Non-Proliferation and Disarmament Papers no.87, Brussels (2024). https://www.nonproliferation.eu/wp-content/uploads/2024/03/EUNPDC-87-CWC-after-RevCon5-2.pdf. Accessed 5 March 2024

Jakob U (2023) The 9th review conference of the biological weapons convention, PRIF blog, Frankfurt. https://blog.prif.org/2023/02/07/the-9th-review-conference-of-the-biological-weapons-convention/. Accessed 23 Feb 2024

Kelle A (2014) Prohibiting chemical and biological weapons. Multilateral regimes and their evolution. Lynne Rienner, Boulder: CO

Kelle A (2023) The CWC at 25: from verification of chemical-weapons destruction to attribution of their use. Nonproliferation Rev 28(4–6):319–336

Kelle A, Siegmann E (2023) Great powers and the norms of the BW prohibition regime, CBWNet Working Paper No. 9, Berlin

Krasner SD (1982) Structural causes and regime consequences: regimes as intervening variables. Int Organ 36(2):185–205

Leitenberg M (1996) Biological weapons arms control, project on rethinking arms control, University of Maryland, PRAC Paper No. 16

Sims NA (2001) The evolution of biological disarmament. Oxford University Press, Oxford

Tucker JB (2002) A farewell to germs: the US renunciation of biological and toxin weapons, 1969–70. Int Secur 27(1):107–148

Ward KD (2004) The BWC protocol: mandate for failure. Nonproliferation Rev 11(2):183–199

Zanders J-P (2023) The biological and toxin weapons convention confronting false allegations and disinformation, EU Non-Proliferation and Disarmament Papers, no. 85, Brussels. Accessed 26 Feb 2024

Zilinskas RA (1999) Cuban allegations of biological warfare by the United States: assessing the evidence. Crit Rev Microbiol 25(3):173–227

Alexander Kelle is senior researcher at the Berlin office of the Institute for Peace Research and Security Policy (IFSH). In the Berlin office, he heads the CBW network to strengthen the norms against chemical and biological weapons (CBWNet). His research focuses on the evolution of the CBW prohibition regimes. He is the author of inter alia *Prohibiting Chemical & Biological Weapons: Multilateral Regimes and Their Evolution* (2014) and, together with A. Ghionis, "The Chemical Weapons Convention After Its Fifth Review Conference: Key Issues for the European Union" (2024).

Chapter 16
Compliance Management: The Verification Dimension

Ralf Trapp

Abstract The verification system of the Chemical Weapons Convention (CWC) is a sophisticated ensemble of systematic and routine on-site inspection measures, and fact-finding mechanisms to address cases of possible non-compliance. Extensive preparations by the OPCW Preparatory Commission, including the development of detailed inspection procedures, the procurement of inspection equipment and the training of the first group of OPCW inspectors ensured an effective start of CWC implementation at the CWC's entry into force in 1997. Since then, verification methods have been refined to optimize resource utilization and adapt to changes in the implementation environment. Although the fact-finding mechanisms of the CWC were never invoked, the OPCW has created ad-hoc fact-finding mechanisms, in collaboration with the United Nations, to investigate cases of chemical weapons use in Syria. Controversy about the use of chemical weapons by the Syrian armed forces has created political pressure, and there are technical challenges to enhance forensic capabilities in support of attribution of responsibility for CWC violations, in particular the use of chemical weapons. The OPCW has responded to these challenges and created new opportunities to strengthen verification as well as international cooperation, including through the establishment of its Center for Chemistry and Technology.

Keywords Chemical weapons convention · Compliance · Verification · Routine inspections · Challenge inspection · Fact finding · Investigation of alleged use

16.1 Introduction

Managing compliance with the obligation and prohibitions of the Chemical Weapons Convention (CWC) involves a range of procedures: provisions relating to national implementation of the CWC (legislation, administrative measures, setting up of a

R. Trapp (✉)
International Disarmament Consultant, Chessenaz, France
e-mail: ralf.trapp@gmail.com
URL: https://ralftrapp.net/

© The Author(s) 2026   189
B. Friedrich et al. (eds.), *Thirty Years of the Chemical Weapons Convention (CWC)*,
https://doi.org/10.1007/978-3-031-98854-7_16

National Authority, etc.) as the basis for domestic compliance enforcement and cooperation with other States Parties, provisions that govern consultation and cooperation among States Parties—directly among them or through the Organisation for the Prohibition of Chemical Weapons (OPCW)—to address and resolve compliance concerns, sanctions to respond to cases of grave violations of treaty obligations including through the UN General Assembly and the UN Security Council, and verification measures to monitor treaty implementation and conduct fact-finding missions to gather factual evidence needed to address and resolve non-compliance concerns.

This paper focuses on the evolution of the verification system of the CWC and its adaptation to challenges and changes in the implementation environment over time. It is the most elaborate procedural system set up under the CWC, accounting for more than half the Convention text. Verification is an essential component of the CWC compliance management system—it is designed to provide a robust and trusted evidential basis for decision masking on compliance and non-compliance matters and on how to respond appropriately to cases of non-compliance.

How does the OPCW system measure up to the expectations of the States that joined the CWC? How has it responded to changes in the implementation environment and to challenges emanating from developments not anticipated by the treaty drafters? And how should it respond to new challenges and the evolving expectations of the international community in the future? The following exposé attempts to give some answers to these questions.

16.2   The Verification System of the CWC

The CWC verification system fulfills several complementary functions: it establishes baselines for safeguarding existing chemical weapons stockpiles and specialized equipment/features of chemical weapons production facilities and for monitoring their destruction in accordance with CWC requirements; it confirms the completeness of destruction operations and compliance with the CWC requirements related thereto; it provides confidence in the legitimacy of the activities of States Parties at chemical industry and other facilities that are relevant to the CWC; it provides tools and procedures to collect and evaluate evidence to support consultative processes about non-compliance concerns, and it provides the tools to investigate allegations that chemical weapons may have been used so as to support actions against such grave treaty violations, as well as provide a factual basis for providing assistance to the victims.

The verification regime for the disarmament of chemical weapons stockpiles and CW products facilities has been designed to ensure full accountability for all declared items, from their initial inspection to their final destruction. It involved the submission of declarations; systematic reporting of destruction plans, operations and results; systematic inspections of CW stockpile locations and former CW production facilities until their destruction or conversion has been completed; systematic inventory controls of critical items and of key steps from initial inspection to completion of destruction; design reviews and the application of safeguards at CW destruction facilities to detect and prevent any diversions of chemical materials; sampling and analysis to confirm the identity of declared weapons fills and the completeness and irreversibility of the destruction operations; and the permanent observation/monitoring of all ongoing destruction activities by inspectors and/or dedicated monitoring instruments. The technical approaches of these types of inspections are fairly straightforward, and have been modeled to a degree on inspection schemes developed during the 1980s in the context of nuclear safeguarding, the Stockholm agreement, the INF Treaty, and other conventional arms control agreements.

The routine verification regimes for the different types of industrial and other facilities (facilities involved with Schedule 1, 2 and 3 chemicals, and other chemical production facilities (OCPFs) that produce discreet organic chemicals) uses a risk-based approach, with systematic verification concepts applied at Schedule 1 facilities, a risk assessment based approach for schedule 2 facilities, and a random selection approach for Schedule 3 facilities and OCPFs (the latter with emphasis on facilities manufacturing organic chemicals that contain any of the elements of phosphor, sulfur or fluorine—reflecting certain structural features of agents known from past CW programs).

Special inspection (fact-finding) mechanisms include challenge inspection—a short notice mandatory inspection that can be invoked against any location or facility of a State Party irrespective of whether it has been declared or not—and investigations of alleged use of chemical weapons (a fact-finding scheme modeled to some degree on investigations of allegations of chemical weapons uses conducted by the United Nations during the 1980s and early 1990s, and formalized in the UN Secretary General's Mechanism to investigate alleged uses of chemical, biological or toxin weapons).

The challenge inspection scheme with its "anytime, anywhere, no right of refusal" concept was a novelty in disarmament treaties. So was the approach developed for the inspection of facilities in the chemical industry—many of them private enterprises rather than State institutions—which poses challenges with regard to such issues as the protection of confidential business information, sampling at privately owned facilities and analysis of such samples at laboratories abroad, and ensuring safety as well as avoiding unnecessary burden or even damage to the facilities as a result of the conduct of international inspections. Many of the details of these types of inspections were only agreed at the very end of the CWC negotiations. The industry verification regime in particular was recognized by the negotiators as needing further refinement based on practical implementation experience. This was the reason why Part X of the Verification Annex foresaw a review of the industry verification regime at the First CWC Review Conference, and why the beginning of OCPF inspections was delayed by several years after the entry into force.

The chair of the final phase of the CWC negotiations, German Ambassador von Wagner, characterized these new achievements in arms control thus[1]:

---

[1] Von Wagner, A.: Conference on Disarmament, Chemical Weapons Final Records (PV), 1992 Session, CD/PV.635, pp. 9–10.

> The political element of challenge inspections reconciles the diverging objectives of maximum assurance against non-compliance, protection of the inspected States Parties' sovereign rights, and the prevention of abuse. Routine verification in industry balances the objectives of reliable confidence-building, simplicity in administration, and non-interference with perfectly legitimate activities in the chemical industry.

It was no surprise that, at the end of the negotiations, many observers praised the CWC verification regime as unprecedented and innovative, and an example for other arms control arrangements. However, what mattered was not merely the soundness of its design, but how it would play out in practice. Already in 1989, with the future industry verification system in mind, Will Carpenter, at the time vice president of Monsanto, observed at the Government-Industry Conference in Canberra[2] that "failure or serious problems in the early days of implementation can lead to an erosion of confidence in the Treaty, misunderstandings amongst states, and a loss of commitment." The preparations for the early phase of treaty implementation fell to the Preparatory Commission, which worked between February 1993 and the entry into force of the CWC on 29 April 1997 in The Hague. The Commission developed a series of guidelines and procedures for the future operations of the OPCW, set up the central OPCW laboratory and equipment store to support on-site inspections and the work of the OPCW's network of designated laboratories, developed and implemented a General Training Scheme for future OPCW inspectors, and procured the initial batch of inspection equipment to be used by OPCW inspectors.[3]

The results of its work were passed on to the First Session of the Conference of the CWC States Parties in April/May 1997, resulting in a number of key decisions relating to the way the verification system would be applied.[4] These included decisions, amongst others, on:

- The OPCW Confidentiality Policy
- The OPCW Health and Safety Policy (including health and safety management during inspection)
- A list of approved inspection equipment together with technical specifications, operational requirements, and a procedure for familiarization with the equipment by National Authorities
- Criteria for the selection of Designated Laboratories to conduct the analysis of authentic verification samples, and a Proficiency Testing Scheme to support the designation of these Member States' laboratories
- Detailed inspection procedures for different types of inspections, which would eventually be compiled by the Technical Secretariat in the OPCW Inspection Manual

---

[2] Department of Foreign Affairs and Trade Canberra of Australia: Final Records of the Government-industry conference against chemical weapons (18–22 September 1989), p. 150.

[3] Preparatory Commission for the OPCW: Final Report. OPCW Homepage, https://www.opcw.org/sites/default/files/documents/PC_series/PC-XVI/en/PCXVI37.pdf, last accessed 2024/03/23.

[4] OPCW: Report of the First Session of the Conference (1997). OPCW Homepage, https://www.opcw.org/sites/default/files/documents/CSP/C-I/en/C-I_9-EN.pdf, last accessed 2024/03/23.

– An initial program of implementation support to help States Parties in the setting
up of functioning National Authorities, including training of personnel, the development and enactment of implementing legislation, and the identification of
declarable facilities.

16.3 Inspector Training

Inspector selection and training was arguably the most important part of preparing
for the start of CWC verification implementation. Without inspectors who had the
right technical/scientific qualifications, thoroughly understood the requirements of
the different verification procedures of the CWC, including their own responsibilities, rights and obligations, could master the operational concepts put in place for
conducting on-site inspections by the OPCW, understood the requirements emanating
from the OPCW's policies with regard to the protection of confidentiality, security,
and health and safety, and had developed the skills and mindsets to function as teams
in the field, CWC inspection could not be conducted effectively and according to the
text and spirit of the treaty.

The inspector training scheme was developed by the Preparatory Commission
with the support of numerous Member States, and also by mobilizing support of
private companies of the chemical industry.[5] Governments and industry associations
made available training centers and facilities, and chemical companies in Europe
and Asia opened their plants for the conduct of mock inspections during the final
module of the training of future inspectors. A first group of inspectors was fully
trained and ready to deploy several weeks after the entry into force, a second group
joined in time to expand the campaign of initial inspections at the end of 1997 when
the Russian Federation and some other countries that had experienced delays in their
ratification processes joined, and additional military and industry facilities needed
to be inspected without delay.

16.4 The Beginning of Verification Operations

The initial inspection campaign can only be described as a heroic effort: The negotiators had "front-loaded" the CWC with what turned out to be unrealistically short
time frames for the initial inspection of chemical weapons stockpiles, CW production facilities, and Schedule 1 facilities. Furthermore, planning assumptions used
by the Preparatory Commission had somewhat underestimated the number of countries with declarable CW facilities (only one other CW possessor State Party was
included in the planning assumptions in addition to the USA and Russia, in fact
there were four), and the number of the facilities that needed to be inspected in the

---

[5] Preparatory Commission for the Organisation for the Prohibition of Chemical Weapons, document
PC-XV/B/10 (1996).

initial inspection campaign had also been underestimated, with many more former CW production facilities that had already been converted for permitted purposes requiring an initial inspection. Most importantly, the Commission had assumed— for political reasons—that a bilateral destruction and verification agreement would be in place between the United States and the Russian Federation (Manley 2007). For such a case, the CWC foresees the option of replacing multilateral verification with bilateral inspections and a simple audit thereof by the OPCW. As a consequence, the size of the inspectorate was designed to audit these bilateral inspections rather than conduct all inspections with full-size OPCW inspection teams.

Luckily, the delay of the Russian ratification created some extra room for the initial inspection campaign. Nevertheless, the Technical Secretariat needed to combine tireless work with finding innovative solutions, which included the conduct of sequential inspections (within a State Party or in a regional context) or the temporary use of draft facility agreements prior to the conclusion of formal agreements by the Executive Council, to ensure that the verification baselines could be established in time.

Despite these challenges, the initial inspection campaign went surprisingly smoothly.[6] Inspectors, verification planners and operations officers gained a vast amount of practical experience in a very short period of time. Equally, a large number of States Parties that had never before experienced international disarmament inspections learned quickly how to receive and escort such inspections. Thorough preparations by the Member States of the Preparatory Commission and the provisional Technical Secretariat, and intensive training of future OPCW inspectors had paid off and ensured that the verification system of the CWC was working well right from the start. The intensity of the early inspection campaign, in addition, helped form a corps of inspectors that was technically competent, conversant in the requirements and provisions of the CWC, and had gelled as a team that was able to function in challenging and, at times, unpredictable circumstances.

16.5   Verification of CW Disarmament

16.5.1   Verification of the Initial Declarations and the Beginning of CW Disarmament

The initial inspection campaign established the baselines for safeguarding the declared chemical weapons at their stockpile facilities (CWSFs) and the items of specialized equipment and structural features, as well as other declarable aspects of former chemical weapons production facilities (CWPFs); it also set up the verification systems for the facilities declared as CWSFs and CWPFs and the only chemical weapons destruction facility (CWDF) operating at the time at Johnston Island in the

---

[6] OPCW Technical Secretariat: Consolidated unclassified verification implementation report April 1997–31 December 2002 (2003). OPCW Homepage, https://www.opcw.org/sites/default/files/doc uments/CSP/RC-1/en/RC-1_S_6-EN.pdf, last accessed 2024/03/23.

Pacific. These baselines are essential for monitoring the progress of CW destruction, destruction or conversion of the CWPFs, and for eventually confirming the completeness of the destruction of all declared stockpiles and facilities.

These early inspections of CW facilities also helped clarify further certain concepts set out in the CWC related to, for example, the definition of chemical weapons (as distinct from similar items used at CWDFs to calibrate aspects of destruction lines), the declaration of underground structures of CWPFs, or the details of how former CWPFs previously converted for legitimate purposes needed to be declared and verified. In addition, the Technical Secretariat was consulted by a number of States Parties about the way in which existing facilities previously association with chemical weapons activities needed to be declared; also, several States Parties offered technical advice and assistance to others to clarify such issues.

The verification concept for chemical weapons facilities and related items is characterized by the CWC as "systematic verification through on-site inspection and monitoring by on-site instruments". This was interpreted to mean that inventory controls of critical items (chemical weapons, specialized equipment and structural features of CWPFs) needed to be full rather than partial, and that each critical step from initial verification to final destruction (CW storage, CW transfer from a CWSF to a CWDF, transfers within a CWDF, destruction, disposal of waste materials) needed to be verified in ways that ensured 100 per cent accountability for each item (Trapp and Walker 2014). That involved multiple controls of each declared item at (and within) different locations by visual inspection and measurements, and systematic efforts to reconcile inventory data emanating from declarations, destruction plans, reports on destruction operations, inventory controls at different CW locations, including transport receipts from CWSFs at CWDFs, and all subsequent operations within a CWDF until the destruction was completed—a labor-intensive undertaking. It also required a detailed engineering review of the layout and technical design of CWDFs to develop and apply a specific verification concept for each destruction facility that would meet the CWC requirements and fit the specifics of the facility in question. It became clear that each destruction facility was unique in some way, and that the verification system for each of them needed to be tailored to the specific task, design and technology used at the facility.

One necessary adjustment related to the confirmation of the identity of the declared weapons fills. The CWC allows for sampling and analysis at all different locations where chemical weapons were being inspected. It became clear, however, that sampling of chemical weapons outside a destruction facility (which is specifically designed to ensure the safe opening and sampling of chemical weapons) would pose significant safety risks, and should therefore be avoided. To ensure that the identity of the agents could be confirmed nevertheless, an approach was implemented that combined physical inspection of the weapons with permanent tagging of a sample of the items, and delayed sampling and analysis at a CWDF only.[7] The use of permanent

---

[7] OPCW Technical Secretariat: Background paper on the conduct of inspections under the Chemical Weapons Convention and related issues (2003). OPCW Homepage, https://www.opcw.org/sites/def ault/files/documents/CSP/RC-1/en/RC-1_S_1-EN.pdf, last accessed 2024/03/23.

tags allowed to track specific items from the initial inspection all the way through to their final destruction, where the chemical fill could be sampled and its chemical composition confirmed. Tagging also allowed the monitoring of movements of stockpile items associated with a tagged item from their original CWSF location to a CWDF and through its aspects to the final destruction unit.

16.5.2 Optimisation of Destruction Verification at CWDFs

The majority of verification resources spent by the OPCW after the entry into force went into the verification of the destruction of chemical weapons at CWDFs. Whilst other aspects of the disarmament process were verified by individual inspections (systematic in nature but limited in terms of time and manpower), the verification of the destruction of chemical weapons was done through permanent presence of inspectors during every period of active destruction operations. This meant deploying teams of inspectors to ensure permanent monitoring of the destruction operations at each operating CWDF large enough to ensure verification in a 24 h seven days a week shift regime (often larger than 20 inspectors per team), throughout each destruction campaign. For longer destruction campaigns, inspection teams would go out for several weeks and then be replaced by a new "rotation" of inspectors at the end of their deployment.

As destruction operations picked up pace in several States Parties, the burden on the OPCW increased accordingly, both in terms of inspector resources committed to this part of the verification enterprise and financially for the OPCW, as well as for possessor States Parties who were required to reimburse part of the verification costs incurred by the OPCW. In response to these challenges, CW possessor States Parties and the Technical Secretariat developed concepts to optimize the utilization of inspector resources, involving a streamlining of verification processes, as well as the use of recording devices to monitor the movement and destruction of CW items in a given CWDF. This allowed a significant reduction of team sizes for the rotations of inspection teams operating at CWDFs during active destruction operation periods.[8]

All in all, the system of systematic verification of the declared chemical weapons and related facilities and their destruction has worked well, albeit at significant cost to the OPCW and the States Parties involved. It provided full accountability for all declared items (chemical weapons in weaponized form as well as bulk storage, specialized equipment and features at CWPFs as well as other items of CWDFs that were subject to destruction) until their ultimate destruction. Systematic verification was also applied at former CWPFs that had been converted to purposes not prohibited, in accordance with the applicable provisions of the CWC. The competence of OPCW inspectors and verification specialists to verify chemical weapons disarmament to

---

[8] OPCW Technical Secretariat: Review of the operation of the Chemical Weapons Convention since the Second Review Conference (2012). OPCW Homepage https://www.opcw.org/sites/default/files/documents/CSP/RC-3/en/wgrc3s01_e_.pdf, last accessed 2024/03023.

the high standards of accountability set forth in the CWC is a key asset of the OPCW that needs to be retained, also after all the declared chemical weapons were verifiably destroyed by 2023.

16.6 Industry Verification

16.6.1 Early Challenges

Routine inspection of facilities and activities not prohibited by the CWC complement the systematic verification of chemical weapons and their disarmament. It provides assurances that no new chemical weapons are being produced by any State Party. Following the initial inspection campaign of chemical weapons and Schedule 1 facilities, the Technical Secretariat included in its inspection planning the first inspections of chemical industry facilities, beginning with Schedule 2 plant sites in late 1997. Schedule 3 facilities followed 1998, and in May 2000 began the inspection of OCPFs.[9]

The OPCW was well prepared for the conduct of industry inspections, supported by the chemical industry: 60 candidate inspectors with industrial backgrounds had been trained at an industry training center in Switzerland, and inspection procedures had been tested in mock inspections at chemical plants in Germany (Hoechst, Bayer, BASF), the Czech Republic (Synthesia Ltd., Semtín), China (Pesticide Complex, Quingdao), and Japan (Daicel Chemical Industries Ltd., Arai, Nippon Soda Ltd., Nihongi).[10] Industry inspections went well, despite earlier concerns of chemical industry about confidentiality, administrative burden, sampling and analysis at their plants, and possible reputational issues. However, there were early concerns about the completeness of industry declarations of a number of States Parties, and some States Parties including the United States lacked the necessary legislative basis to submit industry declarations to the OPCW. When the United States finally submitted its industry declaration in 2000, the OPCW had conducted some 200 industry inspections in 40 countries. A certain implementation practice had been established and was broadly accepted by the Member States.[11]

Industry inspections had to be implemented despite a large number of "unresolved issues" which had been left behind by the Preparatory Commission, or that had emerged during the implementation of the CWC. Some of them affected industry verification. These were early examples to show that provisions of arms control treaties, as well as the way they are being implemented, may have to be adapted when the treaty provides general guidance, but States Parties have not been able to

---

[9] OPCW Technical Secretariat: Consolidated unclassified verification implementation report April 1997–31 December 2002 (2003). OPCW Homepage, https://www.opcw.org/sites/default/files/doc uments/CSP/RC-1/en/RC-1_S_6-EN.pdf, last accessed 2024/03/23.

[10] Ibid, p.6.

[11] Ibid.

agree on specific guidance for how to apply them. Verification methods also need to be adapted to change in response to, for example, advances in science and technology, changes in industry operations, and conditions in the political and security environment that the drafters of a treaty may not have foreseen. Adaptability is also important for arms control treaties to take advantage of new innovative solutions emanating from scientific and technological progress, for example in the area of verification techniques. In the case of the CWC, such adaptations required a close interaction between the Technical Secretariat as the implementer of international verification measures, National Authorities as the partners of the OPCW in verification conduct and at the same time the government bodies responsible for domesticating CWC rules and procedures, and the chemical industry who owns and operates the facilities that are subject to inspections.

When industry inspections started in the USA in 2000, they exposed a number of issues that required adaptation in the way industry verification would be conducted, many of which reflected structural changes in the organization of chemicals manufacturing that had taken place since the adoption of the CWC. These changes affected the way in which chemical manufacturing sites were organized locally, with changing degrees of back-integration within companies, as well as changes in supply chains and local business organization. As a consequence, some definitions of the CWC that govern inspection conduct at industrial sites no longer fitted the reality of how the industry operated (an example being the definition of a plant site versus a plant, which was modelled on industrial structures typical for the 1980s and early 1990s). These CWC concepts were difficult to apply to new types of business organization such as business parks, spin-off companies, the emergence of custom manufacturers and companies renting chemical production plants based on their equipment suit, and other changes in the organization and operations of chemicals manufacturing. This results in different views about access and inspection rights and lead to numerous "unresolved issues", which needed to be resolved through consultations between States Parties, the Technical Secretariat and the chemical industry.

16.6.2 Overcoming Deficiencies in National Implementation

In addition to challenges to the industry verification regime caused by structural and technological changes in the industry as highlighted above, two other issues affected the first decade of CWC industry verification (and to a degree even today):

– Whether all States Parties have identified all declarable industry facilities;
– Persisting inconsistencies between State Party declarations of exports and imports of scheduled chemicals.

These problems are indications of differences between and weaknesses in national implementation systems in different States Parties. National implementation is of course the responsibility of the States Parties, but international verification relies on accurate and timely declarations and on implementation measures that are applied as uniformly as possible by all States Parties. This is not merely a question of effective implementation of verification measures; it also relates to the general requirement to treat all States Parties equally, including with regard to the conduct of the Technical Secretariat in implementing the CWC's verification measures. It was this rationale that prompted the Technical Secretariat to adopt a series of measures to help States Parties overcome deficiencies in their national implementation measures. Key examples included[12]:

- A high-level campaign to assist (in many cases first educate and then persuade) governments to submit initial declarations due 30 days after the date the CWC enters into force for a State Party
- Technical assistance to help States Parties identify declarable industry facilities, using open-source data such as authoritative commercial databases and publications from industry and trade associations, and the development of implementation tools such as an OPCW Handbook on Chemicals and tools for the electronic submission of declarations
- Implementation support for National Authorities, including regional seminars to raise awareness for CWC requirements, training courses for personnel of National Authorities, legislation support (drafting support, commentary on legislative drafts, model legislation, sample legislative texts), and technical support for customs services.

These implementation support measures were reviewed and endorsed by the First CWC Review Conference in 2003. The review conference also decided to request the Executive Council to prepare an Action Plan of Article VII implementation to encourage all States Parties to put in place the necessary measures to ensure full national implementation.[13] The Technical Secretariat was requested to provide the necessary technical support, and States Parties were encouraged to share best practices and help each other in the adoption of effective national implementation measures. The Technical Secretariat and States Parties offering such support closely coordinated their activities under the action plan, and some of the projects were implemented by combined teams of OPCW and State Party experts. The Action Plan lasted for two years; it resulted in a certain improvement with regard to the setting up of National Authorities and the enactment of basic implementing legislation, but at the same time, additional countries joined the CWC in response to a second Action Plan to improve the universal adherence to the CWC, with many new countries

---

[12] OPCW Technical Secretariat: Consolidated unclassified verification implementation report April 1997–31 December 2002 (2003). OPCW Homepage, https://www.opcw.org/sites/default/files/doc uments/CSP/RC-1/en/RC-1_S_6-EN.pdf, last accessed 2024/03/23; OPCW Technical Secretariat: Background paper on international co-operation programmes (2003). OPCW Homepage, https://www.opcw.org/sites/default/files/documents/CSP/RC-1/en/RC-1_S_2-EN.pdf, last accessed 2024/03/23.

[13] OPCW: Decision—Plan of action regarding the implementation of article VII obligations (2003). OPCW Homepage, https://www.opcw.org/sites/default/files/documents/CSP/C-8/en/c8dec16_EN.pdf, last accessed 2024/03/23.

joining the treaty that lacked the necessary experience, capacity and resources to fully implement all the domestic implementation requirements.

Follow-up measures to the original Action Plan were adopted at the end of 2004[14] and the Technical Secretariat continues to this day to offer implementation support to States Parties that need it, and to report on the status of national implementation measures adopted by States Parties and the technical advice and support it had rendered.[15]

16.6.3 Sampling and Analysis in Industry Inspections

The CWC contains provisions for the collection and analysis of samples during routine inspections of chemical industry facilities. The degree of prescriptiveness of the CWC for the employment of this verification method varies from sampling and analysis"shall be undertaken" (Schedule 2 facilities) to "may be undertaken to check for the absence of undeclared scheduled chemicals" (Schedule 3 and OCP facilities). This leaves a degree of flexibility to the Technical Secretariat to acquire samples at chemical industry facilities for chemical analysis. At the same time, the CWC delineates the purpose of this method, focusing it on a check on the presence or absence of undeclared scheduled chemicals rather than on establishing the actual composition of such samples. This provision of the CWC recognizes the need to protect confidential business information (for example the nature and concentration of a catalyst or other ingredient used in the synthesis of a chemical product) that might be exposed should a more detailed analysis of such samples be conducted.

The OPCW recognized the sensitivities in industry about sampling and analysis, and acquired a software package that allows inspectors to perform analyses in "blinded mode". With this software package, the analytical instrument can be set to a mode that does not reveal the actual composition of a sample, but only provides the analytical data for signals that a database match associates with any of the scheduled chemicals contained in the OPCW central analytical database (OCAD).[16] This database is constantly being updated with analytical data (mostly GC retention indices and mass spectra, as well as some additional data such as infrared (IR) spectra), submitted by State Party laboratories, validated by a dedicated

---

[14] OPCW Director-General: Overview of the status of implementation of Article VII of the Chemical Weapons Convention as at 31 July 2022. OPCW Homepage, https://www.opcw.org/sites/default/files/documents/2022/10/ec101dg13a1%20c27dg09%2Ba1%28e%29.pdf, last accessed 2024/03/23.

[15] OPCW Director-General: Overview of the status of implementation of Article VII of the Chemical Weapons Convention as at 31 July 2022. OPCW Homepage, https://www.opcw.org/sites/default/files/documents/2022/10/ec101dg13a1%20c27dg09%2Ba1%28e%29.pdf, last accessed 2024/03/23.

[16] OPCW: Decision—OPCW Central Analytical Database, document C-I/DEC.64 (1997), OPCW Homepage, https://www.opcw.org/sites/default/files/documents/CSP/C-I/en/C-I_DEC.64-EN.pdf, last accessed 2024/03/24.

OPCW Validation Group, and approved for inclusion into the OCAD by the Executive Council. The OCAD is maintained by the OPCW Central Laboratory, shared with National Authorities, and a working copy can be taken to the field by inspection teams along with their suit of analytical instruments (typically a transportable GC–MS instrument).

During the first years of verification implementation, the Technical Secretariat did not deploy sampling and analysis instruments to chemical industry facilities. There were a number of reasons for this, amongst them a focus on the conduct of initial inspections which did not require sampling and analysis but aimed at setting up the country and facility specific arrangements for future inspections and conducting work on risk assessments and facility agreements where applicable, a need to expand the OCAD beyond the initial data sets to ensure reliable results, the need to validate analytical procedures and standards for field use, and a lack of practical experience in deploying and using the OPCW's field analytical equipment at chemical industry facilities. There also would have been resource implications had inspectors wanted to use chemical analysis in the field, which had not been factored into the verification budgets during the early implementation years.

To begin filling this gap and gain practical experience, the Technical Secretariat started a pilot project to test the use of its field sampling and analysis equipment in 2006.[17] Initially, this inspection method was deployed at Schedule 2 facilities where the CWC provisions indicated a more stringent requirement to undertake sampling and analysis, and where the maximum inspection time frames prescribed by the CWC were long enough to easily allow for the setting up and calibration of the equipment and for the running of analyses and controls. Part of the preparation of this inspection technique was a conversation with experts from chemical companies to understand their concerns as well as the conditions that inspectors should expect when attempting to set up their mobile laboratory/equipment at industrial sites. What kind of logistical and technical support would they need and could they expect? What workspace would be available? These were some of the questions that needed an answer before sampling and analysis could be employed at industrial facilities on a more regular basis.

At the OPCW Industry and Protection Forum organized jointly by the OPCW and the International Council of Chemical Associations (ICCA) in 2007, these issues were discussed, and the initial experience of OPCW inspectors with conducting sampling and analysis at chemical industry plants was presented to industry experts.[18] Today, sampling and analysis has become a tool that is regularly deployed in industry inspections. Supporting sampling and analysis remains a challenge for chemical companies, in particular for small and medium-size enterprises. But pragmatic solutions

---

[17] OPCW Technical Secretariat: Support by inspected States Parties for sampling and analysis under Article VI of the Chemical Weapons Convention (2006). OPCW Homepage, https://www.opcw.org/sites/default/files/documents/S_series/2006/en/S_548_2006-EN.pdf, last accessed 2024/03/24.

[18] OPCW Technical Secretariat: The 2007 OPCW Academic Forum and the Industry and Protection Forum in support of comprehensive implementation of the Chemical Weapons Convention. Document S/674/2008 (1 February 2008). OPCW Homepage, https://www.opcw.org/sites/default/files/documents/S_series/2008/en/S_674_2008-EN.pdf, last accessed 2024/03/24.

have been found to enable this important inspection technique in chemical industry inspections.

16.6.4   Adapting to Advances in Science and Technology

The CWC recognizes that implementation and verification will need to be adapted from time to time to take account of changes in the implementation environment caused by advances in science and technology, as well as to take advantage of new scientific and technical methods and instruments that could be used for verification purposes. The CWC provides for a mechanism for the Director-General to establish a Scientific Advisory Board (SAB) which provides advice to him or her, to States Parties and the policymaking organs of the OPCW on advances in science and technology that affect the operations of the CWC. In the early phase of CWC implementation, the SAB was used to provide technical advice on specific issues that had not been resolved in the negotiations during the Preparatory Commission, or that emerged during the implementation of CWC provisions in practice. Many of these issues had a scientific aspect, and in some cases, the SAB's response was instrumental for the OPCW to take regulatory steps, such as with regard to an understanding of what the term "Ricin" in Schedule 1 actually covered, resulting in the exemption from the requirements of Part VI of the Verification Annex of castor oil manufacturers, given that the Ricin contained in the raw material (the seeds of the plant *Ricinus communis*) is never isolated/extracted from the seed pulp but destroyed in the process used to prepare the seeds for oil extraction.[19] Another example was the advice given by the SAB on transfers of small amounts of Saxitoxin for diagnostic purposes, which conflicted with the rules for transfer notifications under Part VI of the Verification Annex; the SAB advice was instrumental for the OPCW to adopt the first amendment to the CWC by inclusion of a paragraph 5bis in Part VI of the Verification Annex that changed the requirement to pre-notify such transfers from 30 days to the day of transfer.[20]

As the OPCW gained experience with implementing the CWC, and in the context of the preparations of the first Review Conference, the SAB took a more comprehensive view and began a form of horizon screening of relevant developments in science and technology, rendering more long-term advice to the OPCW policymakers. To this end, the Director-General, following advice from the SAB, approached the International Union of Pure and Applied Chemistry (IUPAC) with a request for collaboration with the SAB to provide a broader and deeper reach-back into the knowledge base and

---

[19] OPCW: Decision—Reporting on Ricin, document C-V/DEC.17 (2000). OPCW Homepage, https://www.opcw.org/sites/default/files/documents/CSP/C-V/en/C-V_DEC.17-EN.pdf, last accessed 2-24/03/24.

[20] OPCW Executive Council: Decision—Transfers of Saxitoxin for medical/diagnostic purposes. OPCW document EC-XII/DEC.56 (1998); Depositary notification C.N.157.2000.TREATIES-1 (2000). United Nations Homepage, https://treaties.un.org/doc/publication/CN/2000/CN.157.2000-Eng.pdf, last accessed 2024/03/24.

expertise of the international chemistry community.[21] This collaboration between the SAB and IUPAC was an important feature of the preparation of the SAB report to the First CWC Review Conference in 2003, and has become a constant feature of OPCW reach-back into the global scientific community. It has assisted the SAB in preparing comprehensive reports for subsequent Review Conferences on the impact of advances in science and technology on the CWC.

In recent years, this all-around approach of screening the horizon of science and technology has been complemented by taskings of the SAB to review specific implementation areas and provide strategic scientific and technical guidance in areas of importance for the future evolution of the verification system. Examples include a review of relevant technologies and approaches useful in verification, including proposals for a more holistic approach that uses "remote" techniques of information gathering and evaluation,[22] a report on the state of the art in investigative sciences, including on developments in chemical forensics used in investigations of alleged uses of chemical weapons and their attribution to sources and actors involved in such uses,[23] and a report on requirements relate to the analysis of biotoxins.[24]

These examples illustrate that the SAB has become an important source of knowledge and expertise for the OPCW to adapt its verification systems to changes in the implementation environment and new requirements (developments that affect chemicals manufacturing in industry, the discovery of new chemicals of relevance to the CWC, the emergence of new techniques that could enhance OPCW capabilities in verification). This aspect will become more important as the OPCW moves into the post-CW-destruction era and focuses more comprehensively on the prevention of the re-emergence of chemical weapons, including on investigations of cases of suspected misuse of toxic chemicals and their precursors for malicious purposes.

---

[21] Parshall, G. W., Pearson, G. S., Inch, T. D. and Becker, E. D.: Impact of scientific developments on the Chemical Weapons Convention (IUPAC Technical Report). Pure Appl. Chem. 12(74), 2323–2352 (2002).

[22] OPCW Scientific Advisory Board: Verification (2015). OPCW Homepage https://www.opcw.org/sites/default/files/documents/SAB/en/Final_Report_of_SAB_TWG_on_Verification_-_as_presented_to_SAB.pdf, last accessed 2024/03/24.

[23] OPCW Scientific Advisory Board: Investigative science and technology (2019). OPCW Homepage https://www.opcw.org/sites/default/files/documents/2020/11/TWG%20Investigative%20Science%20Final%20Report%20-%20January%202020%20%20%281%29.pdf, last accessed 2024/03/24.

[24] OPCW Scientific Advisory Board: Analysis of biotoxins (2023). OPCW Homepage https://www.opcw.org/sites/default/files/documents/2023/04/Analysis%20of%20Biotoxins%20Final%20Report.pdf, last accessed 2024/03/24.

16.7 Fact-Finding to Resolve Non-Compliance Concerns

16.7.1 Challenge Inspection and Investigation of Alleged CW Use

The challenge inspection mechanism occupies a significant part of the text of the CWC—18 paragraphs of Article IX and an entire Part of the Verification Annex (Part X containing 61 paragraphs) are devoted to this fact-finding mechanism. These provisions can equally be applied to the investigations of suspected uses of chemical weapons, and there are a further 27 paragraphs in Part XI of the Verification Annex that add specifics pertaining to such types of investigation to the factfinding machinery of the CWC.

Except for one of these provisions—paragraph 27 of Part XI of the Verification Annex—none of these provisions has ever been used in practice. The exception was the participation of the OPCW in the UN Secretary General's investigation of allegations of the use of chemical weapons in the Syrian Arab Republic in 2013—an investigation conducted under the authority of the United Nations Secretary General's Mechanism (UNSGM), set up during the 1980s. Based on paragraph 27 of Part XI of the Verification Annex and the Relationship Agreement between the United Nations and the OPCW, the OPCW made a contingent of experienced inspectors available to this UN mission, and OPCW Designated Laboratories supported the analysis of samples collected by the mission.[25] The success of this UN mission demonstrated to the world that the OPCW had created a scientifically sound and operationally robust system of fact-finding that could even be deployed into a conflict zone to investigate successfully incidents of chemical weapons uses. The mission contributed the essential factual basis for subsequent decisions to set up the UN-OPCW Joint Mission to eliminate Syrian chemical weapons in 2013—a process that was supported and verified by OPCW inspectors until the last declared chemical weapon had been removed from the Syrian territory and destroyed by the end of the summer of 2014.[26]

However, the OPCW itself has never been requested by a State Party to execute a challenge inspection or investigation of alleged use. Whilst the Technical Secretariat maintains a state of readiness to implement such special fact-finding missions, and conducts specialized training and field exercises to demonstrate and further improve its competence to conduct such missions,[27] States Parties have been reluctant so far

---

[25] United Nations Office for Disarmament Affairs: The Secretary-General's mechanism for investigation of alleged use of chemical, bacteriological (biological) or toxin weapons—A lessons-learned exercise for the United Nations Mission in the Syrian Arab Republic (2015). United Nations Homepage, https://front.un-arm.org/wp-content/uploads/assets/publications/more/syrian-ll-report/syrian-ll-report-2015.pdf, last accessed 2024/03/24.

[26] Trapp, R: Lesson learned from the OPCW mission in Syria (2015). OPW Homepage, https://www.opcw.org/sites/default/files/documents/PDF/Lessons_learned_from_the_OPCW_Mission_in_Syria.pdf, last accessed 2024/03/24.

[27] OPCW Technical Secretariat: Review of the operation of the Chemical Weapons Convention since the Fourth Review Conference. OPCW document WGRC-5/S/1 RC-5/S/1 (2023).

to invoke these procedures to address compliance concerns. For whatever reasons, States Parties have opted instead for using other mechanisms, including bilateral clarification efforts and unilateral or group sanctions.

16.7.2   Investigations of the Use of Chemical Weapons in Syria

As reports of continuing uses of chemical weapons in Syria (Sarin as well as impro-vised devices using chlorine) emerged in early 2014, the OPCW needed to respond to these reports to protect the norm against chemical weapons use—yet no State Party invoked the formal fact-finding mechanisms of the CWC. In this situation, the Director-General, using the authority vested into his office by the CWC, and subse-quently endorsed by the Executive Council, negotiated with Syria the modalities for the OPCW deploying a Fact-finding Mission (FFM) to investigate these reports.[28] At the same time, the Director-General also set up a Declaration Assessment Team (DAT) to clarify inconsistencies and gaps in the Syrian CW declaration.[29] Both mechanisms are still operational at the time of writing.

The mandate of the FFM was to investigate reports of alleged uses of chemical weapons in Syria by collecting factual evidence about these incidents; it did not have a mandate to investigate who was responsible for these incidents. As the FFM began confirming that chemical weapons had indeed been used in Syria on a number of occasions, with claims and counterclaims appearing about the perpetrators, the UN Security Council agreed to set up a Joint Investigative Mechanism (JIM) of the United Nations and the OPCW to gather factual evidence to attribute responsibility for chemical weapons uses confirmed by the FFM. The mechanism worked between 2015 and 2016, submitting seven reports and identifying the Islamic State and Syrian armed forces as responsible for specific incidents.[30] The JIM's continuation was blocked by a veto by the Russian Federation, which disagreed with the findings attributing certain cases of chemical weapons uses to the Syrian armed forces. This left a gap in the international fact-finding system to investigate responsibility for such acts, and the OPCW took a decision in June 2018[31] that addressed these and other threats of chemical weapons uses. This decision requested the Director-General to

---

OPCW Homepage. https://www.opcw.org/sites/default/files/documents/2023/03/rc5s01%20wgrc 5s01%28e%29.pdf, last accessed 2024/03/24.

[28] OPCW Homepage, https://www.opcw.org/fact-finding-mission, last accessed 2024/03/24.

[29] OPCW Homepage, https://www.opcw.org/declaration-assessment-team, last accessed 2024/03/ 23.

[30] The last JIM report was issued under United Nations: Letter dated 26 October 2017 from the Secretary-General addressed to the President of the Security Council. UN document S/2017/904 (2017).

[31] OPCW: Decision—Addressing the threat from chemical weapons use. OPCW document C-SS-4/ DEC.3 (2018). OPCW Homepage https://www.opcw.org/sites/default/files/documents/CSP/C-SS-4/en/css4dec3_e_.doc.pdf, last accessed 2024/03/24.

establish a dedicated investigative mechanism—known today as the Investigation and Identification Team (IIT) —to collect and analyze factual evidence to allow, if possible, the attribution of responsibility for chemical weapons uses in Syria. The IIT submitted four reports since its inception in 2019. After having provided evidence to conclude on reasonable grounds that the Syrian armed forces had used chemical weapons in March 2017 Ltamenah and April 2018 in Douma, the IIT submitted its fourth report in February 2024. Therein, it identified the Islamic State in Iraq and the Levant (ISIL) as the perpetrator of a sulfur mustard attack in Morea, Aleppo Province in Syria in September 2015.[32]

16.7.3   Responding to the Threat of Chemical Weapons Use

In recent years, the OPCW has taken several of its decisions by vote concerning the use of chemical weapons in Syria and elsewhere (assassination attempts in Malaysia, the United Kingdom, and most recently Russia, using chemical agents that had their origin in State CW programs). This included two decisions on responding to the threat of chemical weapons uses, as well as decisions on sanctions against Syria—a State Party that has been found responsible for flagrant violations of one of the core prohibitions of the CWC.[33] Whilst some observers protested the fact that the OPCW was not able to take these decisions by consensus, it must be pointed out that voting on compliance matters was included in the CWC provisions precisely to ensure that no single power or small group of States Parties could prevent the OPCW from responding appropriately to a confirmed use of chemical weapons. Different from the UN Security Council, where the UN Charter gives each of the five Permanent Members a right to veto decisions—a right that can be, and has been, used to block decisions by the Council when their own interests were at stake—the CWC decision-making rules, although generally in favor of a consensus approach, foresee that the Executive Council vote on issues that require swift action, and in particular when it comes to compliance or non-compliance matters (Krutzsch and Dunworth 2014). The response of the OPCW to the uses of chemical weapons in Syria has underlined the importance of this approach.

The engagement between the OPCW and the United Nations in the context of the different joint missions in Syria has also strengthened the relationship between the

---

[32] The latest IIT report was issued as OPCW Technical Secretariat: Fourth report by the OPCW investigation and identification team pursuant to paragraph 10 of decision C-SS-4/DEC.3 'Addressing the threat from chemical weapons use' Marea (Syrian Arab Republic)—1 September 2015 (2024). OPCW Homepage, https://www.opcw.org/sites/default/files/documents/2024/02/s-2255-2024%28e%29.pdf, last accessed 2024/03/24.

[33] OPCW: Decision—Addressing the threat from chemical weapons use. OPCW document C-SS-4/DEC.3 (2018). OPCW Homepage, https://www.opcw.org/sites/default/files/documents/CSP/C-SS-4/en/css4dec3_e_.doc.pdf, last accessed 2024/03/24; OPCW: Addressing the threat from use of chemical weapons and the threat of future use (2023). OPCW Homepage. https://www.opcw.org/sites/default/files/documents/2023/12/c28dec12%28e%29.pdf, last accessed 2024/03/24.

two organizations, taking the OPCW more firmly into the international system as a recognized partner that brings specialized expertise and capacity to the international community (Krutzsch, Myjer, and Trapp 2014). At the same time, the OPCW decisions about how to respond to the threat of use of chemical weapons have created space for additional engagements between the OPCW and national implementation systems of States Parties, enabling international technical support for national mechanisms in the domain of law enforcement and prosecution of criminal acts and terrorism. At the same time, the activities pursued by the OPCW in the investigative domain open up opportunities for cooperation among States Parties and with the OPCW in such areas as chemical forensics.

16.8 Reflections on the Future

The CWC verification system and the resources put in place by the OPCW to implement these verification procedures (an experienced corps of inspectors, modern inspection equipment, tested procedures, a well-functioning planning and operations support system, an experiences group of verification planners and evaluators, and a network of Designated Laboratories for environmental, biomedical, and now also biotoxin samples) have delivered what the negotiators of the CWC had hoped for: a robust verification system that provides accountability for the destruction of declared chemical weapons and production facilities, and confidence in the non-production of chemical weapons in chemical industry facilities of the States Parties. Although the provisions for challenge inspections and investigations of alleged use have not been invoked by any State Party, the missions conducted in Syria have demonstrated that the OPCW has the capability to investigate possible treaty violations and gather evidence that would allow policy makers to address non-compliance issues and take evidence-based decisions about how to respond to confirmed such cases.

The OPCW has also demonstrated that it is capable of adapting to challenges that were not foreseen by negotiators, such as the conduct of onsite verification and deproliferation measures in conflict zones. It also has been developing competencies and tools to address attribution questions after confirmed cases of CW uses. It will now be important to marshal sufficient political support to build the expertise and techniques developed in Syria back into the regular structure and resources of the OPCW to be able to conduct any future such investigations and to support national investigations, wherever that may become necessary. This will call for strategic investments into technology development in the field of chemical forensics (methods development and validation, the development of certified reference materials and curated and comprehensive databases, chemical and biological informatics tools, and networks with other competent laboratories and scientific centers). The new OPCW Center for Chemistry and Technology[34] can become an important place for international cooperation and coordination of efforts to this end.

---

[34] OPCW Homepage, https://www.opcw.org/media-centre/featured-topics/chemtech-centre, last accessed 2024/03/24.

Over time, the CWC regime will face challenges that the international community is already facing in the context of biological weapons arms control: a growing distance between weapons programs known from the past and current chemical activities in legitimate areas of research, development, manufacturing and application of chemistry, and growing uncertainties about what a new chemical weapons program might actually look like and what types of toxic and precursor chemicals, equipment and facilities it might involve. If CWC verification is to provide assurances against such future possibilities—possibilities that relate to the misuse of emerging technologies rather than a rehash of malicious activities as we know them from the past—then verification concepts and techniques will need to be rethought.[35] This does not mean changing the treaty language, but it would require adapting the way in which provisions of the treaty are being applied in the fast-evolving world of chemistry and technology. At the same time, the OPCW would need to preserve its experiences and competencies gained in past verification conduct—the use of chlorine in Syria was a reminder that even "old" sciences and technology may be used to kill, injure or subdue today.

Finally, it may be worthwhile recalling Fred Iklé's words from his 1961 essay: "detecting violations is not enough. What counts are the political and military consequences of a violation once it has been detected, since these alone will determine whether or not a violator stands to gain in the end" (Iklé 1961). How the verification system of the CWC will evolve and how States Parties will see its role (and its relevance for their security) in the future is not simply a question of how effectively advances in science and technology will be brought to use to enhance the fact-finding and analytical capabilities of the OPCW; even more so, it is a question of political will to base decisions and responses to non-compliance on a technically sound, adaptable, robust, trusted, and sustainable system of fact-finding rather than political preferences.

References

Iklé FC (1961) After detection—what? Foreign Aff 39(2):208–220

Krutzsch W, Dunworth T (2014) Article VIII—The organisation. In: Krutzsch W, Myjer E, Trapp R (eds) The chemical weapons convention—a commentary. Oxford University Press, Oxford, pp 235–296

Krutzsch W, Myjer E, Trapp R (2014) Issues raised by the accession of Syria to the chemical weapons convention. In: Krutzsch W, Myjer E, Trapp R (eds) The chemical weapons convention—a commentary. Oxford University Press, Oxford, pp 689–702

Manley R (2007) Preparing for disarmament: articles III, IV and V. In: Kenyon IR, Feakes D (eds) The creation of the organisation for the prohibition of chemical weapons—a case study in the birth of an intergovernmental organisation. TMC Asser, The Hague, pp 139–178

---

[35] OPCW Director-General: Report of the advisory panel on future priorities of the Organisation for the Prohibition of Chemical Weapons (2011). OPCW Homepage, https://www.opcw.org/sites/def ault/files/documents/S_series/2011/en/Advisory_Group_report_s-951-2011_e_.pdf, last accessed 2024/03/24.

Trapp R, Walker P (2014) Article IV—chemical weapons. In: Krutzsch W, Myjer E, Trapp R (eds) The chemical weapons convention—a commentary. Oxford University Press, Oxford, pp 119–150

Ralf Trapp is a consultant on chemical and biological weapons arms control, providing services, amongst others, to the UN, the OPCW and the EU. A chemist and toxicologist by training, he participated in the negotiations of the Chemical Weapons Convention and worked at the OPCW on issues of verification, international cooperation, and science advice.

Chapter 17
Preventing the Weaponisation of Mid-Spectrum Agents as the Chemical, Life, and Associated Sciences Converge

Michael Crowley

Abstract This paper examines the risks that contemporary chemical, life and associated scientific research and technological developments could be miused for the weaponization of mid-spectrum agents—toxins, bioregulators, and other substances of biological origin, and their synthetic analogues. Drawing on illustrative case studies of dual use research and related activities of potential concern from China, India, Iran, Russia, Syria and the United States, it discusses how States can ensure such research and related activities are not utilized in weapons development. Although midspectrum agents are, in theory, covered by both the Biological and Toxin Weapons Convention (BTWC) and the Chemical Weapons Convention (CWC), this apparent overlap in reality masks a dangerous regulatory gap—with neither Convention implemented effectively to address threats of weaponization of these agents. The paper highlights the potentially damaging consequences of this failure for international peace and security, and proposes realistic routes for action by the BTWC and CWC State Parties to address these challenges.

Keywords Toxin · Bioregulator · CNS-acting chemical agent · Riot control agent · Malodorant

17.1 Introduction

Previous chemical and biological weapons (CBW) programmes of certain States, particularly during the Cold War, attempted to discover (or synthesise) and then weaponise a wide range of potential chemical and biological agents. In the light of these past programmes and in recognition that the list of potential future agents is essentially open-ended, the Biological and Toxin Weapons Convention (BTWC) and

M. Crowley (✉)
Bradford University, Bradford, UK
e-mail: M.J.A.Crowley@bradford.ac.uk

B. Friedrich et al. (eds.), *Thirty Years of the Chemical Weapons Convention (CWC)*,
https://doi.org/10.1007/978-3-031-98854-7_17

211

the Chemical Weapons Convention (CWC) were both constructed to be comprehensive in the substances encompassed and necessarily responsive to scientific and technological developments. The two Conventions both cover and prevent weaponization of toxins, bioregulators, and other substances of biological origin, and their synthetic analogues—and hence such substances have been collectively termed mid-spectrum agents. However, the assumed overlapping protection provided by the CWC and BTWC in reality hides a significant regulatory lacuna, with the danger that neither Convention is being properly implemented so as to effectively prevent and address the development of today's and tomorrow's mid-spectrum weapons. This paper examines contemporary chemical, life and associated scientific research and technological developments of potential concern drawn from illustrative case studies focussing on China, India, Iran, Russia, Syria and the United States. It describes how the CWC and BTWC States Parties have, to date, failed to adequately address these concerns, and provides recommendations as to how both regimes can be strengthened in these areas.[1]

17.2 Mid-Spectrum Agents

The ongoing revolution in the life sciences that has proceeded over the last few decades has resulted in the boundary between chemistry and biology becoming increasingly blurred. This, in turn, has meant that the distinction between certain chemical and biological weapons has become less useful. Rather than thinking of chemical and biological weapons threats as distinct, some analysts[2] have argued that it is more useful to conceptualise such agents as lying along a continuous biochemical threat spectrum from the classical chemical agents on one extreme (i.e. nerve, blood and blister agents), through mid-spectrum agents and on to biological agents (including traditional and genetically modified biological agents). Although the range of substances covered by the term mid-spectrum agents is not fixed, at a minimum they are considered to comprise toxins, and by extension bioregulators, and their synthetic analogues.

---

[1] This paper draws upon longstanding and ongoing joint research undertaken by the author with Professor Malcolm Dando, and specifically utilises material taken from our publication: Crowley, M. and Dando. M, Toxin and Bioregulator Weapons, Preventing the misuse of the chemical and life sciences, Palgrave Macmillan/Springer Nature, Cham, Switzerland, 2022.

[2] Dando, M., Scientific outlook for the development of incapacitants, in Pearson, A. Chevrier, M. & Wheelis, M. (eds) Incapacitating Biochemical Weapons, Lanham: Lexington Books, 2007, p. 125; Aas, P. The Threat of Mid-Spectrum Chemical Warfare Agents, Prehospital and Disaster Medicine, volume 18, number 4, 2003, pp. 306–312.; Davison, N. 'Off the Rocker' and 'On the Floor': The Continued Development of Biochemical Incapacitating Weapons, Bradford Science and Technology Report No. 8, Bradford Disarmament Research Centre, August 2007, pp. 2–4; Pearson, G. Relevant Scientific And Technological Developments For The First CWC Review Conference: The BTWC Review Conference Experience, CWC Review Conference Paper No.1. Department of Peace Studies, University of Bradford. August 2002.

Toxins are toxic substances of natural origin or their synthetic analogues and derivatives that can cause death, permanent or temporary harm or incapacitation to those affected. Certain toxins derived from bacteria are amongst the most poisonous of natural toxins, whilst a range of other toxins—notably those derived from plants and fungi also have significantly high toxicity. As a result of their potency these toxin types were investigated, particularly during the Cold War period, by certain State-level weapons programmes. For example, a series of US military papers highlighting potential agents of concern included abrin, botulinum toxin, conotoxins, saxitoxin, shiga toxin, staphylococcal enterotoxins, T-2 toxin and tetrodotoxin (Nordin 2012, 2013). Although many previous State-level weapons development programmes focussed on toxin weapons that would seriously harm or kill large numbers of people, certain countries also developed 'non-lethal' or 'less lethal' toxin weapons that would supposedly temporarily incapacitate individuals, small groups, or large numbers of individuals (Regis 1999, pp. 201–204). Contemporary advances in the chemical, life and associated sciences and technologies have dramatically widened the range of potential agents available to today's weaponeers by, for example, enabling conjugation of two or more existing toxins or even facilitating the *de novo* design and construction of completely new toxin types (Pitschmann and Hon 2016).

Bioregulators are natural chemicals made by living organisms that operate to facilitate the correct running of core physiological systems in such organisms. In mammals, they regulate wide-ranging body functions including respiration, blood pressure, heart rate, body temperature, consciousness, mood, and immune responses. The structure of bioregulators are extremely varied, ranging from relatively simple, small molecules in some hormones or neurotransmitters to macromolecules such as proteins, polypeptides, or nucleic acids. The physiological impact of bioregulators is not restricted to a single regulatory system. The same bioregulator can be active and have a variety of physiological functions in a number of tissues. In mammals, further complexity also arises as a consequence of the interactions between the endocrine, immune and nervous systems. Thus, for example, changing bioregulator concentration or blocking cell receptor functioning in the nervous system may consequently affect operation of related interacting systems. If introduced in excessively large amounts, these normally benign substances may produce devasting toxic effects, becoming highly potent *de facto* toxins.

To date there appears to be some uncertainty and differing interpretations amongst States Parties as to how toxins and bioregulators should be regulated under relevant arms control and disarmament treaties, notably the BTWC and CWC. This author would argue that this uncertainty has been exacerbated by the lack of clear formal definition of 'toxins' under these instruments and a related lack of clarity with regard to the range of substances (notably bioregulators) that would be covered by this term. Such definitional lacunae may also have limited BTWC and CWC States Parties' effectiveness in regulating a broader spectrum of biological substances (as well as synthesised analogues) that have previously been weaponised or could be weaponised in the future.[3] This raises open questions, firstly as to whether at least some of these additional substances should be classed as toxins (and by extension mid-spectrum agents) and treated as such under the BTWC and/or CWC, and secondly how substances not subsequently deemed to be toxins should be addressed by these two instruments and/or other regulatory measures. Such questions are directly relevant to contemporary development and employment of those so-called 'less lethal' weapons that utilise biological substances or their synthetic analogues, notably:

- *Riot control agents* (RCAs)—which are defined under the CWC "as any chemical not listed in a schedule which can produce sensory irritation or disabling physical effects rapidly in humans and which disappear within a short time following termination of exposure".[4] RCAs include synthetic chemicals such as chloroacetophenone (CN), ortho-chlorobenzylidene malononitrile (CS), and dibenz (b,f)-1,4-oxazepine (CR)—that would not be considered as toxins. However, the term also includes certain biological substances notably the capsaicinoids and the related synthesised analogue, pelargonic acid vanillylamide (PAVA), which the author would argue can be considered to be toxins. RCAs are widely used for law enforcement activities including dispersal of crowds or to incapacitate individuals. However, they are also misused by police and security forces, notably in excessive quantities or in enclosed areas, sometimes resulting in serious detrimental health consequences or fatalities for those affected.[5]
- *Malodorants*—are a diverse range of foul-smelling human-synthesised chemicals and biological substances, the latter category at least the author argues should be considered as toxins. When employed as weapons they are intended to cause strong aversion responses and are notably used for crowd dispersal. They have strong but short-lived physiological effects including nausea, gagging and vomiting.[6]
- *Central nervous system (CNS) acting chemical agents [also referred to as incapacitating chemical agents (ICAs)]*—a diverse range of human-synthesised chemicals and biological substances that effect the CNS (and in the case of ICAs additional

---

[3] In this paper, the term "weapon" is considered to be very broad in scope incorporating any instrument or device [in this case employing toxic chemicals] that is intended to kill, injure or incapacitate. Such weapons could range, for example, from small scale assassination devices employed to kill individuals to wide area dispersal systems designed to effect large numbers of people. The scope incorporates both lethal and "less lethal" devices and instruments.

[4] Organisation for the Prohibition of Chemical Weapons (OPCW) (1993) *Chemical Weapons Convention (CWC)*, 1993, Article II.7 https://www.opcw.org/chemical-weapons-convention.

[5] For an overview of the common riot control agents, their effects under appropriate conditions, and when used inappropriately see: Olajos, E. and Salem, H. Riot control agents: pharmacology, toxicology, biochemistry and chemistry, *Journal of Applied Toxicology*, Volume 21, Issue 5, September/October 2001 Pages 355–391; Crowley, M. *Chemical Control: Regulation of Incapacitating Chemical Agent Weapons, Riot Control Agents and Their Means of Delivery.* Palgrave Macmillan, London, 2016, pp.39–50; Haar, R., Lacopino, V., Ranadive N., Weiser, S. and Dandu, M. Health impacts of chemical irritants used for crowd control: a systematic review of the injuries and deaths caused by tear gas and pepper spray. *BMC Public Health* 17, no. 1, 2017, p. 1–14.

[6] For further discussion see, Crowley, M and Dando, M. R. *Toxin and Bioregulator Weapons: Preventing the Misuse of the Chemical and Life Sciences.* Springer/ Nature, Cham, Switzerland, 2022, pp. 12,162–170.

and/or core physiological systems). When used as weapons, these agents are supposedly intended to cause a long, but non-permanent, incapacitation in those individuals or groups targeted. However, inappropriate doses of these toxic chemicals can cause very serious, long-lasting health effects and can be fatal to those affected.[7]

17.3 Regulation of Mid-Spectrum Agents Under International Arms Control Instruments[8]

The norm against the deliberate use of either poison or disease as weapons of war is ancient. These interconnected taboos were subsequently codified and expanded in the 20th century and continue to this day through three international instruments: the Geneva Protocol (adopted in 1925), the Biological and Toxin Weapons Convention (BTWC) (adopted in 1972) and the Chemical Weapons Convention (CWC) (adopted in 1993). Collectively, these three instruments form two distinct regimes individually prohibiting biological weapons and chemical weapons, which overlap and consequently both prohibit toxin weapons. The inter-relationship between these three instruments and the range of substances they cover within their scope is illustrated in Figure 1.

Geneva Protocol

The Geneva Protocol prohibits the use in armed conflict of biological weapons and chemical weapons, including toxins (and by extension bioregulators). These prohibitions subsequently became accepted as customary international law and are thus applicable to all States regardless of whether they signed and ratified the Protocol; they provided the legal foundations upon which the far more extensive prohibitions and regulations of the BTWC and the CWC could be developed. Because the Geneva Protocol's prohibitions relate solely to use, it cannot be employed to regulate dual-use research, development, manufacture, transfer or stockpiling of toxin and bioregulator weapons.

Biological and Toxin Weapons Convention

The BTWC explicitly prohibits the development, production, stockpiling, acquisition or retention of biological and toxin weapons, and implicitly bans their use. Unfortunately, the Convention does not formally define "toxins" in its text. However, toxins (and it can be argued bioregulators when employed as *de facto* toxins) are encompassed under Article I of the Convention, which declares that:

---

[7] For further discussion see: Crowley, M and Dando, M. R. (2022) *op.cit.*, pp. 12, 124–128, 150–162.

[8] For a more detailed discussion of the regulation of toxins and 0bioregulators see Crowley, M and Dando, M. R. Regulation of toxins, bioregulators and other substances of biological origin under international arms control and disarmament agreements, Chapter 9, pp 197- 235 in Crowley, M and Dando, M. R. (2022) op.cit.

Each State Party to this Convention undertakes never in any circumstances to develop, produce, stockpile or otherwise acquire or retain [*inter alia*]: 1. Microbial or other biological agents, or toxins whatever their origin or method of production, of types and in quantities that have no justification for prophylactic, protective or other peaceful purposes...

Article I's wide and essentially open-ended scope—which makes no restriction with regard to lethality—has become known as the General Purpose Criterion (GPC). The GPC is based on the intent of relevant activities, permitting peaceful activities, but prohibiting those that have "no justification for prophylactic, protective or other peaceful purposes". It has subsequently been regularly confirmed and applied by the BTWC States Parties to the rapidly advancing life sciences and associated technologies through the Additional Understandings adopted as part of Final Documents of repeated BTWC Review Conferences.[9]

The BTWC's provisions—unlike those contained in the CWC—encompass biological agents and toxins acting on plants, as well as animals and human beings. However, there are a number of ambiguities and limitations in the BTWC and its implementation directly related to toxins that, to date, have not been addressed collectively by its States Parties. Firstly, the States Parties have not defined "toxins" nor provided sufficient guidance as to the scope of substances covered by this term, including with regard to bioregulators. Secondly, they have not established whether any forms of toxin (and bioregulator) weapons can be employed for law enforcement purposes, and if such weapons were permitted in principle, what types of substances/delivery mechanisms can be employed, in what specific circumstances and under what restrictions. Thirdly, they have not clarified how dual-use life science and associated research should be controlled so as to ensure that it does not directly facilitate or is not utilised in the development of toxins (and bioregulators) for prohibited purposes. Beyond these toxin-specific limitations, there are a range of broader endemic organisational limitations that further seriously restrict the utility of the Convention to be employed as an effective mechanism for preventing and addressing the multifaceted threats of development and use of toxin and bioregulator weapons. Critically, although the BTWC does contain limited provisions for bilateral and multilateral consultations amongst State Parties to resolve implementation concerns, in practice the BTWC regime suffers from the absence of formal effective and comprehensive verification, transparency and compliance measures. Furthermore, beyond the five yearly review of science and technology undertaken collectively by States and the ISU in the run-up to and during BTWC Review Conferences, there is currently no adequate regular, systematic and comprehensive monitoring and analysis by relevant BTWC bodies of developments in the life and associated sciences and technologies of relevance to the Convention.[10] All of these serious weaknesses and lacunae are further exacerbated by the failure of the BTWC States Parties to

---

[9] ICRC, Treaties, States Parties and Commentaries: *Convention on the Prohibition of the Development, Production and Stockpiling of Bacteriological (Biological) and Toxin Weapons and on Their Destruction (BTWC), 1972.* Available at https://ihl-databases.icrc.org/applic/ihl/ihl.nsf/INTRO/450?OpenDocument, Article 1.

[10] An important step forward in addressing this current lacuna was made by the 9th BTWC Review Conference, in December 2022, which decided to "develop with a view to establishing a mechanism

establish and adequately fund a technical secretariat or similar international organisational body mandated to coordinate these activities and to further encourage and facilitate effective implementation of the Convention by all States Parties.[11]

Chemical Weapons Convention

The CWC, under Article I, explicitly prohibits the development, production, stockpiling, acquisition, retention or use of "chemical weapons".[12] These are consequently defined under Article II:

1. Chemical Weapons' means [*inter alia*]: a. Toxic chemicals and their precursors, except where intended for purposes not prohibited under the Convention, as long as the types and quantities are consistent with such purposes.".[13]

And

2. 'Toxic Chemical' means: Any chemical which through its chemical action on life processes can cause death, temporary incapacitation or permanent harm to humans or animals. This includes all such chemicals, regardless of their origin or of their method of production, and regardless of whether they are produced in facilities, in munitions or elsewhere.[14]

It can be seen that through its interacting definitions of "toxic chemicals" and "chemical weapons", the CWC textual architecture—as with the BTWC—has been designed to incorporate a General Purpose Criterion based on the intent of the activities. In the case of the CWC, this GPC consequently ensures the Convention a comprehensive and essentially 'open-ended' scope of toxic chemicals encompassed which includes all existing toxins and bioregulators, as well as those that will be discovered or synthesised in the future. Once again it should be noted that the Convention makes no explicit reference to lethality as a determinant in coverage so the CWC would encompass toxins, bioregulators and other toxic chemicals of biological origin including those explored as 'less lethal' weapons. Furthermore, although the Convention completely prohibits "chemical weapons", it does not ban the development and employment of certain types of "toxic chemicals"—potentially encompassing certain toxins and bioregulators—for "purposes not prohibited" which specifically include "law enforcement including domestic riot control".[15] However, the use of "toxic chemicals"—potentially including toxins and bioregulators—for

---

to review and assess scientific and technological developments relevant to the Convention and to provide States Parties with relevant advice. In order for this mechanism to be established, the Working Group on the strengthening of the Convention will make appropriate recommendations." [See: United Nations (2022) Final Document of the Ninth Review Conference. BWC/CONF.IX.9. United Nations, Geneva, 21 December 2022, p. 11, paragraph 19].

[11] For further discussion see: Crowley, M. and Dando, M.R. (2022) *op.cit.*, pp. 220–232.

[12] OPCW, CWC, (1993) *op.cit.*, Article I.

[13] OPCW, CWC, (1993) *op.cit.*, Article II.1.

[14] OPCW, CWC, (1993) *op.cit.*, Article II.2.

[15] OPCW, CWC, (1993) *op. cit.*, Article II.9.a-d.

"purposes not prohibited" would only be permitted "as long as the types and quantities are consistent with such purposes."[16]

The applicability of the CWC to toxins is underlined by the inclusion of two named toxins—ricin and saxitoxin—in the Convention's Schedule 1 list, which comprises toxic chemicals that have either been developed as chemical weapons, pose a high risk to the Convention, and/or have little or no use for "purposes not prohibited" under the Convention. Such Scheduled chemicals require specific regulation and are subject to routine verification measures.[17] A further indication that toxins, bioregulators and biological substances more broadly fall within the scope of the Convention is the determination by the Organisation for the Prohibition of Chemical Weapons (OPCW) Scientific Advisory Board (SAB) that six named capsaicinoids and one synthetic analogue fall within the definition of RCAs[18] and should be regulated accordingly, i.e. that they can be used in law enforcement but are banned from being used "as a method of warfare".[19]

However, the implementation of the CWC in a number of key areas of relevance to regulation of toxins, bioregulators and biological substances has not been settled amongst the States Parties, with consequent divergences in interpretation and implementation witnessed. Areas of contestation and uncertainty include determining the boundaries of acceptable dual-use chemical and life science research conducted for "protective purposes", and establishing the legitimacy (or not) of undertaking research and development of, and employing, ICA/CNS-acting weapons for law enforcement purposes. In the latter case, uncertainty continues despite a welcome Decision of the 2021 Conference of States Parties (CSP) which took the form of an "understanding" clarifying that aerosolised use of CNS-acting chemicals for law enforcement purposes was prohibited under the CWC.[20] Additional issues that have not been formally addressed by the CWC States Parties relate to the application of the Convention with regard to 'remote control' and 'wide area' RCA delivery mechanisms, and how maladorants should be regulated under the CWC.

Krutzsch and Trapp, in their Commentary on the Chemical Weapons Convention, underlined the critical "real world" value arising from the inclusion of toxins in the scope of both the BTWC and the CWC:

---

[16] OPCW, CWC, (1993) *op. cit.*, Article II.1.a.

[17] OPCW, CWC (1993) *op. cit.*, Annex on Chemicals, B. Schedules of Chemicals, Schedule 1.

[18] OPCW, Scientific Advisory Board (2017) *Response to the Director General's Request to the Scientific Advisory Board to consider which riot control agents are subject to declaration under the Chemical Weapons Convention.* SAB-25/WP.1. OPCW, The Hague, 27 March 2017. Annex 4 lists the six capsaicinoids as: Oleoresin capsicum (OC); 8-Methyl-N-vanillyl-trans-6-nonemaide (capsaicin);8-Methyl-N-vanillylnomaide (dihydrocapsaicin); N-Vanillyl-9-methyldec-7-(E)-enamide (homocapsaicin); N-Vanillyl-9-methyldecanamide (homodihydocapsaicin); N-Vanillyl-7 methyloctanamide (nordihydrocapasicin); and the synthetic analoge is N-Vanillylnomaide (pseudocapsaicin, PAVA).

[19] OPCW, CWC, (1993) *op. cit.*, Article I.5.

[20] Conference of States Parties (2021) *Decision: Understanding regarding the aerosolised use of central nervous system-acting-chemicals for law enforcement purposes*, C-26/DEC.10. OPCW, The Hague, 1 December 2021.

This overlap recognizes that it is impossible to draw a line between toxins and other toxic chemicals: an increasing number of toxins can be synthesized in laboratories without resorting to organisms which produce them in nature, and a number of low-molecule-weight toxins are at the same time synthetic chemicals manufactured by industry.... At the same time, biotechnological manufacturing methods have increasingly come into use for chemicals which were traditionally produced via chemical synthesis only (Krutzsch and Trapp 2014, p. 85).

Unfortunately, this theoretical double coverage of toxins under both the BTWC and CWC in reality hides a serious potential regulatory lacuna in the actual implementation of the two Conventions by States Parties, with neither instrument being effectively implemented by the relevant States Parties so as to ensure that bioregulators and toxins are not developed and used as weapons. In 2008 Professor Perry Robinson warned of the consequences of such regulatory failure:

Such overlap ought to mean, one might think, that the weapons [including toxins, bioregulators and other bioactive chemicals of biological origin, as well as their synthetic analogues] are well controlled, being subject not just to one but to two international disarmament treaties... (Perry Robinson 2008, p. 2).

But as he continued:

...In the real world, however, that is not the way it is. That overlap seems simply to have given people involved in implementing one of the two treaties opportunity to relinquish, even deny, responsibility for anything also covered by the other treaty. The area of overlap thus risks becoming a gulf into which things disappear. It looks like this has been happening to toxins (Perry Robinson 2008, p. 2).

17.4 Advancing Chemical and Life Sciences

The chemical, life, and associated sciences are undergoing a revolution in capabilities that is allowing scientists to understand and manipulate living systems in unprecedented ways. Whilst capable of producing immense societal and health benefits, these capabilities can be misused for malign purposes. Of particular potential relevance have been developments and convergences in neuroscience, medicinal chemistry, pharmacology, toxicology, immunology, molecular biology, systems biology and synthetic biology, as well as their interactions with nanoscience, artificial intelligence research and computer science. Such overarching processes have significantly impacted and advanced the discovery and study of toxins and bioregulators, their mechanisms of action, the corresponding receptor and sub-receptor sites they interact with, and the broader functioning of the human brain, central nervous system, and other regulatory and physiological systems they affect. Such advances have gone in parallel with the increasing ability to chemically synthesise peptide bioregulators and incorporate chemical modifications resulting in analogues with markedly different physiological properties. Whilst this knowledge will provide significant benefits to society; given its dual or rather multi-faceted applicability, this knowledge and associated technological advances could instead be exploited in the development of toxin and bioregulator weapons or for other malign manipulation of core human physiological systems.

17.5 Dual-Use Toxin and Bioregulator Research Undertaken Purportedly for CBW Defence

One area of continuing concern has been the secrecy surrounding CBW defence establishment research and activities related to toxins, bioregulators, bioregulatory pathways and physiological systems, plus associated measures to facilitate agent dissemination and uptake. The necessity of this kind of defence work is recognised and permitted, under the CWC if it is conducted for "protective purposes",[21] or under the BTWC for "prophylactic, protection and other purposes".[22] However, under Article X of the CWC, there is a general requirement "for the purposes of increasing the transparency of national programmes related to protective purposes, that each State Party provide annually to the Technical Secretariat information on its programme" through CWC Article X Declarations,[23] and likewise BTWC Building Measures.[24] Unfortunately, public transparency in this area in many States is currently very limited and insufficient to address disquiet surrounding the intentions and applications of such research and associated activities conducted by certain State CBW defence establishments.

For example, there has been a long series of concerns raised over decades about US military-funded and military-related institutions reportedly carrying out biological weapons-related research purportedly for defensive purposes, which have been perceived by certain States and civil society observers as coming dangerously near to or actually crossing the line into offensive weapons research. Specific concerns related to the apparent (and questionable) US interpretation that the BTWC allowed development of biological weapons when intended for 'threat assessment' and development of defensive counter-measures. A particular focus of disquiet has been the 2004 establishment of the National Biological Threat Characterization Centre with planned capabilities to investigate *inter alia* "Aerosol Dynamics, Novel Delivery of Threat, Novel Packaging, Simulation & Modelling (Epidemiology), Genetic Engineering, Environmental Stability, work on Bioregulators and Immunomodulators, Genomics/Proteomics/Transcript [and] Red Teaming [i.e. duplication of threat scenarios]…".[25]

---

[21] OPCW, CWC (1993) *op.cit.*, Article X, paragraph 2.

[22] UN, BTWC (1972) *op.cit.*, Article 1.

[23] OPCW, CWC (1993) *op.cit.*, Article X, paragraph 4.

[24] UN, Office of Disarmament Affairs, *Confidence Building Measures*, https://disarmament.unoda.org/biological-weapons/confidence-building-measures/

[25] Korch, G. Leading Edge of Biodefense—The National Biodefense Analysis and Countermeasures Center. Presentation at the DoD Pest Management Workshop, Naval Air Station, Jacksonville, Florida, Feb. Sponsored by the Armed Forces Pest Management Board Office of the Deputy Under

These proposed activities were strongly criticised at the time by Professor Leitenberg, former US Ambassador Leonard (who was previously Head of the United States Delegation to the Biological Weapons Convention Negotiations), and Dr Spertzel. In their opinion:

> Taken together, many of the activities…may constitute development in the guise of threat assessment, and they certainly will be interpreted that way. Development is prohibited by the Biological Weapons Convention. How would these activities differ from their counterparts in the pre-1969 US BW program except for production and stockpiling this time not being envisioned (Leitenberg 2004, pp. 2–3)?

Openly available information demonstrates that a continuing range of contemporary toxin and bioregulator research, of possible dual-use application, has been carried out by the NBTCC and other US military and military-related institutions and researchers. This research included for example investigation of staphylococcal superantigen functioning (Krakauer 2019, p. 178); production of engineered non-toxic botulinum toxins through 'rational design' (Vazquez-Cintron 2017); employment of recombinant technology to develop a non-toxic SEB mutant (Calm et al. 2017); and work on aerosolised ricin and aerosolised botulinum toxin (Roy et al. 2003; Sanford 2010; Boydston 2021).The US has provided sufficient information or allowed public reporting and publication of scientific papers on much of this research indicating their medical or protective purpose. However, given the lack of full public reporting and transparency for certain facilities and programmes, it is not possible to determine the purposes for which all such research has been undertaken or to which it will be applied (Crowley and Dando 2022, pp. 176–186). However even though elements of US CBW defense research are classified, the level of transparency and reporting it undertakes is far greater than many other States.

17.6   The Search for New Toxins to Weaponise

Dual-use toxin research that could be of concern extends well beyond toxins previously explored as potential weapons agents, such as Ricin, SEB, and Botulinum toxin. Military and non-military institutions and scientists have searched out an ever-increasing range of 'novel' toxins that might be weaponised. These include toxins derived, for example, from indigenous poisonous plants, amphibians, reptiles, scorpions and marine animals. Sometimes the intent of the activities was clearly medical or for other "protective purposes". In other examples the purpose was unstated or unclear (Crowley and Dando 2022, pp. 50–56, 94–100, 119–124, 186–190). In certain cases the intention appears to be for toxin weaponization using a low technology route. Scientists from the Indian National Defence Research and Development Organisation (DRDO) and related entities produced a series of papers describing their

---

Secretary of Defense (Installations and Environment). Department of Defense, Washington, D.C. February 2004, Slide 12.

investigations of toxins derived from native Indian stinging and poisonous plants.[26] In 2018, DRDO scientists argued that because the BTWC "banned the use or stockpiling of most of the pathogenic bio-threat agents" this would "necessitate… search [for] some novel natural bio-threat agents from stinging plants that may be used as future bio-weapon for self-defence purposes." Consequently, they conducted research to "identify, characterize and screen the potential of […] stinging plants on the basis of their secondary metabolite contents that may be used for the formulation of novel future bio-threat agents for self-defence" (Gupta et al. 2018).

This DRDO research continued by investigating plants that produce "toxins that have the ability to adversely affect human health in a variety of ways, ranging from relatively mild allergic reactions to serious medical complications, including death." The scientists reported that they had successfully identified "several poisonous plants that can be used for the development of novel multi-system targeted warfare agents for defensive applications." They also identified "poisonous components" from ten possible candidate plants and investigated their "mode of action", that could "have harmful effect on various biological systems like nervous, cardiac, digestive, respiratory, dermal, etc. simultaneously" (Gupta et al. 2018). A number of the toxins investigated as potential agents—such as aconitine—are highly toxic and can be lethal; others while not as poisonous might still potentially be employed as 'less lethal' toxin weapons. Despite repeated requests for further information about this research and its intended purpose, no response was received from the DRDO or Indian authorities with responsibility for implementation of the BTWC and CWC.

17.7 Toxin Production

Concerns are not restricted just to toxin and bioregulator research, but also extend to the intention behind the production of toxins by certain States. Syria was forced to accede to the CWC and undertake a rapid chemical weapons disarmament process after its chemical weapon attacks on its own citizens. On 23 October 2013, Syria formally declared its stockpile of chemical weapons to the OPCW and agreed to facilitate its verification and destruction under OPCW supervision. Yet in its initial declaration, Syria did not provide adequate information concerning its toxin-related activities (Crowley and Dando 2022, pp. 138–145). Reuters reported that "Western officials with access to intelligence about Syria, [highlighted] topics of concern [including] deadly…agent ricin."[27] Necessarily therefore, certain CWC States Parties including the U.S. questioned the "accuracy and completeness of Syria's declaration". In April 2014 the OPCW Director General launched a Declaration Assessment Team [DAT] "[T]o attempt to resolve many of these concerns,

---

[26] See Crowley, M. and Dando, M.R. (2022) *op.cit.,* pp. 79–85, for a detailed discussion together with associated references to relevant Indian scientific research papers.

[27] Charbonneau, L. Western intel suggests Syria can still produce chemical arms. *Reuters*, United Nations, 25 April 2014.

including toxin-relevant activity"[emphasis added].[28] After the work of the OPCW DAT, on 14 July 2014, Syria provided an amendment to its initial declaration and then the OPCW Director General reported that Syria "declared a CWPF [chemical weapons production facility] [The Al-Maliha facility]... a facility for the production of ricin...[T]he newly declared facility is subject to verification and destruction...[A]ccording to [Syria's] amendment, the entire quantity of ricin produced was disposed of prior to the entry into force of the Convention for the Syrian Arab Republic."[29] By July 2018 the OPCW Director General reported that the Technical Secretariat had verified the destruction of all of Syria's declared "chemical weapons production facilities", which included the Al-Maliha facility.[30]

In spite of the verified destruction of this Syrian CWPF, there are still publicly-unresolved areas concerning the purpose and nature of Syria's toxin production activities, which additionally highlight limitations in the OPCW reporting and verification system. For example, if Syria had admitted possessing ricin stockpiles following its accession to the CWC, under the Convention it would have had to declare all locations and quantities of such stockpiles, facilitate OPCW verification of its stockpile declaration through on-site visits and then verification of all stockpile destruction. However, because Syria stated that it had destroyed its ricin stockpiles at some stage before accession, it was not obliged to provide the OPCW with full details of this stockpile nor for it to undergo associated OPCW verification concerning this stockpile.

Additionally, Syria reportedly stated that its ricin production had been for medical purposes, (namely cancer research and treatment).[31] However, this assertion appears to conflict with Syria's formal declaration of the facility to the OPCW as a CWPF, because this designation specifically excludes a "single small-scale facility for production of chemicals listed in Schedule 1 for purposes not prohibited under this Convention."[32] Thus there remains a very important unresolved question: if the ricin production facility was only intended to manufacture this toxin for medical or other legitimate civilian purposes, why didn't Syria simply provide sufficient relevant evidence to the OPCW (and to the public) proving this was actually the case, and continue with this important medical work? Reuters reported scepticism at the time from certain CWC States Parties, regarding Syria's reasons for not including the ricin production facility in its initial declaration. According to one 'diplomatic

---

[28] United States, *Report on Adherence to and Compliance with Arms Control, Nonproliferation, and Disarmament Agreements and Commitments*, Biological and Toxin Weapon Convention, Syria, 5 June 2015, p. 20.

[29] OPCW, Executive Council *Note by the Director-General, Progress in the elimination of the Syrian chemical weapons programme*. EC-M-44/DG.1, Section 4. (b), OPCW, The Hague, 25 July 2014.

[30] OPCW, Executive Council *Note by the Director General, Progress in the elimination of the Syrian chemical weapons programme*. EC-89/DG.1, 24 July 2018.

[31] Zanders, J. P. Gradually making sense of Syria's CW declarations, *The Trench*, 11, August 2014. Available at https://www.the-trench.org/syrias-cw-declarations.

[32] OPCW, *Convention on the Prohibition of the Development, Production, Stockpiling and Use of Chemical Weapons and on their Destruction [Chemical Weapons Convention (CWC)]*, 1993, Article II, 8 (b) (iii). [Text available from https://www.opcw.org/chemical-weapons-convention].

source' "[S]yria will argue that the facilities were not revealed earlier because they were in a rush when they first had to report them .... They had said the ricin was for medical purposes, but we don't believe that's true."[33] In light of the previous, and repeated, obfuscation of its chemical weapons programme, there is still obvious concern as to whether Syria has been completely truthful with regard to its activities at Al-Maliha, and whether any further production facilities or stockpiles of ricin or other potential toxin weapons remain.

17.8   Brain Research Projects

During this century, the technologies available to neuroscientists have developed at an extremely rapid pace and made it more and more possible to understand the neuronal circuits in the central nervous system (CNS) that produce our behaviour, and the roles played by bioregulators in such processes. The benefits of such research in treating people with brain dysfunctions has led certain States in recent years to initiate large-scale brain research projects. In some States, notably China, scientists from military medical institutions and other defence-related facilities are involved in much of the research.[34]

A number of areas of the China Brain Project (CBP) and work of Chinese scientists on neurological systems and associated bioregulators (including noradrenaline, 5-HT, and orexin) in simple animal models—such as fruit flies, zebrafish and mice—have potential 'dual-use' applicability.

In order to better understand more complex human behaviours and the mechanisms that produce them, large-scale work is also being carried out in China on the CNS and responses of non-human primates (NHPs), like macaque monkeys (Quin and Li 2017).[35] This work has potential dual-use implications as previous State-level biological weapons programmes did not limit their activities to manipulation of basic functions of the brain, but also sought to influence human emotions, cognition and behavior. Consequently, as commentators have highlighted, this:

> [R]aises concerns about tacit capabilities… and the yoking of NHP studies and findings to military agendas under programs of dual- or direct-use…Of particular note in such efforts is Junweikejiwei, the newly developed Chinese research agency that conjoins efforts of the CAS [Chinese Academy of Sciences] and China's Ministry of Defense, and which is modelled after the United States' Defense Advanced Research Projects Agency (DARPA), to engage rapid, high-risk/high-return approaches to bioscience and technology" (Palchik et al. 2018).

---

[33] Deutsch, A. Exclusive: Syria reveals more chemical weapons facilities to watchdog—sources. *Reuters*, 17 September 2014.

[34] See Crowley, M. and Dando, M.R. (2022) *op.cit.,* pp. 57–60 (for China Brain Project) and pp. 157–160 (for U.S. brain initiative).

[35] Mu-ming Poo (2016) *China Brain Project and non-human primate research in China.* Presentation Number 25 at *The Brain Forum*, Lausanne. 27 May.

The CBP includes studies of higher functions of the brain using advanced biotechnology with NHP as test subjects. A presentation from one of China's leading researchers in 2016 highlighted the "Goals of non-human primate research" which included "1. Studies of the cognitive functions using non-human primate as the animal model" and "2. Generation of genetically modified [including transgenic] monkeys as animal models of human brain disorders and for basic neurobiology research." This research explores the neural basis of cognition including how cells establish synaptic contacts and generate neural circuit activities.[36] Whilst the purposes of such research may be benign, there are dangers that it could be abused to aid development of bioregulator weapons attacking human cognition.

17.9 'Less Lethal' Weapons

Concerns about harmful use of toxins, bioregulators, and other substances of biological origin and their synthetic analogues, have extended to their development and employment as 'less lethal' weapons—though this term is contested and for certain applications is clearly a misnomer.

ICA/CNS-acting chemical agent weapons

One focus of disquiet amongst key medical and scientific bodies, as well as many States, has been development and use of ICA or CNS-acting chemical agent weapons. In the latter part of the last century the U.S. conducted extensive research on a range of potential ICA/CNS-acting chemicals, including pharmaceutical chemicals and bioregulators, and related bioregulatory pathways, for law enforcement and military purposes. In addition to military and law enforcement funded projects intended to assist development of such weapons. A wider range of potentially relevant dual-use research into related fields such as brain research projects was funded, in part, by U.S. defence bodies. However, there is no evidence that any ICA/CNS-acting weapons were developed or fielded, and in 2013 the U.S. disavowed development, stockpiling and use of CNS-acting weapons. It subsequently reconfirmed this position in November 2021.[37] The significant U.S. ending of such activities must be contrasted with the continuing disturbing actions of Russia and to a lesser extent China, as highlighted in our studies,[38] and the reported activities of Iran.[39]

---

[36] Mu-ming Poo (2016) *op.cit.,* Slide 11.

[37] For further detail on U.S. activities, see Crowley, M. and Dando, M.R. (2022) *op.cit.,* pp. 150–162.

[38] See Crowley, M. and Dando, M.R. (2022) *op.cit.,* pp. 124–128 for Russia's activities and pp. 60–63 for China's activities.

[39] See for example United States, Condition (10) (C) Annual Report on Compliance with the Chemical Weapons Convention (CWC), Bureau of Arms Control, Deterrence and Stability, 18 April 2023. https://www.state.gov/adherence-to-and-compliance-with-arms-control-nonprolif eration-and-disarmament-agreements-and-commitments/

Open-source information indicates that the Soviet Union previously conducted research and attempted to develop ICA/CNS-acting weapons utilising pharmaceutical chemicals and also bioregulators. In the 1980s and 1990s, it explored potential CNS-acting bioregulator weapons including endorphins, enkephalins, and other neuromodulating peptides potentially capable of altering human cognition and emotions. A highly placed whistleblower—Dr Kenneth Alibek former Deputy Director for Science of Bioreparat—stated that "[T]he mood-altering possibilities of regulatory peptides were of particular interest to the KGB" (Alibek and Handelman 1999, p. 164). No evidence of development or use of bioregulator weapons has been found to date, but it is unclear whether all such activities terminated following the collapse of the Soviet Union. In 2002 Russian security forces employed aerosolised ICA/CNS-acting pharmaceutical chemicals (i.e., fentanyl derivatives) in a largescale anti-terrorist operation to rescue over 900 hostages held in a Moscow theatre. While most of the hostages were freed, more than 120 were killed by the chemical agents, and many more suffered long-term health problems. More recent Russian 'dual use' research into pharmaceutical chemicals with potential ICA/CNS weapons application has also been reported.[40] Russia has stated that law enforcement use of CNS-acting weapons is not regulated by the CWC and rejected the legitimacy of the 2021 CWC Conference of State Parties "understanding" that clarified that law enforcement use of aerosolized CNS-acting weapons is not regulated by the CWC and rejected the legitimacy of the 2021 CWC Conference of State Parties "understanding" that clarified that law enforcement use of aerosolised CNS acting chemicals was prohibited under the Convention.

Chinese State-owned companies—including China North Industries Group Corporation (NORINCO) and the State 9616 Plant developed and/or promoted ICA/CNS-acting weapons employing dart guns for employment against specific individuals rather than groups,[41] and such weapons have been demonstrated to be in the possession of the Chinese military in 2011 in Hong Kong.[42] The ICA/CNS-acting chemicals utilised in these Chinese weapons cannot be identified and there is no publicly available information as to whether development of ICA/CNS-acting weapons employing bioregulators has occurred. It is notable that China also rejected the legitimacy of the 2021 CWC CSP "understanding".

Malodorant weapons

Another area of potential concern relates to research and attempted development of malodorant weapons. Within the US, for example, projects undertaken or funded by the Department of Defense and associated bodies explored utilising malodorants—both naturally occurring substances and synthetic chemicals—as weapons.

---

[40] Crowley, M., and Dando, M. *Down the Slippery Slope: A Study of Contemporary Dual-Use Chemical and Life Science Research Potentially Applicable to Incapacitating Chemical Agent Weapons.* University of Bradford and University of Bath. 2014, Available online at: https://biochemsec2030d otorg.files.wordpress.com/2013/08/down-the-slippery-slope-final-web.pdf.

[41] For further details and details of original sources see: Crowley, M., and Dando, M. (2014) *op.cit.,* pp. 14–20.

[42] Arthur, G. New Equipment in Hong Kong, *Defence Review Asia, 2012.*

The US military also conducted or funded and oversaw projects to develop a range of malodorant dispersal and delivery mechanisms.[43] From 2004, a US commercial company/military partnership sought to develop a 155 mm malodorant artillery round. This munition encased multiple submunitions which were intended to be released above the target area, falling to the ground on parachutes and dispersing their malodorant payloads as they fell. It had a range of at least of 20 kms and covered a minimum of 5,000 square metres. Although the project was suspended in 2008, subsequent US military interest is certainly indicated by solicitations made in 2011, 2014 and 2021 requesting industry proposals for projects to develop malodorants and dispersal mechanisms.[44]

Riot control agents and related means of delivery

An issue of widespread concern to civil society that has not been adequately addressed to date by either the CWC or BTWC States Parties is the use and misuse of riot control agents (RCAs). Although the CWC explicitly prohibits the use of RCAs as a method of warfare,[45] it permits their use for "law enforcement including domestic riot control purposes,"[46] provided they are used in "types and quantities" consistent with such purposes.[47] However, RCAs have been frequently misused by law enforcement officials for serious human rights violations, most commonly in non-custodial settings to restrict, intimidate, or punish those participating in public protest the world over; and also in the prisons, detention centres or police stations of certain countries to ill-treat individuals.[48] A recurring medical concern has been their use in excessive quantities in the open air or in confined spaces where the targeted individuals cannot disperse. In such situations, serious injury or death can result including from the toxic properties of the chemical agents or from asphyxiation.[49] A particularly shocking case has been the inappropriate use of excessive amounts of tear gas by Indonesian police to quell disturbances at a football match on 1 October 2022. The use of tear gas,

---

[43] For further details and details of original sources see: Crowley, M., and Dando, M.R. (2022) *op.cit.,* pp. 162–171.

[44] For further details and details of original sources see: Crowley, M., and Dando, M.R. (2022) *op.cit.,* pp. 163–168.

[45] OPCW, Chemical Weapons Convention, 1993, Article I.5.

[46] OPCW, Chemical Weapons Convention, 1993, Article II.9.

[47] OPCW, Chemical Weapons Convention, 1993, Article II.1.a.

[48] For illustrative cases, see Amnesty International, *Tear Gas: An Investigation*, n.d., https://teargas.amnesty.org/#top (accessed 5 June 2023); Crowley, M. *Chemical Control: Regulation of Incapacitating Chemical Agent Weapons, Riot Control Agents and Their Means of Delivery*. Palgrave Macmillan, London. 2016, pp. 50–80.

[49] See Amnesty International, *Tear Gas: An Investigation*, n.d., https://teargas.amnesty.org/#top (accessed 5 June 2023); Crowley, M. *Chemical Control: Regulation of Incapacitating Chemical Agent Weapons, Riot Control Agents and Their Means of Delivery*. Palgrave Macmillan, London, 2016, pp. 48–49 and pp. 72–75.

the resulting crowd panic and stampede, compounded by overcrowding and locked exits, led to the deaths of over 130 people with hundreds more injured.[50]

The present situation could further deteriorate because of contemporary manufacture, promotion, trade and subsequent use of mechanisms capable of dispersing large quantities of RCAs over wide areas or extended distances. Besides the potential misuse for collective punishment of crowds, such 'wide-area' RCA delivery mechanisms might also be used as 'force multipliers' in conjunction with firearms, making lethal force deadlier on a large scale. While perhaps developed for law enforcement purposes, they could also be incorporated into military arsenals, and employed in warfare in contravention of the CWC and in the case of the capsaicinoids, the BTWC. Academic and non-governmental organisation researchers have provided evidence of manufacture and marketing of 'wide area' RCA delivery mechanisms, including internal dispersal systems, water cannon, multiple barrel launchers, large calibre projectiles, and delivery systems mounted upon unmanned aerial vehicles, unmanned ground vehicles and remote weapons systems.[51] Likewise, the OPCW SAB has repeatedly warned of the dangers in this area, most notably in its 2023 report to the 5th CWC Review Conference where it stated that:

> The continued development, testing, production, and promotion of diverse munition systems capable of disseminating RCAs on a large scale or over a long distance remain areas of concern. A review of developments and specifications of commercial products shows that a large variety of delivery systems and ammunition carrying chemical payloads are now available. *Moreover, the capabilities being developed increasingly resemble military equipment. These systems could be repurposed and filled with other chemicals including CWAs [chemical warfare agents], CNS-acting chemicals, and bioregulators.*[52]

In recent years proliferation, use and misuse of certain 'wide area' RCA delivery mechanisms has begun. In 2018 the Israeli security forces employed unmanned aerial vehicles to drop tear gas projectiles upon protesting crowds, including reportedly against peaceful protestors, bystanders, journalists and field medical facilities.[53] In 2022 Israel authorised further employment of RCA unmanned aerial vehicles for

---

[50] See for example Lamb, K. and Tereisa, A. Tear gas, locked gates led to Indonesian soccer stampede, spectators say. *Reuters*, 4 October 2022.

[51] Crowley, M. *Drawing the Line: Regulation of 'Wide Area' Riot Control Agent Delivery Mechanisms under the Chemical Weapons Convention*, University of Bradford/Omega Research Foundation, 2013; Crowley, M. *Tear Gassing by Remote Control: The Development and Promotion of Remotely Operated Means of Delivering or Dispersing Riot Control Agents*, University of Bradford/Omega Research Foundation/Remote Control Project, 2015; Crowley, M. *Chemical Control: Regulation of Incapacitating Chemical Agent Weapons, Riot Control Agents and Their Means of Delivery*. Palgrave MacMillan, London, 2016.

[52] OPCW, Director-General *Report of the Scientific Advisory Board on Developments in Science and Technology to the Fifth Special Session of the Conference of the States Parties to Review the Operation of the Chemical Weapons Convention*. RC-5/DG.1, 22 February 2023 paragraph 80.

[53] UN Human Rights Council Report of the Detailed Findings of the Independent International Commission of Inquiry on the Protests in the Occupied Palestinian Territory, A/HRC/40/CRP.2, 18 March 2019; Greenwood, F. and Zaqqout, O. Drones Don't Wear Uniforms. They Should. Foreign Policy, 22 May 2018; Al Jazeera Israeli Drone Targets Journalists, 12 November 2018.

law enforcement purposes.[54] According to human rights organisations and the media, in 2019[55] and 2021[56] respectively, security forces in Lebanon and Colombia inappropriately utilised multiple barrel launchers to fire batches of tear gas projectiles simultaneously into protesting crowds.

The dangerous consequences of the unregulated proliferation and use of wide area means of delivering and dispersing RCAs have been underlined by reports that from late 2022 to at least April 2024, Russian military forces have used RCAs against Ukrainian military forces during their ongoing invasion of Ukraine, as part of what they call a "Special Military Operation". This included repeated reported use of drones to drop K-51, VOH, RH-VO and KS-23 Cheremukha ("Bird Cherry") grenades filled with the RCAs CS[57] and CN[58] onto Ukrainian troops embedded in fortified positions, in order to dislodge them and enable subsequent military engagement.[59] In a video broadcast on 2 May 2023, on the Russian State-controlled Channel One, soldiers assigned to Russia's 810th Naval Infantry Brigade openly admitted to dropping RCA grenades into Ukrainian trenches "in order to smoke them out from fortified positions."[60] In December 2023, Russia's 810th Naval Infantry Brigade stated on its Telegram channel that the brigade is using a "radical change in tactics" by dropping K-514 grenades on Ukrainian forces.[61] In April 2024, Colonel Serhii Pakhomov, acting head of the Ukranian military's Atomic, Biological and Chemical Defence Forces, claimed that Kyiv had recorded "around 900 uses of riot control agents by Russia in the past six months out of over 1,400 since the Feb[ruary] 2022 invasion".[62] The issue was raised at the 5th CWC Review Conference in May 2023

---

[54] Breiner, J. Israel Using Drones to Tear Gas Palestinian Demonstrators in West Bank. *Haaretz*, 28 April 2021.

[55] Amnesty International, *Suppressing Protests: French Less-Lethal Weapons Used In Lebanon*, January 2021; Amnesty International *Lebanon: New evidence reveals French law enforcement equipment unlawfully used to crush protests*, 28 January 2021.

[56] Human Rights Watch *Colombia: Egregious Police Abuses Against Protesters*, 9 June 2021; Amnesty International *Cali: In the epicentre of repression: human rights violations during the 2021 national strike in Cali, Colombia* AMR 23/4405/2021, 30 July 2021.

[57] 2-Chlorobenzylidemalonitrile.

[58] Dibenz[b,f][1,4]oxazepine.

[59] Phillips, L. and Crouch, D. Have Chemical Weapons been Used in Ukraine? Royal United Services Institute, 20 June 2023; Hambling, D. Russia's Tear Gas Bombings In Ukraine May Be First Step In Dangerous Chemical Escalation, Forbes, 1 November 2022; Hoske,F., Malenko, A. and Gatilova, S. Ukraine says Russia steps up illegal use of tear gas to clear trenches, Associated Press, 17 April 2024.

[60] Russian TV, Channel 1, "Time" programme, 2 May 2023, https://www.1tv.ru/news/issue/2023-05-02/21:00#6.

[61] See United States, Condition (10)(C) Annual Report on Compliance with the Chemical Weapons Convention (CWC), Bureau of Arms Control, Deterrence, and Stability, 4 April 2024.

[62] Hoske,F., Malenko, A. and Gatilova, S. (17 April 2024) *op.cit.*

by Ukraine[63] and also by the UK.[64] Subsequently additional CWC States Parties, notably Canada, France, Germany, Italy, Japan, the UK, and the US[65] have repeatedly requested further information from Russia. Russia has denied using RCAs in armed conflict, instead accusing Ukraine of conducting such activities with the assistance of the US.[66] On 1 May 2024, the US State Department declared "Russia has used riot control agents as a method of warfare in Ukraine, … in violation of the CWC.[67] Although a growing number of States have raised this issue within the OPCW, the facts of the case are contested and the States Parties have not, at the time of writing, enacted formal OPCW procedures to resolve the issue. Instead, allegations of Russian misuse of RCAs in Ukraine continue, with potentially damaging consequences for the stability of the OPCW and associated weakening of the norm prohibiting chemical weapons.

17.10   Conclusions and Recommendations

It is apparent that the textual architecture of both the BTWC and the CWC contains critical ambiguities, limitations and weaknesses that have led to diverging patterns of implementation by the States Parties with regard to regulation of toxins, bioregulators and other substances of biological origin. Unfortunately, these long-standing challenges, which will only increase with the advance in the life and associated sciences, have not been addressed by the States Parties. Consequently, the theoretical double coverage of such mid-spectrum agents by both the BTWC and the CWC regulatory regimes in practice hides a dangerous and potentially widening lacuna, with neither Convention being effectively implemented by the States Parties so as to ensure that these substances are not weaponised.

Comparing both Conventions, it is clear that despite its limitations, the CWC and its attendant regulatory regime incorporates a number of existing measures that can be further strengthened and implemented to more effectively regulate toxins and bioregulators, but only if States Parties were so willing. Consequently, in the medium

---

[63] OPCW, Statement by Dr Kateryna Bila, Head of the Secretariat of the Chemical Weapons Convention of the Ministry of Foreign Affairs of Ukraine at the 5th Special Session of the Conference of the States Parties to review the operation of the Chemical Weapons Convention, RC-5/NAT.13, 15 May 2023.

[64] OPCW, UK National Statement to the Fifth Review Conference of the Chemical Weapons Convention by Her Excellency Joanna Roper, Permanent Representative of the United Kingdom of Great Britain and Northern Ireland.

[65] See for example, OPCW, Note Verbale from the Permanent Representation of the Federal Republic of Germany to the Technical Secretariat of the OPCW, Executive Council, EC-104/NAT.6, 6 October 2023.

[66] Note Verbale from the Permanent Representation of the Russian Federation to the Technical Secretariat of the OPCW Dated 12 October 2023.

[67] US State Department, Imposing New Measures on Russia for its Full-Scale War and Use of Chemical Weapons Against Ukraine, Office of the Spokesperson, 1 May 2024.

term (i.e., the next five to ten years) the CWC and the OPCW provide the best routes for controlling 'dual-use' bioregulator and toxin research and associated activities of potential concern so as to ensure that they are not, and are not (mis)perceived to be, misused for malign purposes. And the following recommendations recognise this.[68]

1. *Strengthening CWC States Parties/OPCW regulation*

- Reaffirm the importance of the GPC as a vital safeguard ensuring the Convention's comprehensive scope and future-proofed prohibition *inter alia* of all naturally occurring and synthetic toxin and bioregulator weapons. Establish a consultative process to develop guidelines on how the 'types and quantities' principle of the GPC should be applied in practice. The consultative process should explore specific challenges to the GPC arising from contested interpretation as to the range of toxic chemicals (including toxins and bioregulators) and associated delivery mechanisms that could be legitimately employed for law enforcement, and the nature of what constitutes legitimate use. It should specifically address:

  - *Riot control agents*: Clarifying nature and scope of activities consistent with 'law enforcement including domestic riot control'. Building upon previous work of the SAB identifying chemicals (including those of biological origin) that fulfil the definition of RCA,[69] guidance should be developed as to quantities of identified RCAs that can legitimately be employed in law enforcement. Such guidance should acknowledge obligations under relevant international human rights law, so as to ensure that this use of RCAs is targeted, proportionate, necessary, and does not endanger the health or the life of those targeted or of bystanders.
  - *Malodrants*: Examining the properties of those malodorants that were previously and are currently being researched and/or developed as weapons by States, and determine whether these substances fulfil the definitions of 'toxic chemicals' and/or 'RCAs' and/ under the Convention. Consequently, provide guidance to States Parties as to whether these malodorants can legitimately be employed for law enforcement purposes, and if so under what circumstances and with what restrictions.
  - *Dispersal and delivery mechanisms*: Determining which dispersal and delivery mechanisms that are purportedly intended for RCAs (including biological substances and their human synthesised analogues) and that could alternatively be used to disperse other toxic chemicals, are inappropriate for law enforcement and consequently breach Article II.1 of

---

[68] For discussion of these recommendations and for further associated technical recommendations see Crowley, M., and Dando, M.R. (2022) *op.cit.,* pp. 272–286.

[69] OPCW, Office of Strategy and Policy *Note by the Technical Secretariat, Declaration of riot control agents: advice from the Scientific Advisory Board.* S/1177/2014, 2014; See also OPCW, Scientific Advisory Board *Response to the Director-General's Request to the Scientific Advisory Board to consider which riot control agents are subject to declaration under the Chemical Weapons Convention.* SAB-25/WP.1., OPCW, The Hague, 27 March 2017.

the CWC. Such prohibited means of dispersal and delivery should at a minimum include aerial-delivered bombs, artillery shells, mortar munitions, cluster munitions, and other delivery mechanisms that are indiscriminate or that disseminate quantities of RCAs that are likely to cause serious injury or death to those targeted. Appropriate guidance should be given as to which RCA means of dispersal and delivery can legitimately be utilised for law enforcement purposes, and if so under what circumstances and with what restrictions.

– *ICA/CNS-acting chemical agent weapons*: Following adoption of the 'understanding' by the 26th Conference of CWC States Parties, aerosolised use of CNS-acting chemical agents for law enforcement purposes is effectively prohibited.[70] Further guidance is, however, needed to define 'CNS-acting chemicals' and the range of chemicals that would be covered by the 'understanding'. A potential forum for informing development of such guidance could be the *TWG to study current developments concerning CNS-acting chemicals*, as raised by the SAB in its Report to the 5th CWC Review Conference.[71] Additional guidance is also required in order to ensure that any existing or future law enforcement weapons that use toxic chemicals, including toxins and bioregulators, which act on other core human physiological processes beyond the CNS (i.e. ICA weapons) are also prohibited. Guidance should further clarify that not only aerosolised but all weaponised use of such toxic chemicals for law enforcement purposes, no matter how they could be delivered, should be prohibited.

– Update and strengthen routine OPCW industry monitoring and verification measures applicable to toxins and bioregulators, as well as analytical methods and databases available for challenge inspections and investigations of alleged use. Consideration could be given to updating the CWC Schedules by adding further toxins that have been or may be employed in weapons programmes, notably including botulinum toxin and SEB, and additionally incorporate indicators of new forms of toxic chemical (and their precursors) of potential concern, notably bioregulators such as Substance P.[72] An additional approach to consider is modification of existing verification provisions relating to 'Other Chemical Production

---

[70] OPCW, Conference of States Parties *Decision: Understanding regarding the aerosolised use of central nervous system-acting-chemicals for law enforcement purposes*, C-26/DEC.10., OPCW, The Hague 1 December 2021.

[71] OPCW, Director-General *Report of the Scientific Advisory Board on Developments in Science and Technology to the Fifth Special Session of the Conference of the States Parties to Review the Operations of the Chemical Weapons Convention.* RC-5/DG.1, OPCW, The Hague, 22 February 2023, p. 8, paragraph 31.

[72] See for example research conducted by the Swedish Defense Research Establishment on the effects of the aerosolised bioregulator, Substance P. Koch, B., Edvinsson, A and Koskinen, L. Inhalation of substance P and thiophan: acute toxicity and effects on respiration in conscious guinea pigs, *Journal of Applied Toxicology*, vol. 19, no.1, 1999, pp. 19–23.

Facilities' (OCPFs) so as to capture OCPFs that could support State-level bioregulator weapons production.[73]

2. *Strengthening BTWC States Parties regulation*

- Reaffirm the importance of the GPC as a vital safeguard ensuring the Convention's comprehensive scope and future-proofed prohibition *inter alia* of all naturally occurring and synthetic toxin and bioregulator weapons.
- Establish an appropriate process to collectively examine and subsequently define the terms 'toxins' and 'other biological agents' as they are employed under the BTWC, and also to clarify the nature and scope of substances covered by both of those terms, notably with regard to bioregulators. This process should specifically attempt to clarify the implementation of the Convention with regard to certain relevant substances promoted as 'less lethal' weapons notably those riot control agents, malodorants and CNS-acting chemical agents of biological origin and their synthetic analogues.
- Establish a body of experts that is functionally similar to the OPCW's Scientific Advisory Board, which can provide independent scientific advice, including regular analysis of relevant life and associated science and technology (S&T) developments, to States Parties. This S&T review body needs, for its proper functioning, to be guaranteed long-term financial, technical, and human resources, and this must include a permanent S&T staff position drawn from the ISU.

3. *Addressing the regulatory gap between the BTWC and CWC*

The BTWC-CWC regulatory deficiencies have been further exacerbated by significant differences regarding location, resources, size and mandates of the two international organisations (BTWC ISU and the OPCW) responsible for facilitating implementation of the respective Conventions. Consequently, we would argue that the BTWC and CWC States Parties should now individually and collectively examine how the two regulatory regimes and the institutional organisations can best cooperate. Particular consideration should be given to regulating those substances—notably bioregulators, toxins and other substances of biological origin or their synthesised analogues such as certain RCAs, malodorants and ICAs/CNS-acting chemicals—that are covered by both Conventions.

Some positive interactions have already occurred at the technical and scientific levels, and these have been greatly facilitated by the OPCW Technical Secretariat, notably through the actions of the SAB. A particularly important process facilitating such interactions was the SAB Temporary Working Group study and subsequent report on convergence of chemistry and biology, published in 2014. This SAB TWG provided a unique international forum for interaction between experts drawn from the BTWC and CWC and associated life and chemical scientific communities.

---

[73] This was originally proposed by the late Dr Jonathan Tucker. See: Tucker J. The body's own bioweapons, *Bulletin of Atomic Scientists*, Volume 64, number 1, March/April 2008, pp. 16–22 & 56–57.

The OPCW should now establish a similar forum, possibly as a TWG or ideally a standing arrangement under the SAB. Such a body could encourage and facilitate further engagement and explore more effective coordination and cooperation on technical issues where both organisations have undertaken parallel activities, for example monitoring and analysing chemical and life science developments of relevance to both Conventions, and awareness raising and outreach to the chemical and life science communities.

At the institutional level, the BTWC ISU and the OPCW TS should examine the effectiveness of current information exchange, cooperation, and collaboration measures with regard to overlapping issues, notably regulation of bioregulators, toxins and other biological substances of concern, and how such measures should be improved in response to continuing convergence of the chemical and life sciences. The OPCW TS and the BTWC ISU should examine the benefits and challenges of negotiating more formal organisational cooperation, potentially through a Relationship Agreement or Memorandum of Understanding (MOU). Such an RA or MOU could more clearly determine how the BTWC ISU and OPCW TS would cooperate in certain 'real world' situations where both organisations had a mandate for action, e.g. if a State declares or is accused of developing bioregulator or toxin weapons production facilities or of possessing such stockpiles of such weapons, or clarify how both organisations could collectively and most effectively respond and facilitate assistance provision to a State that has been threatened with or has been attacked by bioregulator or toxin weapons.

Bibliography

Alibek K, Handelman S (1999) Biohazard: the chilling true story of the largest covert biological weapons program in the world. Random House, New York

Army Edgewood Chemical Biological Center Aberdeen Proving Ground, United States (2017). http://www.dtic.mil/dtic/tr/fulltext/u2/1039416.pd

Boydston J et al (2021) Infuence of aerodynamic particle size on botulinum neurotoxin potency in mice. Inhalation Toxicol 33(1):1–7

Calm A et al (2017) Pilot scale production and testing of a recombinant staphylococcal enterotoxin SEB triple mutant, ECBC-TR1471

Crowley M, Dando M (2022) Toxin and bioregulator weapons, preventing the misuse of the chemical and life sciences. Palgrave Macmillan/Springer Nature, Cham, Switzerland

Gupta SM et al (2018) Phytochemical analysis of Indian stinging plants: an initiative towards development of future novel biothreat agents for self-defence. Proc Natl Acad Sci India Sect B Biol Sci 88(2):819–825

Krakauer T (2019) Staphylococcal superantigens: pyrogenic toxins induce toxic shock. Toxins 11(3):178

Krutzsch W, Trapp R (2014) Article II: definitions and criteria. In: Krutzsch W, Myjer E, Trapp R (eds) The chemical weapons convention: a commentary. Oxford, Oxford University Press

Leitenberg M et al (2004) Biodefense crossing the line. Politics Life Sci 22(2):2–3

Nordin JS (2012–2013) Biotoxins used as warfare agents. Combat WMD J Part I(8):3–7, Part 2(9):28–35 and Part 3(10):11–16

Palchik G et al (2018) Monkey business? Development, infuence, and ethics of potentially dual-use brain science on the world stage. Neuroethics 11:111–114

Perry Robinson J (2008) Bringing the CBW conventions together. CBW Conv Bull 80

Pitschmann V, Hon Z (2016) Military importance of natural toxins and their analogues. Molecules 21(556):1–20

Quin Z, Li X (2017) Non-Human primate models for brain diseases-towards genetic manipulations via innovative technology. Neurosci Bull 33(2):238–250

Regis E (1999) The biology of doom: the history of America's secret germ warfare project. Henry Holt, New York

Roy C et al (2003) Impact of inhalation exposure modality and particle size on the respiratory deposition of ricin in BALB/c mice. Inhalation Toxicol 15(6):619–638

Sanford D et al (2010) Inhalational botulism in rhesus macaques exposed to botulinum neurotoxin complex serotypes A1 and B1. Clin Vaccine Immunol 17(9):1293–1304

Vazquez-Cintron EJ et al (2017) Engineering botulinum neurotoxin C1 as a molecular vehicle for intra-neuronal drug delivery. Nat Sci Rep 7(42923)

Michael Crowley is a Honorary Visiting Senior Research Research Fellow at Bradford University, UK and Research Associate with the Omega Research Foundation, UK. Hus current research explores contemporary development and use of 'less lethal' weapons including riot control agents, malodorants, and CNS-acting agents, as well as related means of delivery. He co-authored (with Malcolm Dando) *Toxin and Bioregulator Weapons: Preventing the Misuse of the Chemical and Life Sciences* (2022).

Chapter 18
Three Decades of Chemical Weapons Destruction: Successes, Failures, and Lessons Learned

Paul Walker

Abstract Over thirty years ago the world began the destruction of chemical weapons stockpiles, one of the major goals of the new Chemical Weapons Convention (CWC). The author, very involved in stockpile demilitarization since at least the first US on-site inspection of the easternmost Russian stockpile in Shchuch'ye in 1994, describes the challenges, successes, and failures of these years, many unforeseen by negotiators who set a ten-year deadline for chemical weapons destruction. Despite these many hurdles—technical, financial, and political—the strength of the CWC and its 193 States Parties shines through as it verifies the complete destruction of stockpiles in eight officially declared possessor states by 2023. The article concludes with primary and interrelated goals for further strengthening the CWC, the most comprehensive, multilateral disarmament treaty to date.

Keywords Chemical weapons · Demilitarisation · Citizen involvement · Toxic chemicals

18.1 Introduction

The year 2023 was marked by the last chemical weapon destroyed at the Blue Grass Army Depot in Kentucky, USA. This historic event marked completion of over 72,000 metric tons of chemical weapons in eight countries under verification auspices of the Organization for the Prohibition of Chemical Weapons (OPCW) in The Hague. The July 7, 2023 destruction was the last of 55,000 M55 rockets, a two-meter-long projectile carrying eleven pounds of GB nerve agent. It also marked the end of one of three separate campaigns to destroy 523.4 US tons of chemical agents at Blue Grass, and the last of nine US chemical weapons stockpiles declared to the OPCW in 1997. Dr. Candace Coyle, Blue Grass site project manager, stated in a press release: "It is a historic moment not only for us in Kentucky, but the United States and across

P. Walker (✉)
Chemical Weapons Convention Coalition and Arms Control Association, Washington, DC, USA
e-mail: pwalker@armscontrol.org

© The Author(s) 2026
B. Friedrich et al. (eds.), *Thirty Years of the Chemical Weapons Convention (CWC)*,
https://doi.org/10.1007/978-3-031-98854-7_18

the world. With this last munition, the Chemical Weapons Convention mandate to eliminate all declared chemical weapons stockpiles is now globally achieved."[1]

At a December 13, 2023 commemoration of the US accomplishment in Washington, DC, OPCW Director-General Fernando Arias stated: "I wish to wholeheartedly congratulate you for this historic achievement. It marks the destruction of all declared stockpiles worldwide, by all former possessor states, fulfilling one of the main goals of the Chemical Weapons Convention." He went on to comment: "In the contemporary degraded international security environment, this monumental accomplishment we commemorate today should serve as a strong message for peace".[2]

These celebrations recognized a major goal, i.e. the safe and verified demilitarization of all declared chemical weapons stockpiles in eight countries, but they also represented the long, arduous, and costly course of stockpile destruction in each country over the past thirty-three or more years. The time frame which the Chemical Weapons Convention negotiators had envisioned as taking 10–15 years had lasted more than double that time, and the effort and cost many times more than diplomats had assumed.

This paper will offer a brief survey of the destruction of chemical weapons over the last three decades—successes and challenges involving planning, funding, technology, arms control, and other related issues.

18.2   The Chemical Weapons Convention

An international agreement to ban chemical weapons had been discussed in multilateral fora since the end of the nineteenth century, but the first major accomplishment came after World War I when toxic chemicals were introduced on the battlefield at Ieper, Belgium on April 22, 1915. Although the Germans had used tear gas earlier in the war, Ieper was the first time that deadly chlorine gas was used in massive quantities, this time against allied soldiers—Algerian, British, Canadian, French, and Senegalese. The world was horrified by the thousands of battlefield injuries and deaths at Ieper and later WWI battles, and by the fact that allies responded subsequently with deadly chemicals themselves. Indiscriminate chemical warfare had shown its deadly face.

The 1925 Geneva Protocol, an effort to ban the atrocities prevalent in WWI, prohibited the use of chemical and biological weapons in war. It entered into force in 1928,[3] but did not prohibit the research, development, production, and stockpiling of such weapons which most major powers pursued at the time. Fortunately, World War II witnessed no major use of chemical weapons on the battlefield, although several countries had them stockpiled and ready for use. Formal negotiations started

---

[1] Blue Grass Press Release (2023).

[2] OPCW News Release (2023).

[3] United Nations, Office for Disarmament (1925).

in 1968 in Geneva on the abolition of biological and chemical weapons, leading to the 1972 Biological Weapons Convention; after many years of a "rolling text," the Chemical Weapons Convention eventually emerged and was opened for signature in January, 1993.[4] It had been propelled forward in the late 1980s by the use of chemical weapons by Saddam Hussein in his eight-year war against Iran and in Halabja against the Kurds, and by the bilateral negotiations between the US and Soviet Union in which the two superpowers agreed to unilateral and reciprocal destruction of their enormous chemical weapons stockpiles comprising 95% of the world's remaining arsenals.

The CWC collected 130 signatures from countries during its three-day signing in Paris and expanded from there. With widespread public support from civil society, chemical industry, and foreign ministries around the world, the CWC quickly expanded its signatories and ratifications. Needing 65 ratifications 180 days prior to entry-into-force (EIF), Hungary became the 65th country to ratify in late October 1996, establishing entry-into-force as April 29, 1997. In the meantime, the signatories established a Preparatory Commission or "PrepCom" to begin the work of the CWC and clarify the many issues which remained.

Fortunately, the United States and Russia, with chemical stockpiles totaling 28,600 and 40,000 metric tons respectively, ratified in April (before EIF) and December (after EIF), 1997, after contentious political battles in both Washington and Moscow.

Today the CWC has 193 States Parties representing 98% of the world's population and is the largest multilateral arms control treaty with verification and on-site inspections.[5] Situated in The Hague, The Netherlands, the Organization for the Prohibition of Chemical Weapons (OPCW), the CWC's implementing agency, organizes an annual Conference of States Parties (CSP, 29 to date) and a Five-Year CWC Review Conference (RevCon, five to date). The treaty also allows for Special Sessions (SS, four to date) to be called when necessary by the Conference, Executive Council, or a third of the States Parties.[6]

The CWC requires all States Parties to declare their past and existing chemical weapons programs and stockpiles upon joining, and to submit a follow-up declaration annually.[7] It also requires four specific deadlines for destruction of declared chemical weapons stockpiles, explained in Annex IV(C) of the Convention: 1% within three years of EIF, 20% with five years, 45% within seven years, and 100% within ten years.[8] A five-year extension would also be possible. It was generally thought during negotiations that any stockpile could be destroyed and verified within ten years after entry-into-force, which would be several years after signature opening. And

---

[4] For a concise history, see OPCW, "Looking back helps us look forward," https://www.opcw.org/about-us/history.

[5] See OPCW, https://www.opcw.org/

[6] For organizational procedures, see Article VIII of the CWC, https://www.opcw.org/chemical-weapons-convention/articles/article-viii-organization.

[7] See Article III of the CWC, https://www.opcw.org/chemical-weapons-convention/articles/article-iii-declarations.

[8] See Annex IV of the CWC, https://www.opcw.org/chemical-weapons-convention/annexes/verification-annex/part-iva-destruction-chemical-weapons-and.

with a possible five-year extension, this provided more than enough time—almost twenty years—for a possessor state to comply. So, the treaty, with EIF in April 1997, established a 2007 deadline with an extension possible to April, 2012, over 19 years since it was opened for signature. Unfortunately, none of the declared possessor states met all of their treaty-required deadlines.

18.3 Chemical Weapons Destruction

18.3.1 United States

The United States was the first possessor states to unilaterally and publicly start chemical weapons destruction. In 1990, three years before the CWC was opened for signature, the newly-designed prototype incinerator began operating on Johnston Atoll (nicknamed JACADS) in the Pacific Ocean,[9] about 860 miles southwest of Hawaii. The 1,842 metric tons of mustard and nerve agent weapons had been secretly moved to Johnston Atoll in 1990 from Germany and 1971 from Okinawa where they had been forward deployed. By 1997 when the CWC entered into force, the US had completed destruction of 1,202 metric tons, about 65%, at JACADS and would complete the remaining 35% by 2000.

Incinerator testing at Johnston Atoll demonstrated that chemical weapons could be completely burned in the furnace, but not without some atmospheric risks of live agent being released along with heavy metals (mercury was discovered in some mustard weapons). The US Army downplayed these risks, but in the continental US, where plans were to build eight more "baseline incinerators" at each stockpile, communities and state regulatory agencies started raising concerns.

The US went forward with construction of four additional incinerators at Tooele, Utah; Umatilla, Oregon; Anniston, Alabama; and Pine Bluff, Arkansas. Environmental and public health activists and officials eventually agreed to another process at four stockpiles: Aberdeen, Maryland; Newport, Indiana; Pueblo, Colorado; and Blue Grass, Kentucky. The incinerator construction had started in the mid-1990s with little public input. This forward momentum, partly to meet CWC deadlines but also to invest in what thermal engineers considered the most mature and tested technology, led to communities eventually agreeing to the incinerator process. But the four communities in the late 1990s without early construction decided to use a more controllable process, at least in their eyes, which resulted from a national dialogue on destroying chemical weapons.

---

[9] Johnston Atoll was named "Kalama Atoll" by the Hawaiians in 1856. In the 20th century it was used as a runway for military forces and for nuclear testing, eventually contaminated with plutonium. It was also used for biological weapons testing in the 1960s, to store Agent Orange during the Vietnam War, and finally for chemical weapons storage from 1971 to 2000. Today the atoll has returned to a National Wildlife Refuge with considerable contamination remaining. https:// en.wikipedia.org/wiki/Johnston_Atoll

Neutralisation is a wet chemistry process whereby the munition is drained of its liquid, toxic chemical, which is mixed with hot water and another chemical, often sodium hydroxide, thereby destroying its toxicity. This process was chosen from a dozen alternative proposals in the 1997 Assembled Chemical Weapons Assessment (ACWA) program, which included a national dialogue of stakeholders—local and state officials and regulators, vendors, military officials, and environmental and public health experts—and lasted about five years until 2002. Meeting every two–three months, this dialogue proved to be very productive in building consensus, involving community stakeholders, and vetting industrial processes for destroying chemical weapons and high toxic waste.[10]

Mr. Michael Parker, the first director of the ACWA Program and Dialogue, reported to the US Congress in December 1997, that "[t]he task of identifying and demonstrating alternative technologies is not only challenging technically, but also is challenging logistically when attempting to include stakeholders in the full range of activities being conducted by the Program to include the procurement process. By utilizing innovative methods, both DoD and the Dialogue worked hard to establish a means by which the Dialogue could be fully involved."[11]

The congressional report went on to describe the rationale behind ACWA: "The Dialogue on Assembled Chemical Weapons Assessment was formed in an effort to effectively address the charge of Public Law 104–201 and Public Law 104–208 to demonstrate not less than two alternatives to the baseline incineration process. DoD and others interested in this issue stated the need to integrate the input of communities, regulators, and other concerned parties into the process of developing criteria and assessing alternative technologies so that decisions are technically sound and publicly acceptable."[12]

Tooele, just west of Salt Lake City in Utah, was the largest stockpile of the nine US declared arsenals—12,353 metric tons. It began destruction in the mid-1990s, so had destroyed 232 metric tons before CWC EIF in 1997. The other three incinerator sites all began operation in the late 1990s or early 2000s, and completed operations before both Pueblo, Colorado and Blue Grass, Kentucky began their neutralization process. None of these massive efforts were simple, however. The US Army and the Program Manager for Chemical Demilitarization (PMCD) had begun public outreach efforts in local communities and sought to adhere to state environmental and public health regulations. The ACWA dialogue process was also impactful on all sites, and after 1997 the OPCW managed round-the-clock inspection and verification operations at every site.

Because the US was the first to start large demilitarization operations, it was able to meet the 1 and 20% deadlines of the OPCW in 2000 and 2002. But the 45% deadline in 2004 was another story and the US asked the OPCW for an extension until April 2007. In early 2007 the Chemical Material Agency (CMA, successor to

---

[10] The author was part of the ACWA process, serving on the National Dialogue and its subcommittee, the CATT (Citizens Advisory Technical Team). See https://www.peoacwa.army.mil/.

[11] US Department of Defense (1997).

[12] Ibid, p. iii.

PMCD, Program Manager for Chemical Demilitarization) realized that they would not make the April 2007 extended deadline and made a decision to change the second-stage destruction process at Newport, Indiana. The Newport stockpile community and state had previously agreed to use neutralization rather than incineration for the first stage and had also agreed to use a new post-neutralization process, Super Critical Water Oxidation (SCWO), for the second stage of destruction.

SCWO was one of the technologies tested in the ACWA process. It involved shooting neutralized agent and other contaminated materials into a long tube-like reactor under very high pressure and temperature; this very quickly disassociated all materials and produced water, salts, and heavy metals after a few seconds. Newport was planned to be the prototype test site for this process, in preparation for later use at Blue Grass, Kentucky. However, the US Army, to meet the 45% CWC deadline, decided to secretly ship all waste via tanker trucks to Port Arthur, Texas, a waste incinerator not far from Houston used by the military to burn toxic wastes.

This truck shipment, over 1,500 miles crossing several state borders, started late one night without formal notice to state and local authorities, and met with heavy criticism from regulatory authorities and local advocates. People felt deceived by the Army after an agreement had been reached to utilize neutralization and SCWO for the processing of 1,152 metric tons of VX nerve agent. They were also concerned over safety and security of the many truck trips required to carry at least 11,000 metric tons of neutralized nerve agent, and the fact that they would be sending toxic waste to an African-American community where the Veolia waste incinerator was located. Port Arthur citizens had protested many times with the sign, "No VX in TX," and were worried over the safety and security of high toxic waste processing in their town.

The Army argued that the shipments would be perfectly safe, heavily protected with global positioning systems (GPS) in each truck, and the SCWO system would have added years to secondary processing. In the end, the tanker truck shipments went forward without any reported incidents and Newport completed its processing in 2007. What planners did not address was the impact on Blue Grass, Kentucky, which was assuming early testing of SCWO at Newport, Indiana before use at Blue Grass.

Aberdeen, Maryland was home to 1,422 metric tons of mustard agent. Very little attention was paid to this stockpile, one of several open-air arsenals, just a few miles north of Baltimore and not far from the major north–south highway, Route 93. The Army had planned to build an incinerator here as well, against objections from the local community, but the terrorist attacks of September 11, 2001 changed planning. The military had not focused on the possibility of aerial attacks, especially against open-air stockpiles, but immediately after the September attacks all existing chemical weapon stockpiles became potentially primary targets for further terrorist attack.

The Maryland stockpile was not only vulnerable from the air but sat on the edge of Chesapeake Bay with little-to-no protection. After 911, security was greatly improved and a year later the decision was made to neutralize the mustard on-site and ship the neutralized agent by truck north to Dupont Chambers Works in Delaware. Such a move would have been opposed by local groups and communities but, considering the perceived risks to the stockpile after 911, this went forward relatively quickly and easily.

In 2006 at the OPCW's CSP-11, the annual conference of States Parties, the United States requested a five-year extension to 2012 for its stockpile destruction program, knowing full well that they wouldn't make that extended deadline. The CWC allows for this final extension of five years in its Article IV and its Annexes, Part IV(C), paragraph 24.[13]

The US Ambassador Eric Javits stated at CSP-11 before the States Parties: "Experience has shown that the task of eliminating the legacy of chemical weapons stocks has proven more difficult than any of us imagined. All but one of the declared possessor States have had to request extensions to the 100% percent destruction deadline. While there are great challenges, the commitment to complete destruction of all CW stocks is very clear." He went on to say: "…the US remains absolutely and irreversibly committed to destroying 100% of its stockpile and will make every effort to do so safely and as soon as possible under its federal, state, and local laws, and under the watchful and continuing scrutiny of the OPCW's inspection regime."[14]

It was obvious at CSP-11 that no country would meet its deadlines, and the final report of the conference illustrated that very clearly. The United States, contrary to the objections of Russia and a few other countries, obtained its five-year extension for complete destruction to 2012. Russia also obtained extensions for its 45% deadline to 2009 and for its 100% deadline to 2012. Under Sub-item 9(c) of the CSP-11 final report, the conference also granted extensions to Albania, India, Libya, and South Korea ("A State Party"), with the request that possessor states report annually on progress toward the CWC deadlines.[15]

The final two stockpiles in the United States, under management of the Assembled Chemical Weapons Assessment (ACWA) Program, were Pueblo, Colorado and Blue Grass, Kentucky. Pueblo had 2,369 metric tons and Blue Grass 475 metric tons, combined about 10% of the total US volume. From early in the process of determining the destruction technology, both Pueblo and Blue Grass were very skeptical of incineration as the first stage technology and two leaders—Irene Kornelly at Pueblo and Craig William at Blue Grass—were instrumental is rallying local communities for a more controllable process. They and many other community leaders formed the Chemical Weapons Working Group (CCWG)[16] and, after years of dialogue and deliberation, agreed to neutralization as a first-stage technology. Pueblo also agreed to bioremediation as a second-stage technology and as noted earlier, super-critical water oxidation (SCWO) as a second-stage technology at Blue Grass.

---

[13] OPCW, https://www.opcw.org/chemical-weapons-convention/annexes/verification-annex/part-iva-destruction-chemical-weapons-and.

[14] Ambassador Eric Javits at CSP-11 https://www.opcw.org/sites/default/files/documents/CSP/C-11/national_statements/US-en-Javits.pdf.

[15] See Agenda Item 9 (2006).

[16] See Kentucky Environmental Foundation for more information, https://www.kyenvironmental foundation.org/chemical-weapons/.

Pueblo, Colorado was fortunate to have only mustard agent and its choice for neutralization for first-stage destruction went well. Also, its second-stage process, bioremediation, went very smoothly. Certain weapons, however, presented problems such as hardening of the old mustard agent, so two "bang boxes," Static Detonation Chambers (SDCs), were brought in to handle these difficult weapons. Destruction operations began in March 2015 and were finalized in June 2023, requiring slightly more than eight years to destroy over 780,000 artillery and mortar rounds holding over 2,600 US tons of mustard agent.[17]

The longtime chair of the Colorado Chemical Demilitarization Citizens' Advisory Commission, Ms. Irene Kornelly, stated that "[t]he destruction of all the chemical weapons stored for so many years at the Pueblo Chemical Depot is an incredible achievement. This could not have been done without the excellent Pueblo workforce committed to doing their job and doing it well. The support of the Pueblo community, the State of Colorado and the Department of Defense has shown the world what can happen when we all work together. We have made the world a safer place!".[18]

Blue Grass, Kentucky was the smallest but most complex of the nine stockpiles, holding both mustard and nerve agents and a number of weapon types including M-55 rockets considered to be the most vulnerable for self-igniting. Its 475 metric tons and over 101,000 weapons would prove to be challenging, requiring five separate campaigns and a change in the second-stage destruction technology. Destruction began in June 2019 and finished July 7, 2023, having taken just over four years. Most of the weapons were drained of their agents which were neutralized, and SDCs were installed as a second-stage technology to replace SCWO which had been rejected for technical reasons. Non-contaminated parts were shipped off-site for treatment, and contaminated waste was treated in the SDCs. The final weapon to be destroyed was an M-55 rocket with GB nerve agent, more than 50% of the original stockpile at Blue Grass.[19]

Although these two stockpiles represented only about 10% of the overall US stockpile, they were the last to be destroyed but did so in a very different way than the prior 90%. Having rejected incineration as the prime technology, the communities chose neutralization and demonstrated (along the Aberdeen, Maryland and Newport, Indiana) that it could be effectively and safely used, not only with bulk agent but with complex weapons, in a timely and cost-competitive way. Five chemical weapon stockpiles had been destroyed with incineration and four with neutralization, but this mix might have looked very different had it taken place five years later.

"Having advocated on behalf of the community's interest for almost 40 years, it is thrilling to witness the completion of this monumental effort," said Craig Williams, Kentucky Chemical Demilitarization Citizens' Advisory Commission and Chemical Destruction Community Advisory Board co-chair. "Kudos to everyone

---

[17] See the Pueblo Press (2023).

[18] Ibid

[19] For more information on Blue Grass, see https://www.peoacwa.army.mil/2023/07/07/press-rel ease-last-chemical-weapon-destroyed-at-blue-grass-army-depot/.

involved in what can only be seen as a model of government, contractor and citizen cooperation to accomplish such an important objective."[20]

18.3.2   Russian Federation

It was apparent that Russia was very interested in destroying its enormous chemical weapons stockpiles given their agreement in the late 1980s with the United States for reciprocal and unilateral reductions and inspections. In 1994 the US undertook an inspection of the easternmost site, Shchuch'ye, which was the only one of seven declared sites east of the Urals. This site contained 5,400 metric tons of nerve agents in thousands of artillery shells and missile warheads. The official on-site inspection[21] was helpful in realizing that the Russian Army was very interested in getting rid of its stockpiles of munitions, but had little idea how to do it.

The Shchuch'ye inspection also brought home the fact that a public outreach and education campaign would be a critical part of the Russian demilitarization effort, but urgent security upgrades were necessary in the meantime. Shchuch'ye was one of two Russian stockpiles which were man-portable, i.e., included munitions which could be carried out in a backpack or gym bag; five of the seven Russian sites were large munitions or bulk storage tanks, but both Shchuch'ye and Kizner in the Udmurt Republic had thousands of modest-sized artillery or weapons which could be moved off-base fairly easily.

The US inspection team included an Assistant Secretary of Defense who offered to construct an incinerator at Shchuch'ye to burn the stockpile, similar to five of the US stockpiles, but the Russians in Shchuch'ye and later in Moscow refused, saying that this technology was too difficult to construct and maintain, and would have unnecessary environmental and public health impacts. They preferred to do a study on alternative technologies available in Russia which became the "Joint Research and Evaluation Project" (JREP) to settle on one or more technologies of destruction.

Meanwhile the US requested $20 million to upgrade security at the two worrisome sites as part of the Cooperative Threat Reduction (CTR or Nunn-Lugar) Program, which would fund a major part of the Russian chemical weapons destruction program in coming years. These security upgrades—upgraded fencing, nightlights, guard training, video and other surveillance systems, and guard towers—were installed at Kizner and Shchuch'ye, but unfortunately took eight years until 2002 to finalize. These intervening years were very worrisome given the breakup of the Soviet Union, the location of Shchuch'ye along the Kazakhstan border, and the lack of any proper inventory of weapons (at least until OPCW inspectors were allowed access after Russian ratification of the CWC in December 1997) (Walker 2003).

---

[20] Ibid

[21] This author took part in the Shchuch'ye inspection on behalf of the Committee on Armed Services, US House of Representatives.

The Russian chemical weapons stockpile of 40,000 metric tons of agent, the largest in the globe (about 55%), consisted largely of nerve agents –VX, Sarin, and Soman—with some older stocks of mustard, lewisite, and phosgene. The neutralization program got started in Gorny in the Saratov Oblast in 2002, twelve years after the United States, likely because Gorny was the smallest Russian stockpile with 1,150 metric tons of mustard and lewisite, which were the easiest chemicals to neutralize. They were the second declared possessor State to begin demilitarization after the US, but the next to last to finish, five and a half years before the US.

All the sites used neutralization and were much easier to demilitarize due to the lack of propellent and explosives in the stored munitions. The larger aerial bombs could in fact be neutralized intact by adding neutralent (often hot water and sodium hydroxide) directly to the air space in the bomb and letting them neutralize ("cook") over a month or two. But the second-stage process was more varied. The proposal in Shchuch'ye was to mix the liquid hydrolysate, at least ten times the volume of the original agent after neutralization, with asphalt and utilize it for road pavement. The US and other Western funders tested this mixture and found it carcinogenic, so Russia finally settled on filling barrels with the toxic product and storing it in bunkers for later decision-making. This was unsettling for the residents of the Kurgan Oblast where Shchuch'ye was located because it appeared like an "unfunded mandate" on the local government. To this day, years after processing at Shchuch'ye was finished, at least 50,000 metric tons of this toxic asphalt still sits near the destruction complex. The other man-portable site, Kizner, used a foreign-made incinerator for the second stage, apparently burning the hydrolysate. A large lewisite stockpile, over 6,300 metric tons, at Kambarka was neutralized but the toxic hydrolysate, over 63,000 metric tons, was also stored to apparently extract the arsenic included in the waste.

All seven Russian sites established local Public Outreach and Information Offices (POIOs) managed by Green Cross Russia (GCR), the national affiliate of Green Cross International (GCI)[22] and was internationally managed by GCI's "Legacy of the Cold War Program."[23] This was later changed to the "Environmental Security and Sustainability (ESS) Program" and included Green Cross Switzerland (GCCH) and Global Green USA in addition to GCR and GCI. This outreach program was funded by the US Cooperative Threat Reduction (CTR) Program and was subsequently shut down by the Putin Administration about five years before Russia finished its processing in 2017.

Russia's chemical weapons destruction program was very late in starting, but once going it turned out to be very successful. Started in 2002 and ending in 2017, it took only fifteen years to complete the demilitarization of 40,000 metric tons, an average of over 2,600 metric tons per year. This is over three times the speed of the US program, about 866 metric tons per year. But Russia had several advantages— much of the Russian stockpile was stored in bulk; the weaponized portion did not have explosives or propellent; and a good portion of program costs were covered by US and Western governments. The United States alone spent over $2 billion for

---

[22] Founded in the mid-1990s by Mikhail Gorbachev. See Gorbachev (2000).

[23] The author was the director of this GCI program.

Russian chemical weapons destruction, largely focused on Shchuch'ye and Kizner, with Britain, Germany, Italy, and other countries covering at least $1 billion. This accounted for perhaps 25% of Russia's costs, although no cost figures for the total amount appear available.

The largest downside of the Russian program is the very large amount of toxic waste, mostly liquid, remaining for later processing. This will someday be an enormous cost, risking environmental and public health impacts in the years ahead. The head of the Russian delegation to the OPCW, G.V. Kalamanov, Deputy Minister of Industry and Trade, stated in 2017: "On 27 September 2017, the last of our country's chemical munitions was destroyed at the Kizner facility in the Udmurt Republic. This is without a doubt an important step on the path to a more sustainable and stable world. The elimination of chemical weapons in Russia was carried out under strict international control and in full compliance with the provisions of the Convention. We voice our gratitude to the OPCW Technical Secretariat and the States Parties for many years of close cooperation. We look forward to continued effective cooperation with this Organization. We call upon States in possession of chemical weapons to follow our example and make every effort to complete, in the soonest possible time, the destruction of their remaining stockpiles."[24]

18.3.3   Albania

Albania joined the CWC as a non-possessor State Party in January 1993 with the 130 original signatories; it ratified the Treaty 17 months later in May 1994, still claiming no chemical weapons. However, in December 2002 it discovered a cache of chemicals in the mountains outside of Tirana and asked the OPCW for help in evaluating the apparent chemical arsenal. With the help of several States Parties, Albania verified that they were in possession of 16.6 metric tons of mustard and lewisite and officially declared this to the OPCW in March 2023. Markings on the barrels were all in Chinese, so the assumption was that China had supplied the chemical agent, likely before the CWC entered into force.

Although Albania's chemical stockpile was the smallest declared to date, it would be a challenge to destroy given its location in a remote mountain location prohibiting any major industrial construction or transportation. A German firm built a small incinerator with an afterburner which was able to be moved in pieces to the stockpile location and eventual start-up in 2007. The plan was to be the first State Party to meet the CWC ten-year deadline by April 2007. However, when the first of 600 canisters was placed into the furnace for the estimated twenty-minute burn process, it literally exploded in twenty seconds, burning a hole in the furnace and burning out the afterburner.

This surprising accident fortunately did not kill any workers but it was a major problem to repair the furnace and afterburner. After a couple of months of repairs,

---

[24] OPCW (2017a, b).

the process restarted and the stockpile was destroyed safely by July 2007, a few months after the CWC deadline. On July 12, 2007, the OPCW announced that "Albania is the first nation completely and verifiably to destroy all of its chemical weapons by eliminating in total 16,678 kg of chemical warfare agent. The Albanian stockpile included mustard, lewisite, mustard/lewisite mixture, adamsite, and chloroacetophenone agents."[25]

OPCW Director-General Rogelio Pfirter also stated: "[I welcome] Albania's completion of this campaign, which had required them to overcome considerable technical challenges associated with chemical disarmament. He also extended his appreciation to OPCW Member States Greece, Italy, Switzerland and the United States for the support they had provided in Albania's destruction effort."[26] Pfirter went on to add that approximately 33% of the 71,000 metric tons of chemical warfare agent declared by June 2007 had been verified destroyed—over 23,000 tons primarily from the US and Russian stockpiles.

The following day the US Department of State issued a statement congratulating Albania and noting that the effort was funded by the United States for about $48 million.[27] Not mentioned was the fact that the weapons were destroyed, but all the waste produced by the incinerator was left at the site and not taken care of. This would come back to haunt the US seven years later when they dealt with the removal of Syria's chemical weapons stockpile.

18.3.4 South Korea

The Republic of Korea (hereinafter referred to as South Korea) signed the CWC in January 1993 as one of the 130 original signatories and ratified the Convention in April 1997 before the CWC's entry-into-force. It also declared a stockpile of chemical weapons, estimated by this author as 600 metric tons, but decided not to make this declaration publicly known. It is because of this secrecy that the OPCW calls it the "other state party" or "another state party" in any listing of possessor states.

Pfirter addressed the Tenth Conference of States Parties on November 7, 2005, and stated: "Good progress has also been made by another State Party, which has destroyed approximately 65% of its stockpile. I think that this State Party can take legitimate pride in this achievement."[28] One can surmise that South Korea had been working on its demilitarization program for a couple of years by then, and it completed destruction by 2008. The Thirteenth Conference of States Parties in December 2008 announced the following: "Albania and A State Party completed the

---

[25] OPCW (2007).

[26] Ibid.

[27] US Department of State (2007).

[28] OPCW (2005).

destruction of their entire chemical weapons stockpiles on 4 July 2007 and 10 July 2008, respectively."[29]

South Korea is thought to have had a binary chemical weapons capability, much like the newest US chemical weapons, and this no doubt made its destruction process easier, but no validation of technology used and how their destruction process proceeded is available in public literature. One can only speculate that because of the large and undeclared North Korean chemical weapons stockpile, estimated at 5,000 metric tons or more, South Korea is very sensitive to discussion of their own unilateral disarmament program. Hopefully in the near future we will have better information thereon. It indeed should be a very positive story for the OPCW to tell.

18.3.5   India

India signed the CWC also in January 1993 and ratified it in September 1996, thus joining the Convention before it entered into force in April 1997. It declared a chemical weapon stockpile (or stockpiles) of about 1,000 metric tons but is reluctant to discuss publicly anything about the weapons or how they were destroyed. The stockpile was largely or totally a mustard agent arsenal and was incinerated, according to off-the-record discussions over the years at the OPCW, but India is reluctant to discuss any relevant facts about location or challenges.

At the Fourteenth Conference of States Parties in December 2009 an OPCW report stated:

> A State Party [South Korea] completed the destruction of all its chemical weapons in 2008, leaving four remaining possessor States at the end of the year. By 31 December 2008, India had destroyed 99%, the Libyan Arab Jamahiriya 2%, the Russian Federation 30%, and the United States of America 57% of their declared stockpiles of chemical weapons.[30]

Thus, early in 2009 India became the third possessor state to complete its destruction, behind Albania and South Korea, and continued the line of declared possessor states to make steady but slow progress toward meeting their CWC Article VI obligations.

18.3.6   Libya

Libya acceded to the CWC on January 6, 2004, eleven years after the Convention opened for signature, and after its leader, Muammar Gaddafi, decided to improve his standing with the global community and multilateral treaty regimes. The CWC entered into force for Libya in February 2004, a month after its accession.

---

[29] OPCW (2008).
[30] OPCW (2009).

Libya declared about 35.2 metric tons of chemical agent, largely bulk mustard, to the OPCW in 2004 and later, but the challenge was to find the financing and technologies to accomplish the mission. By the end of 2004 all 3,300 empty munitions had been destroyed, but the bulk agent stored in the middle of the desert was a longer-term project.

Fortunately, Germany, Canada, and the United States were able to help begin destruction by 2010 and by the end of 2013 all bulk mustard had been neutralized.[31] In February 2014 OPCW Director-General Ahmet Uzumcu shook hands with Libyan Foreign Minister Mohamed Abdulaziz in Tripoli to announce the destruction of the remaining mustard artillery shells and bombs: "The destruction of these munitions was a major undertaking in arduous, technically challenging circumstances, as we saw first-hand earlier today at the remote Ruwagha Chemical Weapon Destruction Facility. (…) From start to finish, meeting these challenges was the product of close cooperation between Libya, the OPCW Technical Secretariat and other States Parties."[32]

By 2016 and 2017 the Libyan National Authority had destroyed 114 metric tons of isopropanol at Ruwagha and 19 tons of pinacol alcohol at a second site, Bir al-Osta Milad, near Tripoli. These were all Schedule 2 chemicals. It was decided at the 52nd Executive Council meeting in July 2016 that all remaining chemicals should be removed from Libya and destroyed in Germany's capable facilities at Münster. With the support of Germany and other States Parties, these chemicals were shipped from the port of Misrata and subsequently destroyed under OPCW auspices.[33]

Libya thus became the fourth possessor state to complete destruction of its chemical weapons stockpile, all beyond official Treaty deadlines but within several extended deadlines provided by the States Parties. It offers an excellent example of mutual support extended by the United States, Canada, Germany, Cyprus, New Zealand, Finland, Tunisia, the European Union, and other States Parties to undertake the task in difficult technical and political circumstances with little environmental and public health impacts.

18.3.7 Iraq

Iraq joined the CWC on January 13, 2009, sixteen years after the Chemical Weapons Convention was opened for signature. The Convention entered into force for Iraq one month later, February 12, 2009. Iraq had used chemical weapons during the long Iran-Iraq War in the 1980s, killing and injuring thousands of Iranians, and attacking its own Kurdish population in northern Iraq in 1988. The remnants of Saddam Hussein's chemical warfare efforts—weapons, agents, precursor chemicals, and laboratory equipment—were collected and stored in two large bunkers in

---

[31] OPCW (2013).

[32] OPCW (2014). See also OPCW (2012).

[33] OPCW (2017a, b).

the desert by United Nations inspectors in the early 1990s. These bunkers et al.-Muthanna, numbered 13 and 41, were struck by large US bombs in the 2003 war, one of which reportedly was unexploded inside one bunker. These two bunkers were large, concrete, multistory buildings which were temporarily sealed in 1994 by the UN inspectors.

With Iraq's accession to the CWC, it was required to declare its possession of these bunkers and its toxic contents to the OPCW. But the question for several years was how best to access the storage site without risking the lives of workers. No one was anxious to enter the bunkers.

The Iraqi National Authority established a committee to study "liquidation" of the bunkers in 2010 but soon realized that it would be very difficult, risky, and expensive to destroy the bunkers and their contents.[34] It was therefore decided between Iraq and States Parties to further seal the bunkers with rebar and concrete and live with this toxic legacy of war for the indefinite future. Some States Parties argued that this was not "destruction" of the existing detritus of Saddam Hussein, but the OPCW decided it was the best and safest option at the time. In 2018 the OPCW declared that Iraq had met its obligations under the Convention for destruction of chemical weapons.

18.3.8 Syria

Syria was not a member of the CWC until 2013, acceding to the Convention a month after it allegedly used chemical weapons in an attack on Ghouta, a suburb of Damascus, killing an estimated 1,400 civilians. Chemical weapons use had been alleged earlier in several attacks since late 2012, but the Ghouta-area attacks on August 21, 2013, were particularly brutal, indiscriminate, and deadly using Sarin nerve agent.

Syrian President Bashar Al-Assad denied use of chemical weapons, and Syria had acceded to the 1925 Geneva Accords which banned the use of biological and chemical weapons. But the pressure from Russia and the United States to join the CWC was high, as it was from the OPCW and United Nations after the UN Mission to investigate chemical weapons attacks was fired upon and ran into roadside bombs in Syria. This was the first time that OPCW inspectors had ever come under fire on an inspection mission, and bullet-proof vests became standard equipment. On September 14, 2013, Syria acceded to the CWC and agreed to destroy all chemical weapons stockpiles and production facilities. Its entry-into-force was one month later on October 14, 2013. Two months later in December the UN fact-finding mission, with OPCW and World Health Organization components, delivered its final report on Ghouta to US Secretary-General Ban Ki-moon. It investigated sixteen alleged

---

[34] OPCW (2010).

chemical incidents from October 17, 2012, to August 25, 2013, and recommended further investigation and analysis of seven including Ghouta on August 21, 2013.[35]

The OPCW decided that it would be safer to remove the declared chemical agents and precursors from Syria than to try to destroy them in country while civil war raged on. But several questions remained—how they could be removed, and where would they go to be destroyed. The story of the Syrian operation is a complex one, with much written about it, but the declared 1,304 metric tons of chemicals were loaded onto a foreign transport ship from the port of Latakia and transferred to the US Merchant Marine ship, Cape Ray, which was modified to neutralize a good portion of them while at sea in the Mediterranean. This was a controversial plan, with many protesters from the fishing and tourist industries making their voices heard in Athens, Cyprus, Italy, and throughout the northern Mediterranean.

The Syrian government would not allow the Cape Ray into local waters, so a transfer port had to be found; most Mediterranean ports refused access due to the risk involved, but finally a southern Italian port offered its services, conditioned that none of the chemicals would land on local territory. The transfer was made successfully by keeping the chemicals on board the docks at all times for the transfer between ships.

It was initially hoped that the chemicals could be off-loaded in Albania for destruction, but Albania refused to participate in the effort, reportedly because the US had still not cleaned up the former chemical agent storage site where they had destroyed Albania's stockpile in 2007. The Cape Ray, with NATO naval units surrounding it in the Mediterranean, neutralized a portion of the Syrian stockpile, and then sailed to Germany, Finland, the UK, and the US for further treatment of the chemicals. These were largely treated in thermal systems, but Veolia Environmental in Port Arthur, Texas required overpacking (reportedly an additional $20 million) before they could incinerate the corroded canisters in their possession.[36]

Syrian-declared chemicals, except for a small portion which was successfully neutralized in Syria, were all removed by the end of 2014 and destroyed by 2015. The OPCW however notified Syria of the inadequacy of their official declaration and established the Declaration Assessment Team (DAT), the Fact-Finding Mission (FFM), and the Investigation and Identification Team (IIT) to further investigate ongoing allegations of Syrian (and ISIS) use of chemical weapons.

The DAT was established in April 2014 "… to engage with relevant authorities of the Syrian Arab Republic on gaps, inconsistencies, and discrepancies identified by the Technical Secretariat in the Syrian initial declaration (2013) of its chemical weapons program."[37] The Director-General in May 2024 submitted its 128th monthly report on Syria and determined that "[c]onsidering the identified gaps, inconsistencies,

---

[35] For further information on the OPCW-UN Joint Mission, see https://opcw.unmissions.org/.

[36] See Walker (2014).

[37] OPCW, "Declaration Assessment Team," https://www.opcw.org/declaration-assessment-team.

and discrepancies that remain unresolved, the Secretariat assesses that the declaration submitted by the Syrian Arab Republic still cannot be considered accurate and complete in accordance with the Convention…".[38]

The FFM was also established in April 2014 after "…repeated allegations of the use of toxic chemicals for hostile purposes at a number of locations in the Syrian Arab Republic…" Over the last decade, the FFM has issued 21 reports covering 74 instances of alleged chemical weapons use in Syria, interviewed over 600 people, and collected over 450 samples. "The FFM concluded that chemical weapons were used or likely used in 20 instances: in 14 cases the chemical used was chlorine, in three cases the chemical used was Sarin, and in three cases the chemical used was mustard agent".[39]

The IIT was established in June 2018 by a Special Session of the Conference of States Parties after many countries showed growing concern over chemical weapons allegations in Syria with no resolution of who might have committed such CWC violations. The IIT was mandated to identify the perpetrators of specific instances of chemical weapons use in the Syrian Arab Republic.

"The IIT is responsible for investigating only those instances in which the Fact-Finding Mission (FFM) has determined that use or likely use of chemical weapons in Syria has occurred, as well as cases for which the now-expired OPCW-United Nations Joint Investigative Mechanism (JIM) did not issue a report."[40] To date the IIT has identified five instances of chemical weapons use by the Syrian military: three in Ltamenah in March 2017, one in Saraqib in February 2018, and one in Douma in April 2018.

Syria would have been the fourth State Party to complete stockpile destruction, after Albania, South Korea, and India, if there were not further allegations after 2015, but many alleged instances of chemical weapons use raise serious concerns about meeting its treaty obligations. Therefore today, we have seven declared possessor states having now completed their Article VII obligations (chemical weapons program destruction), but we still await the latest reports of the OPCW on Syria.

18.4 Conclusions

Over the past three decades the multiple efforts to eliminate chemical weapons stockpiles and programs have been technically complex, very expensive, politically contentious, labor intensive, and unfinished, but they represent an extraordinary challenge to fully destroy a whole class of weapons of mass destruction. The goals of the 1925 Geneva Protocol have been partially realized in the Chemical Weapons Convention over seventy years later and, although the diplomats may not have predicted

---

[38] OPCW, "Fact-Finding Mission," https://www.opcw.org/fact-finding-mission.

[39] OPCW, "Fact-Finding Mission," https://www.opcw.org/fact-finding-mission.

[40] OPCW, "Investigation and Identification Team," https://www.opcw.org/iit.

timelines and costs correctly, they certainly made major progress in creating a safer and secure world.

There are many lessons learned which could be enumerated here but we will focus on six, all interrelated but not all-inclusive:

Universality—With 193 States Parties, the Chemical Weapons Convention is almost universal with 98% of the global population under the treaty regime. The last country to join the CWC was Syria in 2013, and we still have four countries—Egypt, Israel, North Korea, and South Sudan—outside the treaty. Israel signed the treaty in 1993, attends the annual meeting as an observer, but refuses to accede to the CWC. South Sudan has voiced strong interest in acceding but has yet to follow through on their promises. Egypt refuses to sign until Israel signs, and North Korea refuses to engage in any negotiations. None of these excuses is acceptable; no country should have chemical weapons, and the OPCW and States Parties should continue to take every opportunity to convince these "final four" to accede.

National Implementation—The CWC requires every country to establish a National Authority, typically residing in the foreign ministry and chaired by the foreign minister, and to pass legislation banning use of chemical weapons. This allows all States Parties to fully implement the treaty and hold individuals account-able for using chemical weapons. Yet, to date only about two-thirds of countries have fully implemented the CWC. The OPCW should press forward with helping those outstanding countries to complete implementation in order to reach 100% by the Sixth Five-Year Review Conference in 2028.

Nonproliferation—Syria is an excellent example of diplomatically pressing a country to meet its obligations under the CWC. Monthly reports to the Executive Council, condemnation of any use of chemical weapons, trade restrictions on precursor chemicals, and OPCW sanctions have all been effective in apparently pressing Syria to stop chemical weapons attacks in the last couple of years. Russia is also a good case in point with the illegal use of Novichok in assassination attempts on Sergei Skripal in 2018 in the United Kingdom, and on Alexei Navalny in Russia in 2020.

While assuring States Parties abide by their CWC obligations, its also very impor-tant to recognize that non-states parties or terrorist groups not mentioned in the CWC have also violated CWC rules. Daesh, ISIS, or the Islamic State used chem-ical weapons while operating in Iraq, according to OPCW inspectors, so nonpro-liferation—stopping chemical weapons from re-emerging anywhere—remains very important. This will require continually maintaining a strong inspectorate, ready to launch a challenge or other inspection on short notice. It also requires excellent national responses to stop such inhumane behavior by States Parties.

Transparency and Inclusivity—Much effort has been committed by Director-Generals, the OPCW Technical Secretariat, and States Parties in recent years to promote transparency in operations and inclusion of civil society, industry, academia, and interested stakeholders. Yet there are still roadblocks, not surprising given that the world still has many non-democratic countries who do not recognize large parts of civil society outside of government. States Parties should resist efforts to narrowly define the goals of the CWC as only "technical" and only "destruction of existing stockpiles." By including civil society, including industry, academia, and interested stakeholders, the OPCW strengthens public outreach and education and thereby the CWC.

The CWC Coalition, founded the 2009–10 and supported by the OPCW, States Parties, and Director-Generals, has been very effective in increasing the number of NGOs and experts from civil society to participate in annual meetings, five-year review conferences, and other opportunities. This effort must be further nurtured and become a regular feature of annual meetings. The Advisory Board on Education and Outreach (ABEO) was also established in 2016 and can be another effective engagement vehicle.[41] Yet more always needs to be done, as the recent establishment of an OPCW working group on education and outreach has already shown.

Chemical Weapons Victims—The world has witnessed very little use of chemical weapons since World War I, but the horrible attacks by Saddam Hussein in the 1980s against his own people and against Iran illustrated once again that these inhumane and indiscriminate weapons need to be eliminated from world conflict. This left thousands of Iranians and Kurds as a legacy to chemical weapons, in addition to the more recent victims in Syria and Russian assassination attempts.

The OPCW has a memorial next to headquarters in The Hague and organizes an annual event during the yearly conference. It also has a Victims' Trust Fund, but not much attention has been paid to victims except for these annual meetings. More needs to be done to help all victims, particularly with ongoing medical care, be they from Iran, Iraq, and Syria, or legacy issues from Morrocco, Yemen, Ethiopia, or elsewhere where chemical weapons were used in WWI.

Science and Technology—The Chemical Weapons Convention bans a discrete number of toxic chemicals in its Schedule 1 in the treaty annexes. Schedule 2 deals with dual-use toxic chemicals which have commercial use and require import and export regulations. After the recent Russian assassination attacks with a Novichok nerve agent, the schedules were updated.

But the science of chemistry moves forward every year with thousands of new chemicals produced, including now Generative Artificial Intelligence options. Fortunately, the Science Advisory Board (SAB) has been helpful in this regard, and the announcement of a working conference on AI and Chemistry in 2024 helps the process by staying informed on how modern chemistry may impact and endanger the CWC and global security.[42]

With these six and other related issues impacting the Chemical Weapons Convention, we must recognize the treaty regime as a living document. We must see the recent elimination of all declared chemical weapons as an important step toward building a more safe and secure world, but also one of many steps to come to enforce the CWC, identify perpetrators, and refer them to the United Nations or other global legal venues for prosecution. Also, the OPCW must offer more assistance and training to all States Parties in an effort to provide excellent monitoring and surveillance to prevent toxic

---

[41] OPCW (2016).

[42] OPCW (2024a, b).

chemicals from re-emerging and being used to inflict illegal, indiscriminate, and inhumane harm.
/newpage

Bibliography

Walker P (2003) Russian chemical weapons demilitarization: successes and challenges. In: Einhorn RJ, Flournoy MA (eds) Protecting against the spread of nuclear, biological, and chemical weapons: an action agenda for the global partnership. CSIS, Washington DC, pp 53–69

Gorbachev M (2000) Ridding Russia of its chemical weaponry requires U.S. help. Los Angeles Times

Webography

Agenda Item 9 (2006) Status of the implementation of the convention. In: OPCW, report of the eleventh session of the conference of states parties, C-11/5. https://www.opcw.org/sites/default/files/documents/CSP/C-11/en/C-11_5-EN.pdf. Accessed 8 Dec 2006

Ambassador Eric Javits at CSP-11. https://www.opcw.org/sites/default/files/documents/CSP/C-11/national_statements/US-en-Javits.pdf.

Annex IV of the CWC. https://www.opcw.org/chemical-weapons-convention/annexes/verification-annex/part-iva-destruction-chemical-weapons-and

Article III of the CWC. https://www.opcw.org/chemical-weapons-convention/articles/article-iii-declarations

Article VIII of the CWC. https://www.opcw.org/chemical-weapons-convention/articles/article-viii-organization.

Blue Grass Press Release (2023) Last chemical weapon destroyed at blue grass army depot. https://www.peoacwa.army.mil/2023/07/07/press-release-last-chemical-weapon-destroyed-at-blue-grass-army-depot/. Accessed 7 July 2023

CATT (Citizens Advisory Technical Team). See https://www.peoacwa.army.mil/

Kentucky Environmental Foundation for more information. https://www.kyenvironmentalfoundation.org/chemical-weapons/

OPCW (2007) Albania the first country to destroy all its chemical weapons. PR73/2007. https://www.opcw.org/media-centre/news/2007/07/albania-first-country-destroy-all-its-chemical-weapons. Accessed 12 June 2007

OPCW (2017b) Statement by G.V. Kalamanov, Deputy Minister of industry and trade, head of the delegation of the russian federation at the twenty-second session of the conference of states parties. C-22/NAT.28. https://www.opcw.org/sites/default/files/documents/CSP/C-22/national_statements/c22nat28_e_.pdf. Accessed 28 Nov 2017

OPCW (2005) Opening statement of the director-general to the conference of states parties at its tenth session. C-10/DG.11, p. 6, para. 37. https://www.opcw.org/sites/default/files/documents/CSP/C-10/en/C-10_DG.11.pdf. Accessed 7 Nov 2005

OPCW. https://www.opcw.org/chemical-weapons-convention/annexes/verification-annex/part-iva-destruction-chemical-weapons-and

OPCW Looking back helps us look forward. https://www.opcw.org/about-us/history

OPCW (2008) Note by the Director-General: status report on the progress made by those states parties that have been granted extensions of deadlines for the destruction of their category 1 chemical weapons. C-13/DG.7, p 2, para. 5. https://www.opcw.org/sites/default/files/documents/CSP/C-13/en/c13dg07.pdf. Accessed 14 Nov 2008

OPCW (2009) Report of the OPCW on the implementation of the convention on the prohibition of the development, production, stockpiling, and use of chemical weapons and on their destruction in 2008. C-14/4, p 1, para. 3. https://www.opcw.org/sites/default/files/documents/CSP/C-14/en/c1404_en.pdf. Accessed 2 Dec 2009

OPCW (2013) Libya completes destruction of its bulk sulphur mustard stockpile. https://www.opcw.org/media-centre/news/2013/05/libya-completes-destruction-its-bulk-sulfur-mustard-stockpile. Accessed 6 May 2013

OPCW (2014) Libya completes destruction of its category 1 chemical weapons. https://www.opcw.org/media-centre/news/2014/02/libya-completes-destruction-its-category-1-chemical-weapons. Accessed 4 Feb 2014

OPCW (2012) Libya: review of the progress made toward destruction of chemical weapons in Libya. Statement by The Libyan delegation at the seventieth session of the executive council. EC-70/NAT.18. https://www.opcw.org/sites/default/files/documents/EC/70/en/ns/ec70nat18_e_.pdf. Accessed 25 Sep 2012

OPCW (2017a) Libya: Report to the Executive Council at The Eighty-Fifth Session on the Progress Achieved Towards Complete Destruction of The Remaining Chemical Weapons Stockpile (31 May 2017). EC-85/NAT.4. https://www.opcw.org/sites/default/files/documents/EC/85/en/ec85nat04_e_.pdf. Accessed 15 June 2017

OPCW (2010) Republic of Iraq: Statement by H.E. Ambassador Muhammad Abddullah Alhumaimidi, head of the department of international organizations and cooperation in the Iraqi Ministry of foreign affairs at the fifteenth session of the conference of states parties. C-15/NAT.18. https://www.opcw.org/sites/default/files/documents/CSP/C-15/national-statements/CSP15_Iraq_en.pdf. Accessed 30 Nov 2010

OPCW Declaration Assessment Team. https://www.opcw.org/declaration-assessment-team

OPCW (2024b) Report by the director-general: progress in the elimination of the syrian chemical weapons program. EC-106/DG. https://www.opcw.org/sites/default/files/documents/2024/05/ec106dg07%28e%29.pdf. Accessed 24 May 2024

OPCW Fact-Finding Mission. https://www.opcw.org/fact-finding-mission

OPCW investigation and identification team. https://www.opcw.org/iit

OPCW (2016) Report of the first session of the advisory board on education and outreach. ABEO-1/1. https://www.opcw.org/sites/default/files/documents/ABEO/abeo-1-01_e_.pdf. Accessed 29 Apr 2016

OPCW (2024a) Germany provides €65,000 to support OPCW conference on role of AI in chemical weapons convention implementation. https://www.opcw.org/media-centre/news/2024/07/germany-provides-eu65000-support-opcw-conference-role-ai-chemical-weapons

OPCW News Release (2023) OPCW Director-General attends celebratory ceremony on end of destruction of U.S. chemical weapons stockpile. https://www.opcw.org/media-centre/news/2023/12/opcw-director-general-attends-celebratory-ceremony-end-destruction-us. Accessed 20 Dec 2023

Pueblo Press (2023) https://www.peoacwa.army.mil/2023/06/22/press-release-pueblo-plant-team-completes-colorado-chemical-weapons-stockpile-destruction/

OPCW-UN Joint Mission. https://opcw.unmissions.org/

Walker PF (2014) Syrian chemical weapons destruction: taking stock and looking ahead. Arms Control Today. https://www.armscontrol.org/act/2014-12/features/syrian-chemical-weapons-destruction-taking-stock-and-looking-ahead

United Nations, Office for Disarmament (1925) 1925 Protocol for the Prohibition of the Use of Asphyxiating, Poisonous or Other Gases, and of Bacteriological Methods of Warfare, commonly known as the 1925 Geneva Protocol. https://disarmament.unoda.org/wmd/bio/1925-geneva-protocol/

US Department of State (2007) Media note. Albania—first country to destroy all of its chemical weapons. https://2001-2009.state.gov/r/pa/prs/ps/2007/88378.html. Accessed 13 July 2007

US Department of Defense (1997) Assembled chemical weapons assessment program: annual report to congress. page 1 https://www.peoacwa.army.mil/wp-content/uploads/1997_supplem ental_rtc.pdf. Accessed Dec 1997

Paul Walker coordinates the Chemical Weapons Convention (CWC) Coalition, is Vice Chair of the Arms Control Association, and is a member of the US Department of State Inter-national Security Advisory Board. He has worked on chemical weapons demilitarization since undertaking the first US on-site inspection of a Russian CW stockpile in 1994 when he was a Professional Staff Member of the US House of Representatives Armed Services Committee. He led the Green Cross International Program on Environmental Security and Sustainability for 20 years and was awarded the Right Livelihood Award in 2013 "for working tirelessly to rid the world of chemical weapons." Walker holds an MA from Johns Hopkins SAIS and a PhD in international security from MIT; he is also a US Army Vietnam-era veteran.

Part V
Chemical Weapons and Terrorism: Ongoing and Future Challenges

Chapter 19
The Chemical Weapons Convention as a Tool in the Global Fight Against Terrorism

Alexander Kelle and Yasmin Cürük

Abstract States parties to the Chemical Weapons Convention (CWC) have agreed among themselves to prohibit the development, production, and use of chemical weapons (CW) and ensure their destruction. During its first 25 years of operation the focus of the CWC and the Organization for the Prohibition of Chemical Weapons (OPCW), which has been set up to oversee its implementation, has been on the destruction of state-level CW arsenals. Yet, ever since the September 11 attacks in 2001, the specter of chemical terrorism has influenced the OPCW in the performance of several of its functions. Given its origins at the end of the Cold War, with a concomitant focus on state-level CW programs, we ask what role the CWC has played in the global fight against terrorism. To this end, the first section of the paper outlines key provisions of the CWC, functions of the OPCW, and their relevance in the fight against CW terrorism. We then address the mandate of the United Nations for fighting terrorism, as contained in its Global Counterterrorism Strategy and Security Council Resolution 1540, and its relationship agreement with the OPCW. Subsequent sections analyze practical cooperation measures in the areas of assistance, protection and capacity building, and in verifying compliance with the CWC. The final section draws together the argument and provides an assessment of CWC and OPCW in supporting the global fight against terrorism.

Keywords Chemical weapons convention · Terrorism · United Nations

A. Kelle (✉) · Y. Cürük
Institute for Peace Research and Security Policy (IFSH), University of Hamburg, Berlin, Germany
e-mail: kelle@ifsh.de

© The Author(s) 2026
B. Friedrich et al. (eds.), *Thirty Years of the Chemical Weapons Convention (CWC)*,
https://doi.org/10.1007/978-3-031-98854-7_19

261

19.1 The Chemical Weapons Convention, the Organization for the Prohibition of Chemical Weapons, and Chemical Terrorism

19.1.1 Key Provisions of the Chemical Weapons Convention and Their Applicability to Chemical Terrorism

Key provisions of the CWC[1] are contained in Article I of the treaty and encompass several norms, understood here as "standards of behavior in terms of rights and obligations" (Krasner 1982) to guide state party behavior. More specifically, according to Article I of the Convention states possessing chemical weapons (CW) undertake to destroy their stockpiles and all states parties equally agree to refrain from the development, production, acquisition, stockpiling, transfer and use of chemical weapons. In order to put the CW destruction norm into effect, CWC states parties agree in Article III on detailed declarations regarding chemical weapons and CW-related facilities. States parties have to submit these declarations to the Organization for the Prohibition of Chemical Weapons (OPCW) where its Technical Secretariat (TS) will process them. Articles IV and V contain the standards of behavior for CW possessor states concerning chemical weapons and CW production facilities, including their declaration and inspection obligations. Articles IX and—to some extent—Article X of the Convention express the investigation norm, for whose realization challenge inspections and investigations of alleged use of CW are the key instruments foreseen in the treaty.

The CWC does not explicitly mention terrorism with chemical weapons. However, neither does the treaty exclude an extension of the prohibitions for states parties to non-state actors, such as terrorist groups. Quite to the contrary, CWC Article VII explicitly requires states parties to internalize obligations stemming from the treaty to their domestic legal systems by way of implementing legislation and other measures. As a result, the CWC prohibits the development, acquisition and use of CW not only for states but also via the internalization norm expressed in CWC Articles VI and VII for non-state actors. In addition to the applicability of the CWC to chemical terrorism deriving from the treaty text, the Director General of the OPCW submitted to its Executive Council a concept paper shortly after the 9–11 attacks and the anthrax mailings through the US postal system. This outlined the contribution that full and effective CWC implementation could make in the global struggle against terrorism.[2] The paper identified CWC universality, the full implementation of the CW destruction norm (CWC Articles IV and V), Article VI (on activities not prohibited), the internalization norm expressed in Articles VI and VII, the assistance and protection provisions under Article X, and the international cooperation norm

---

[1] The treaty text of the Chemical Weapons Convention is available at https://www.opcw.org/chemical-weapons-convention.

[2] OPCW, *Note by the Director-General. The OPCW and the Global Struggle against Terrorism*, The Hague: OPCW, document EC-XXVII/DG.3, 9 November 2001.

in Article XI as contributing to global anti-terrorist efforts. Based on the paper by the Director General the OPCW Executive Council in December 2001 decided "the full and effective implementation of all provisions of the Convention is in itself a contribution to global anti-terrorist efforts".[3] The decision stressed that the contribution of the OPCW should focus on practically all of the above-mentioned Articles of the CWC—with the exception of the inspection norm contained in Article IX—and established an Open-Ended Working Group on Terrorism (OEWG-T) as a subsidiary organ to the Executive Council. The OEWG-T would subsequently serve as an important venue for the discussion of chemical terrorism-related matters at OPCW.

19.1.2   Key Functions of the Organization for the Prohibition of Chemical Weapons in the Global Fight Against Terrorism

The drafters of the CWC created the OPCW under Article VIII of the convention. The organization consists of the two so-called policy making organs, which are the Executive Council and the Conference of the States Parties, as well as the Technical Secretariat. Together, they ensure CWC implementation and seek to maintain a fit-for-purpose organization in a changing global security environment. The threat of chemical terrorism has increased in salience since the CWC entered into force in April 1997. Key functions that the OPCW has performed over this period with respect to chemical terrorism generally fall into three areas: (1) classification and attribution of meaning (2) capacity building and (3) investigation of reported terrorist CW use.

Classification and attribution of meaning. According to Michael Barnett and Martha Finnemore, all international organizations "classify the world, creating categories of problems, actors and action" and "fix meanings in ways that orient action and establish boundaries for acceptable action" (Barnett and Finnemore 2004, pp. 31–2). The aforementioned note by the OPCW Director-General and related Executive Council action in the fall of 2001 clearly fall into this category. Through this, the OPCW established the relevance of the CWC for the global fight against terrorism and its own role in this context. In addition, with the OEWG-T, the Executive Council created a new actor with a targeted responsibility in the fight against terrorism.

The OEWG-T was later complemented by a sub-working group on non-state actors in order "to discuss in more detail, and on the basis of three discussion papers issued by the Secretariat, the topics of: the legal accountability of non-State actors under the Convention …; measures to prevent the hostile use of toxic chemicals …;

---

[3] OPCW, Executive Council. *Decision – The OPCW's Contribution to Global Anti-Terrorist Efforts*, document EX-XXVII/DEC.5, The Hague, 7 December 2001, quote on p. 1.

and ensuring an effective response to the hostile use of a toxic chemical …".[4] In general terms, the sub-working group was more focused on "supporting the OEWG-T in making practical recommendations on how the OPCW could further contribute to global anti-terrorism efforts."[5] The three discussion papers were complemented by a fourth one on the utility of CWC Article VI in the context of states parties' efforts to address chemical terrorism.[6] In sum, the discussion papers further established the OPCW's role in the classification and organization of meaning concerning how CWC states parties could best implement the provisions of the treaty in the global fight against chemical terrorism. Classification and attribution activities during this period culminated in the first ever conference on countering chemical terrorism held by the OPCW in June 2018, which brought together representatives from states parties, relevant international and non-governmental organizations, and academia.[7]

Capacity building. In some areas of CWC implementation in a broader sense, which are also relevant for being able to counter chemical terrorism, such as national implementing legislation or protective measures, it was evident since the early phase of CWC implementation that states parties would benefit from capacity building measures undertaken by the OPCW. The national implementation action plan agreed upon by the Conference of the States Parties in October 2003 represents an early indication in this regard. At that time, about 60% of states parties had submitted a declaration to the OPCW under CWC Article VII (5) and about a third indicated that they had legislation in place covering all key areas of CWC implementation. Through a number of diverse OPCW capacity-building activities over the subsequent 20 years, these figures have improved significantly. As of July 2023, 192 of 193 CWC states parties had notified the OPCW of their national authority for CWC implementation and 179 states parties had submitted their Article VII (5) declaration. However, only 128 states parties reported having implementing legislation in place that covers all initial measures for effective CWC implementation.[8]

The provision of assistance to increase levels of preparedness and capabilities to respond to a terrorist use of chemical weapons has also received sustained attention by the OPCW. Over the years, the organization's Technical Secretariat has conducted numerous courses, workshops and trainings to assist its member states in their efforts to protect themselves from chemical attacks, including by terrorists. OPCW activities in this regard complement those of CWC states parties under Article X, which include

---

[4] OPCW, *Report of the Implementation of the Convention in 2015*, document C-21/4, The Hague: OPCW, 30 November 2016, p. 30.

[5] OPCW, *Report of the Implementation of the Convention in 2015*, document C-21/4, The Hague: OPCW, 30 November 2016.

[6] OPCW, *Note by the Technical Secretariat. The Contribution of Article VI to States Parties' Efforts to Counter Terrorism*, document S/1387/2016, The Hague: OPCW, 19 May 2016.

[7] OPCW, *Note by the Technical Secretariat. Summary of the Conference on Countering Chemical Terrorism. OPCW Headquarters, The Hague, The Netherlands, 7-8 June 2018*, document S/1652/2018, 16 July 2018.

[8] OPCW, *Report by the Director-General: Overview of the Status of Implementation of Article VII of the Chemical Weapons Convention as at 31 July 2023*, document C-28/DG.7, The Hague, 6 September 2023.

the provision of emergency assistance and protection to any state party that is exposed to or threatened with chemical weapons. Importantly, Art. X (8) does not distinguish between a state and a non-state perpetrator. Assistance provided by states parties may take the form of a financial contribution to a voluntary fund managed by the OPCW. Member states could also donate certain items of chemical defense equipment, either on a bilateral basis or through the organisation.

Investigation of use. A special form of assistance that has less to do with capacity building and more with verification of treaty compliance is also foreseen under CWC Article X. It stipulates that within 24 h after the receipt of a request for assistance in case of a suspected CW use, the OPCW Director-General must launch an investigation of the alleged chemical attack, irrespective of the type of perpetrator, to establish relevant facts that would enable further action. After another 72 h, he has to submit a report to the Executive Council, which has then to deal with the issue. Although no state party has ever requested the investigation of alleged use of CW following CWC provisions, other mechanisms to investigate the reported use of CW have been established and used on a regular basis, with regard to chemical weapons use in Syria by both state and non-state actors alike.

The so-called Fact-Finding Mission (FFM) was established by then Director-General Ahmet Üzümcü in 2014. Over the past decade, the FFM has investigated over 70 cases of alleged CW use in Syria and confirmed the use of chemical weapons in 20 of them.[9] While it was the FFM's task to determine whether CW had been used, a separate mechanism, the OPCW-UN Joint Investigative Mechanism (JIM), was created in August 2015 by UN Security Council Resolution 2235 (2015) to identify the perpetrators of such use. The JIM identified both the Syrian government and the so-called ISIL/Da'esh as perpetrators of CW use (see Sect. 19.2.3).

Underlying OPCW activities in these three areas is a further function that Barnett and Finnemore have identified in their extensive analysis of international organizations: the diffusion of norms, understood here as standards of behavior, defined in terms of rights and obligations. Most notably, this has affected the internalization norm expressed most prominently in Article VII of the CWC, but has also been relevant in relation to the international cooperation, assistance and investigation norms. Before analyzing the OPCW's pursuit of these normative guideposts in the UN-led global fight against terrorism, the following section will briefly discuss the mandate and role of the United Nations in this context.

---

[9] OPCW, *Note by the Technical Secretariat. Accession of the Syrian Arab Republic to the Chemical Weapons Convention: Ten Years On*, document S/2213/2023, The Hague (2023).

19.2 The OPCW's Contribution to the Global Fight Against Terrorism

19.2.1 The United Nations and Its Relationship with the OPCW in the Fight Against Chemical Terrorism

The United Nations is the premier international organization tasked with securing international peace and security. This applies mostly to the UN Security Council, but also to the General Assembly and the Secretary General. UN activities in the global fight against terrorism have their basis in the UN Global Counter-Terrorism Strategy, adopted by the General Assembly in Resolution 60/288 on 8 September 2006.[10] The plan of action annexed to the resolution *inter alia* encouraged the UN Secretary General to update his mechanism (UNSGM) for the investigation of alleged use of chemical and biological weapons (Littlewood 2006), and called on the OPCW to continue its capacity building measures for CWC states parties to prevent terrorists from accessing chemical materials, to improve chemical security at relevant facilities and to respond effectively to chemical attacks by terrorists.[11] With this, the UN General Assembly strengthened the investigation and international cooperation and assistance norms expressed in the CWC.

Since then, the General Assembly has undertaken regular reviews of the UN Global Counter-Terrorism Strategy. The most recent, eighth biennial review reiterated previous references to United Nations Security Council Resolution 1540 (see below). It also called upon member states to continue to "support international efforts under the auspices of the United Nations to prevent terrorists from acquiring weapons of mass destruction and their means of delivery, and urges all Member States to take and strengthen national measures, as appropriate, to prevent terrorists from acquiring weapons of mass destruction."[12] Thus, the UN Global Counter-Terrorism Strategy and its periodic reviews express an expectation for states to support the treaty-based norms to internalize international obligations against terrorist use of weapons of mass destruction and to cooperate and assist in preparing and responding to such terrorist incidents.

The UN Security Council adopted Resolution 1540 on 28 April 2004. It provides WMD terrorism-related guidance that is more detailed than the broad-based UN

---

[10] United Nations, *Resolution adopted by the General Assembly on 8 September 2006. 60/288. The United Nations Global Counter-Terrorism Strategy*, document A/RES/60/288, New York: United Nations, 20 September 2006.

[11] *UN Counter-Terrorism Strategy*, p. 8.

[12] United Nations, *Resolution adopted by the General Assembly on 22 June 2023, The United Nations Global Counter-Terrorism Strategy: eighth review*, document A/RES/77/298, New York, quote on p. 20.

counter-terrorism strategy.[13] One observer has succinctly described it as a piece of law that is "mandatory for all United Nations Member States, brings together obligations under numerous single-technology focused treaties and agreements, focusses attention on the activities of non-state actors, and requires Member States to go beyond mere declarations of support for nonproliferation" (Gahlaut 2019, p. 53f). After its adoption UNSC Resolution 1540 was criticized for its legislative nature, and for directing UN Member States to take steps against an abstract threat. Nonetheless, the operative paragraphs of the resolution strengthen several of the core norms of the CW prohibition regime, such as the ones on non-acquisition, non-transfer and non-use of CW, albeit with a focus on non-state actors such as terrorist groups. To implement the resolution, the Security Council set up a Committee and established a group of experts.[14] While their mandate was initially limited to a two-year period, it quickly became obvious that the full and effective implementation of the resolution would be a longer-term undertaking. Hence, the UN Security Council extended the mandate of the Committee and the group of experts several times, most recently in November 2022 for a period of ten years, following the third comprehensive review of resolution 1540.[15]

The practical cooperation between the UN and OPCW envisaged under Resolution 1540—or at least implied by it—has as its foundation the relationship agreement that the two international organizations had already concluded in 2000. The OPCW Conference of the States Parties approved the agreement in May 2001, which entered into force later that year. It distinguishes five main areas of cooperation: first, the OPCW Executive Council—in line with paragraph 36 of CWC Article VIII—shall bring cases of non-compliance with the CWC of particular gravity to the attention of both UN General Assembly and the Security Council. Second, the OPCW Conference of the States Parties—in reference to paragraph 4 of CWC Article XII—shall bring cases of non-compliance of particular gravity to the attention of both UN General Assembly and the Security Council. Third, according to paragraph 27 of Part XI of the CWC Verification Annex, the OPCW shall closely cooperate with the UN Secretary General in case of CW use by a non-CWC state party, and put its resources at the Secretary General's disposal. Thus, the first three of the areas identified for cooperation between UN and OPCW are already codified in the CWC treaty text itself. The latter of those would become relevant in the initial investigation of reported CW use in Syria in early 2013, before that country became a state party to the CWC. The remaining two areas identified by the UN-OPCW relationship agreement concern cooperation in the provision of assistance to a CWC state party that has suffered from the threat or actual use of CW—in relation to which the wording of CWC Article X is somewhat vague, making reference only to "relevant international

---

[13] United Nations Security Council, *Resolution 1540 (2004) Adopted by the Security Council at its 4956th meeting*, document S/RES/1540 (2004), New York, 28 April 2004.

[14] For a discussion of Resolution 1540, criticism voiced and its early implementation, see Alexander Kelle (2014), pp. 208–213.

[15] United Nations Security Council, *Resolution 2663 (2022) Adopted by the Security Council at its 9205th meeting*, document S/RES/2663 (2022), New York, 30 November 2022.

organizations"[16]—and to foster international cooperation for peaceful purposes in the field of chemistry. Before addressing practical UN-OPCW cooperation in ensuring CWC compliance in Sect. 19.2.3, we will first highlight some aspects of cooperation in the fields of assistance and capacity building and its evolution over time.

19.2.2 Practical OPCW-UN Cooperation in the Areas of Assistance, Protection, and Capacity Building

Through its extensive work in the realm of assistance, protection, and capacity building, the OPCW plays a pivotal role in aiding member states to prepare for and respond to incidents of chemical terrorism. In relation to the OPCW's contribution to the global fight against terrorism, the Third Review Conference emphasized the "need to explore further cooperation ... and build on existing work with relevant international organisations and international bodies that deal with the potential threats of chemical terrorism."[17] Due to this, and given the necessity of a coordinated response spanning various international organizations in the event of a terrorist attack with a toxic chemical, the organization places a large emphasis on its cooperation with different UN sub-actors and other relevant international organizations. With this, it aims to ensure the building of comprehensive capacity against chemical threats on the global, regional, and national levels, in order to create synergies, eliminate the duplication of efforts, and improve efficiency.

Collaborative Initiatives and Interagency Coordination. Unlike nuclear and radiological terrorism, there is no single agency responsible for chemical terrorism response. Instead, multiple UN and international entities share overlapping responsibilities. Due to this complex network of responsibilities, the UN Counterterrorism Implementation Task Force (UNCTITF) Working Group on Preventing and Responding to WMD Terrorist Attacks, which the OPCW has co-chaired alongside the International Atomic Energy Agency (IAEA), has held regular inter-agency meetings. In 2011, the working group scoped out the corresponding responsibilities of each relevant international actor in the event of a terrorist attack using chemical or biological weapons.[18] The OPCW takes on a key role in the response efforts following a chemical terrorist attack through its unique technical expertise, and can be involved in coordinating emergency relief efforts alongside the United Nations Disaster Assessment and Coordination (UNDAC) mechanism and other relevant organizations to manage the immediate response and mitigate the impact on affected

---

[16] See the CWC treaty text at https://www.opcw.org/chemical-weapons-convention.

[17] OPCW, *Report of the Third Special Session of the Conference of the States Parties to Review the Operation of the Chemical Weapons Convention*, document RC-3/3*, The Hague, 19 April 2013, p. 29.

[18] UNCTITF, *Report of the Working Group on Preventing and Responding to Weapons of Mass Destruction Attacks: Interagency Coordination in the Event of a Terrorist Attack using Chemical or Biological Weapons or Materials*, New York, August 2011.

populations. The number of other agencies involved is too great to name here, but include the International Maritime Organization (IMO), the World Organization for Animal Health (OIE), the United Nations Office on Drugs and Crime (UNODC), and the United Nations Office for the Coordination of Humanitarian Affairs (UNOCHA), among others.[19]

Moreover, as a key participant and vice-chair in the UN Global Counter-Terrorism Coordination Compact's Working Group on Emerging Threats and Critical Infrastructure Protection, led by the International Criminal Police Organization (INTERPOL) and including the United Nations Office for Disarmament Affairs (UNODA) and the United Nations Interregional Crime and Justice Research Institute (UNICRI), the OPCW works on projects enhancing inter-agency interoperability and public communication during chemical incidents. The OPCW's commitment to inter-agency coordination extends to partnerships with the United Nations Office of Counter-Terrorism (UNOCT), UNOCHA, the World Health Organization (WHO), INTERPOL, UNICRI, and the BWC ISU.[20] These collaborative efforts seek to ensure a cohesive and effective international response to chemical terrorism.

Legal and Policy Assistance. Ever since 2001, the OPCW's focus has remained on the full implementation of the Convention as the main contribution of the CWC to the global fight against terrorism. In 2017, the Executive Council recognized the pivotal role of assistance and cooperation in achieving full national implementation of the CWC and stressed the importance of the criminalization of activities prohibited by the CWC, as required by Article VII.[21] To support states with the implementation of the Convention, the Secretariat collaborates with external partners such as the Expert Group of the 1540 Committee, UNODC, and UNODA. They jointly seek to enhance States Parties' awareness of their obligations and provide practical assistance, particularly in areas where the Convention and Resolution 1540 overlap (e.g., national legislation, customs and border control, and chemical security).[22]

The Secretariat also actively contributes to workshops of its external partners, such as workshops aimed at strengthening the implementation of Resolution 1540, underscoring the relevant obligations of each States Party under the Convention that coincide with the resolution. Similarly, these external partners support activities organized by the OPCW. To name just some examples, in 2023 UNODA regional coordinators and experts from the 1540 Committee joined regional meetings in Africa and Asia, as well as the annual meeting of National Authorities, which aims to enhance

---

[19] Ibid.

[20] OPCW, *Report of the OPCW on the Implementation of the Convention on the Prohibition of the Development, Production, Stockpiling and Use of Chemical Weapons on Their Destruction in 2020*, document C-26/3, The Hague, 1 December 2021.

[21] OPCW, *Decision: Addressing the Threat Posed by the Use of Chemical Weapons by Non-State Actors*, document EC-86/DEC.9, The Hague, 13 October 2017.

[22] OPCW, *Note by the Director-General: Status of the OPCW's Contribution to Global Anti-Terrorism Efforts*, document EC-105/DG.10, The Hague, 16 February 2024, p. 2.

National Authorities' capacity to comply with their obligations under the Convention.[23] Additionally, a recent regional workshop on Best Practices in the Development of Legislative and Regulatory Framework on Chemical Security for African countries was supported by experts from INTERPOL, the World Customs Organization (WCO), and the chemical industry. Even more recently, in May 2024 the OPCW held a pilot workshop on the role of implementing legislation in addressing threats arising from non-state actors, including participation from UNODC, INTERPOL, the 1540 Committee Group of Experts, and the WCO.[24]

The issue of the legal accountability of non-state actors and strengthening legal frameworks through collaboration with other organizations has also been a key focus of the OEWG-T and the sub-working group on non-state actors, especially since 2016. This was highlighted in a 2020 meeting where the UNODC held a presentation, and attendees discussed various international counter-terrorism instruments, the role of the UNODC in their implementation, and its cooperation with the OPCW.[25]

Preventing the Use of Toxic Chemicals by Non-State Actors. Another important concern in countering chemical terrorism is preventing access to dual-use chemicals. In this context, the Executive Council in 2017 emphasized the "obligation of States Parties under paragraph 2, Article VI of the Convention to 'adopt the necessary measures to ensure that toxic chemicals and their precursors are only developed, produced, otherwise acquired, retained, transferred, or used' for purposes not prohibited under the Convention".[26] To aid States Parties in achieving this goal, the Secretariat hosts an extensive array of workshops, training programs, and regional seminars each year, focusing on chemical safety and security management, custom controls and transportation issues, and others.

These are usually held in collaboration with relevant UN entities including UNOCT, UNICRI, and UNODC, as well as international organizations such as INTERPOL and the WCO. The long-standing collaboration with the WCO, in particular, is very relevant to transportation issues. Here, a focus has been a train-the-trainer course on the technical aspects of the transfer regime. The aim of this partnership is to enhance customs officials' knowledge of the Convention's transfer regime and the tools to identify scheduled chemicals.[27]

Ensuring an Effective Response. Enhancing the global response capacity to the hostile use of chemicals, including use by a non-state actor, has been another OPCW

---

[23] Ibid.

[24] OPCW, *OPCW strengthens legislative and regulatory framework on chemical security in Africa*, available at https://www.opcw.org/media-centre/news/2024/05/opcw-strengthens-legislative-and-regulatory-framework-chemical-security, last accessed 2024/05/21.

[25] OPCW, *Note by the Director-General: Status of the OPCW's Contribution to Global Anti-Terrorism Efforts*, document EC-96/DG.10, The Hague, 18 February 2021, p. 3.

[26] OPCW, *Decision: Addressing the Threat Posed by the Use of Chemical Weapons by Non-State Actors*, document EC-86/DEC.9, The Hague, 13 October 2017.

[27] OPCW, *Report of the OPCW: On the Implementation of the Convention on the Prohibition of the Development, Production, Stockpiling and Use of Chemical Weapons and on their Destruction in 2015*, document C-21/4, The Hague, 30 November 2016, p. 16.

priority. The Secretariat's focus on collaborations in this area includes building national and regional response capacities, increasing the organization's own capacity to respond and provide assistance during an attack with a hostile chemical, and improving the collective global response in coordination with the UN and other relevant international organizations.

Tabletop exercises have been a particularly important tool in assessing and improving states' response capabilities. Due to the necessity for a coordinated response in the event of an attack, exercises are often held in cooperation with relevant agencies, e.g., ASSISTEX 3 in 2010 in cooperation with UNOCHA.[28] Furthermore, the OPCW's commitment to capacity building at the national and regional levels includes conducting specialized courses to enhance emergency response capabilities to chemical attacks. These initiatives are crucial for training first responders, improving laboratory capacities, and facilitating international cooperation in emergency responses, ensuring a coordinated and efficient global response to chemical threats.

In May 2016, the OPCW created the Rapid Response and Assistance Mission (RRAM) to be deployed upon a member state's request to provide emergency assistance in the event of a chemical attack. This explicitly includes attacks by non-state actors; in fact, the first field exercise, including participation from INTERPOL and WHO in December 2017, aimed to test RRAM competencies required for addressing an incident of alleged CW use by a non-state actor.[29] The RRAM can also support a state party in its efforts to coordinate a response with other international organizations, contributing to the role of the OPCW in inter-agency coordination in the event of a terrorist attack.

More recently, the OPCW conducted a chemical emergency response capacity building exercise, dubbed CHEMEX-Africa, from 23 September to 5 October 2023 in Algiers, Algeria.[30] The multi-faceted training exercise involved the training of more than 80 first responders from African countries, by African instructors, who had previously participated in OPCW capacity building courses. CHEMEX-Africa was conducted with funding from the Canadian government and sought to "evaluate and analyze existing gaps in chemical emergency response capabilities in Africa."[31] This event is indicative of both the continuing chemical security concerns among African CWC states parties and a broader and well-established OPCW focus on supporting its African member states.

---

[28] OPCW, *Note by the Technical Secretariat: Final Exercise Instructions for Exercise ASSISTEX 3*, document S/866/2010, The Hague, 16 September 2010.

[29] OPCW, *Field Exercise in Romania to Improve OPCW's Rapid Response and Assistance Capabilities*, available at https://www.opcw.org/media-centre/news/2017/12/field-exercise-romania-improve-opcws-rapid-response-and-assistance, last accessed 2024/05/21.

[30] OPCW, *Note by the Technical Secretariat: CHEMEX Africa (Multi-Component Chemical Emergency Response Capacity-Building Exercise for the Africa Region), Algiers, Algeria, 23 September –5 October 2023*, document S/2185/2023, The Hague 27 June 2023.

[31] OPCW, *CHEMEX Africa 2023*, available at, https://www.opcw.org/media-centre/featured-topics/chemex-africa-2023 last accessed 2024/05/17.

Africa Programme. This focus is exemplified through the long-standing "Programme to Strengthen Cooperation with Africa on the Chemical Weapons Convention", more commonly referred to as the "Africa programme". This initiative aims to streamline the CWC implementation in Africa and address the continent's needs in relation to the Convention, with one focus being bolstering the resilience and response capacities of African states parties against chemical terrorism.[32]

The program is based on three-year phases and is currently in its sixth phase, which covers the period from 2023 to 2025. Reports on the OPCW's counter-terrorism activities indicate that since its inception in 2007, there has been an increasing emphasis on the counter-terrorism aspect of the program. Beginning with the 2020 report and including all subsequent reports to date, there is an acknowledgment that Africa faces specific challenges related to terrorist threats and attacks. The focus of capacity building activities under the program has evolved over time and broadened in scope. It has shifted from general chemical safety and compliance with the Convention in its early stages, for example through legislative assistance, to a larger emphasis on national protection programs and response training targeting chemical terrorism prevention and response more directly, with CHEMEX-Africa being a recent example of that trend.[33]

To achieve the programs' objectives, the OPCW collaborates with several international and regional organizations in its implementation. These include African Regional Economic Communities, the African Union (AU) Commission, the European Union (through the CBRN Centres of Excellence based in Africa), WCO, and various OPCW Member States and institutions that host activities under the program. Together, these collaborations facilitate targeted capacity building and technical assistance, crucial for enhancing chemical safety and security in Africa.[34]

19.2.3 Practical OPCW-UN Cooperation in Ensuring CWC Compliance

Concrete practical cooperation between the OPCW and the UN in the area of investigating possible non-compliance with the CWC by terrorist groups started in late 2012 and early 2013 when first reports about the use of CW in Syria surfaced. While Western governments suspected the Syrian government to be the perpetrator, the latter always asserted that terrorist groups inside the country were to blame for the

---

[32] OPCW, *Africa Programme*, available at https://www.opcw.org/resources/capacity-building/international-cooperation-programmes/africa-programme, last accessed 2024/05/21.

[33] OPCW, *Report of the OPCW On the Implementation of the Convention on the Prohibition of the Development, Production, Stockpiling and use of Chemical Weapons and on their Destruction*, document C-16/4, The Hague, 30 November 2011, p. 19; OPCW, *Note by the Director-General: Status of the OPCW's Contribution to Global Anti-Terrorism Efforts*, document EC-105/DG.10, The Hague, 16 February 2024, p. 11.

[34] OPCW, Africa Programme, available at https://www.opcw.org/resources/capacity-building/international-cooperation-programmes/africa-programme, last accessed 2024/05/21.

CW attacks. Although not specifically designed to confirm terrorist CW use, the 2013 UN Mission to Investigate Allegations of the Use of Chemical Weapons in the Syrian Arab Republic marked the first substantive example of cooperation in this area. Syria acceded to the CWC only in September 2013, and therefore the above-mentioned UN Secretary General's mechanism had to be utilized for investigating the first documented CW use cases in that country. As noted in the report of this UN mission, which focused on the alleged CW use in the Ghouta area of the Syrian capital Damascus, "[f]or the purpose of ascertaining the facts related to the allegations of use of chemical weapons, gathering relevant data and undertaking the necessary analyses, the [UN] Secretary General has requested the Organisation for the Prohibition of Chemical Weapons ("OPCW") to put its resources at his disposal, including providing a team of experts to conduct fact-finding activities."[35] OPCW inspectors and other experts constituted the bulk of the mission team. Evidence collected and analysed included surface-to surface rockets found on site and still containing residues of the nerve agent Sarin, biomedical and environmental sample analysis, and more than fifty interviews with survivors of the attack. The report concluded that "on 21 August 2013, chemical weapons have been used in the ongoing conflict between the parties in the Syrian Arab Republic, also against civilians, including children, on a relatively large scale."[36] It is noteworthy that the identification of the perpetrators behind the attack was not part of the UN mission's mandate. Concerning the nature of the cooperation in this mission, Makdisi and Pison Hindawi have concluded that while it was based on the UN Secretary General's mechanism and the UN-OPCW relationship agreement, collaboration "effectively remained ad hoc rather than integrated" (Makdisi and Hindawi 2019, p. 545).

The OPCW-UN Joint Mission. The Syria-related verification work continued and expanded with the OPCW-UN Joint Mission on the elimination of Syrian chemical weapons in late 2013 and 2014. After Syria joined the CWC on 14 September 2013 as a CW possessor state, it had to declare its CW stockpile and past CW program. In order to destroy the Syrian CW arsenal without delay, the OPCW Executive Council on 27 September decided to establish strict deadlines and stipulated special procedures for the swift and verifiable destruction of Syria's CW program.[37] The UN Security Council endorsed this decision on the same day with its Resolution 2118 (2013). The resolution also authorized an advance team of UN personnel to assist the OPCW in Syria, and requested the Secretary General to develop recommendations regarding the role of the United Nations in eliminating the Syrian CW program.[38]

---

[35] United Nations, *Report of the United Nations Mission to Investigate Allegations of the Use of Chemical Weapons in the Syrian Arab Republic on the alleged use of chemical weapons in the Ghouta area of Damascus on 21 August 2013*, document A/67/997—S/2013/553, New York, 16 September 2013, quote on p. 4.

[36] Ibid., p. 8.

[37] OPCW, *Decision. Destruction of Syrian Chemical Weapons*, document EC-M-33/DEC.1, The Hague, 27 September 2013.

[38] United Nations Security Council, *Resolution 2118 (2013) Adopted by the Security Council at its 7038th meeting*, document S/RES/2118 (2013), New York, 27 September 2013.

Collectively these decisions, in combination with intense informal consultations,[39] enabled the UN Secretary General to establish the OPCW-UN Joint Mission, which began its operation on 16 October 2013. After the Joint Mission had successfully supervised the destruction or removal of declared CW stocks and related facilities, the collaborative effort ended on 30 September 2014, after which the remaining work related to the Syrian CW program reverted to the OPCW.

The OPCW-UN Joint Investigative Mechanism. Notwithstanding the destruction of all declared Syrian chemical weapons, the repeated use of CW, mainly chlorine, in Syria in 2014 continued. This led the UN Security Council to establish the OPCW UN Joint Investigative Mechanism (JIM) on 7 August 2015 when it adopted Resolution 2235. This resolution tasked the JIM to "identify to the greatest extent feasible individuals, entities, groups, or governments who were perpetrators, organisers, sponsors or otherwise involved in the use of chemicals as weapons, including chlorine or any other toxic chemical."[40] As with the Joint Mission before, the OPCW contributed the CW expertise to this joint endeavor. This new role of attributing responsibility represented a significant change in the verification of CW use in Syria and beyond,[41] and resulted in the identification of both state and non-state perpetrators. Concerning the latter, the JIM concluded that the Islamic State in Iraq and the Levant (ISIL) was responsible for using sulfur mustard in an attack on Marea on 21 August 2015 and at Umm Hawsh on 15 and 16 September 2016 (Kelle 2023). As noted in the seventh JIM report, the leadership of the mechanism "has continued to be supported by three components: the Investigative Office, the Political Office and the Planning and Operations Support Office. The Investigative Office is based in The Hague, Netherlands, and comprises two units: the Information Collection Unit and the Analysis and Corroboration Unit."[42] Thus, there was a clear division of labour between OPCW and UN with the former leveraging its technical investigative capabilities, while the latter brought its political weight and planning and operational assets to the table.

During the fall of 2017, Russia used its veto power in the UN Security Council several times to prevent a further extension of the JIM's mandate. The last such veto in mid-November de facto terminated OPCW-UN cooperation in jointly investigating CW use in Syria.[43] This subsequently led some CWC states parties to call for the transfer of the attribution function of the JIM to the OPCW Technical Secretariat. In June 2018 a special session of the OPCW Conference of the States Parties adopted

---

[39] See Makdisi and Pison Hindawi 2019, pp. 548–550.

[40] United Nations Security Council, *Resolution 2235 (2015)*, document S/RES/2235, New York, 7 August 2015.

[41] United Nations Security Council, *Letter dated 24 August 2016 from the Secretary-General addressed to the President of the Security Council*, document S/2016/738, New York; and *Letter dated 26 October 2017 from the Secretary-General addressed to the President of the Security Council*, document S/2017/904, New York.

[42] *Letter dated 26 October 2017 from the Secretary-General*, document S/2017/904, p. 3.

[43] United Nations Security Council Meetings Coverage, "Security Council Fails to Adopt 2 Draft Resolutions on Extending Mandate of Joint Mechanism Investigating Chemical Weapons Attacks in Syria", 8105th meeting, SC/13072, New York, 16 November 2017.

a decision entitled "Addressing the Threat from Chemical Weapons Use". This *inter alia* "condemns the use of chemical weapons by State and by non-State actors", including the cases reported on by the JIM.[44] The Conference also decided "that the Secretariat shall put in place arrangements to identify the perpetrators of the use of chemical weapons in the Syrian Arab Republic by identifying and reporting on all information potentially relevant to the origin of those chemical weapons".[45]

The Investigation and Identification Team. The so-called Investigation and Identification Team (IIT), which was established subsequently within the OPCW Technical Secretariat, has since issued four reports on suspected CW use cases. In three of them, the IIT found the Syrian armed forces responsible for the chemical attacks on its civilian population. The latest report covers another instance of CW use in Marea, where the IIT concluded that on 1 September 2015 "units of the Islamic State in Iraq and the Levant (ISIL or Islamic State) deployed sulfur mustard, using one or more artillery guns."[46] Ongoing work and findings of the IIT have been a regular topic on the agenda of the OPCW Policy Making Organs. Thus, while the investigative aspects of addressing CW use have remained with the OPCW since the OPCW UN Joint Investigative Mechanism, the political dimension formerly addressed by the UN Security Council has been brought under the purview of the organization. This leaves the UN's operational and planning capabilities still involved in the OPCW's investigative work on suspected CW use cases in Syria, including those committed by terrorists.

19.3 Summary and Conclusions

The Chemical Weapons Convention is a treaty that seeks to regulate behavior of its states parties vis-à-vis one another. As such, it is not an obvious contributor to the global fight against terrorism. Hence, in this chapter we have briefly presented the provisions of the CWC and the functions of the OPCW that are relevant in the context of chemical terrorism, discussed the relationship of the OPCW with the United Nations in this issue area, and analyzed practical OPCW-UN cooperation in the areas of assistance, protection and capacity building, as well as ensuring compliance with the CWC via verification measures. In sum, the OPCW over the past quarter century has evolved into a sector-specific, i.e. focused on chemical terrorism, but nonetheless key actor in the UN-led global fight against terrorism.

The broad nature of the prohibitions contained in the CWC, in combination with the obligation of states parties to internalize these normative guideposts into their

---

[44] OPCW, *Decision: Addressing the Threat from Chemical Weapons Use*, document C-SS-4/DEC.3, The Hague, 27 June 2018, p. 2.

[45] Ibid, p. 3.

[46] OPCW, *Report by the Director General: Progress in the Implementation of Decision C-SS-4/ DEC.3 on Addressing the Threat from Chemical Weapons Use*, document EC-105/DG.14, The Hague, 27 February 2024, p. 2.

domestic legal systems, makes them applicable to terrorism with chemical weapons, too. Both the OPCW Director-General at the time and the Executive Council reaffirmed this understanding shortly after the terrorist attacks in the USA in the fall of 2001, established the OPCW's role in the global fight against terrorism, and provided guidance for its future involvement in this issue area. The classifications and attribution of meaning undertaken back then were further supported by the creation of the Open-Ended Working Group on Terrorism as a subsidiary organ to the Executive Council, and reinforced over time through conceptual work undertaken by the OPCW Technical Secretariat.

Preventing or responding to chemical terrorism are key elements in the broader global fight against terrorism, in which the United Nations plays the leading role among international organizations. As the CWC contains references to the UN and its organs in some of its treaty provisions, both organisations had already begun to negotiate a relationship agreement by the time the salience of transnational terrorism increased significantly in 2001 and thereafter. The OPCW UN Relationship Agreement entered into force in late 2001 and foresaw cooperation in case of particular grave instances of non-compliance with the CWC, in case of CW use by a non-state party to the CWC, and with a view to providing assistance and fostering international cooperation in the peaceful uses of chemistry. This basic agreement was subsequently complemented by UN Security Council Resolution 1540 (2004), which contains WMD terrorism related obligations for UN member states and through its operative paragraphs supports the implementation of several regime norms expressed in the CWC. Similarly, the UN Global Counter-Terrorism Strategy, adopted by the General Assembly in September 2006 called on the OPCW to continue to provide capacity-building measures for its states parties in the areas of prevention, response and investigations. The strategy as well as its periodic reviews continue to demonstrate the international community's expectation that in its area of expertise the OPCW should use the normative framework provided by the CWC to contribute to the global fight against terrorism.

Over the past quarter century, the OPCW has undertaken extensive work in relation to assistance, protection and capacity building. Concerning chemical terrorism, CWC states parties have long acknowledged that the OPCW needs to collaborate with other international organizations in the UN system and beyond. One of its most prominent collaborations centered on counter-terrorism has been the OPCW's role in co-chairing the UNCTITF Working Group on WMD Terrorist Attacks together with the IAEA, focusing on inter-agency coordination in mitigation and response efforts. The OCPW has also engaged in work with entities like the 1540 Committee, UNODC, and UNODA to promote and enhance national compliance with the Convention and strengthen legal frameworks. The OPCW regularly conducts workshops and training programs on the prevention of chemical terrorism, working closely with organizations like the WCO to improve knowledge on chemical security and the prevention of terrorists' access to dual-use chemicals. It also works to enhance states parties' response capacity, including through practical exercises like ASSISTEX and CHEMEX-Africa, and stands ready to provide assistance and coordinate a response in the event of an attack through its own RRAM, established in 2016. Last, but not least, the Africa Programme, now in its sixth phase, has been developed to strengthen CWC implementation and resilience against chemical terrorism in Africa, collaborating with regional and international organizations for targeted capacity building on the continent. Through these collaborative efforts, the OPCW aims to create a coordinated and effective international response to the global threat of chemical terrorism, improving prevention, preparedness, and response capabilities.

While the OPCW's contribution to the global fight against terrorism through capacity building and assistance and protection measures has largely followed a path that was predetermined by the CWC treaty text and relevant normative guidance promulgated by the United Nations, the same cannot be said for ensuring compliance with the CWC through investigations of CW use by terrorists. Although the CWC's provisions for the investigation of alleged CW use in principle are applicable to terrorist use, no CWC state party ever made such a request. Instead, the first joint OPCW-UN investigation of CW use in Syria in 2013 was based on the UN Secretary General's mechanism for this purpose, with the OPCW providing the chemical weapons expertise necessary to successfully conduct the investigation. The subsequent OPCW-UN Joint Mission (2013–2014) to disarm the Syrian CW program and destroy the country's declared CW stockpile represents the highest level of integration of a collaborative CW-related mission between OPCW and UN. Based on these experiences, the two organizations could continue their collaboration in the context of the UN Security Council-mandated Joint Investigative Mechanism (JIM) from 2015 to 2017. The JIM's investigations concluded that ISIL was responsible for using sulfur mustard in an attack on Marea in August 2015 and Umm Hawsh in September 2016. After Russia had vetoed the extension of the JIM's mandate in the fall of 2017, the attribution function conferred to it was transferred to the OPCW with the Investigation and Identification Team (IIT) taking up its work in 2018. The IIT also issued a report in which it identified ISIL as the perpetrator in a sulfur mustard attack in Marea in September 2015. Over time, the OPCW-UN collaboration in ensuring CWC compliance by investigating CW use has evolved from reliance on the Secretary General's Mechanism to a highly integrated Joint Mission to disarm Syria, to the Security Council mandated JIM. Compared to these efforts, today's cooperation between the two organizations in this area is significantly reduced after the Russian vetoes in the Security Council prevented the JIM from continuing its work and the OPCW Conference of the States Parties transferring the attribution function to its technical secretariat.

Bibliography

Barnett M, Finnemore M (2004) Rules for the world. International organizations in global politics. Cornell University Press, Ithaca and London

Gahlaut S (2019) United Nations security council resolution 1540 implementation: more of the same or brave new world. Strat Trade Rev 5(7):53–66

Kelle A (2014) Prohibiting chemical and biological weapons. Multilateral regimes and their evolution. Lynne Rienner, Boulder: CO

Kelle A (2023) The CWC at 25: from verification of chemical-weapons destruction to attribution of their use. Nonprolifer Rev 28(4–6):319–336

Krasner SD (1982) Structural causes and regime consequences: regimes as intervening variables. Int Organ 36(2):185–205

Littlewood J (2006) Investigating allegations of CBW use: reviving the UN Secretary-General's mechanism. Compliance Chronicles Number 3. Canada: Canadian Centre for Treaty Compliance, Ottawa

Makdisi K, Pison Hindawi C (2019) Exploring the UN and OPCW partnership in Syrian chemical weapons disarmament. Interorganizational cooperation and autonomy. Glob Gov 25(4):535–562

Alexander Kelle is senior researcher at the Berlin office of the Institute for Peace Research and Security Policy (IFSH). In the Berlin office, he heads the CBW network to strengthen the norms against chemical and biological weapons (CBWNet). His research focuses on the evolution of the CBW prohibition regimes. He is the author of inter alia Prohibiting *Chemical & Biological Weapons: Multilateral Regimes and Their Evolution (2014)* and, together with A. Ghionis, "The Chemical Weapons Convention After Its Fifth Review Conference: Key Issues for the European Union" (2024).

Yasmin Cürük holds a BSc in Economics from the University of Bonn. From September 2023 to July 2025 she worked at the Berlin Office of the Institute for Peace Research and Security Policy (IFSH), where she supported the CBW network to strengthen the norms against chemical and biological weapons (CBWNet) as a student employee.

Chapter 20
Defence Against Terror Weapons: Why Are Chemical Weapons Still Used and How Do We Defend Against Them?

Lincoln Sheff

Abstract Despite chemical weapons being prohibited in 1997 when the Chemical Weapons Convention (CWC) came into force, states still spend considerable resources towards Chemical Warfare Defence (CWD). The chapter argues against claims that chemical weapons fundamentally lack utility in warfare. Such claims miss the point about perceptions of the weapons' ability to create terror to explore the terror aspect around chemical weapons, this chapter combines literature around chemical weapons' ability to cause psychological terror with taboo literature to support its argument. However, this chapter argues against conventional taboo approaches. This is because analysis of the taboo primarily focuses on explanations of non-use. This chapter puts forward the case that social and psychological perceptions of chemical weapons can be studied in order to explain why the weapons are used. It is psychologically taxing to defend against chemical weapons. This chapter then uses two examples of states' defences against chemical weapons, the First World War, and the Iran-Iraq War, finding some key similarities between them. Following this, the chapter uses informal discussions with CWD experts to understand perceptions of chemical weapons, whilst also understanding CWD institutions' broader social and normative place in history.

Keywords Warfare · Chemical weapons · CWC · Defence · Terror

20.1 Introduction

Within the history of chemical weapons there is a debate around the weapons' utility. This centres around whether they are ineffective as battlefield weapons and can be defended against with the right equipment and training (Spiers 2020, p. 221). A proponent of this argument is United States Army Chemical Corps Officer Dan Kaszeta who argues that in the First World War, "The outcome of very few battles

L. Sheff (✉)
University of Bath, Bath, UK
e-mail: lcs58@bath.ac.uk

© The Author(s) 2026
B. Friedrich et al. (eds.), *Thirty Years of the Chemical Weapons Convention (CWC)*,
https://doi.org/10.1007/978-3-031-98854-7_20

was determined by chemical weapons" (2020, p. 8). He also claims that one type of chemical weapon, nerve agents, does not invoke fear, suggesting that "as I learned more about nerve agents, the less I feared them" (2020, p. xi). He further argues that; "Nerve agents never won any battles. Nor did they win any wars" (2021, p. 258). This chapter stands against this perspective, arguing that chemical weapons have utility in that they are terror weapons.

To support this argument, this chapter focuses on the psychological social dimension surrounding chemical weapons. Chemical weapons are perceived to be an "agent of terror" or a "weapon of terror" (Phillips and Crouch 2023, Palmer 2004). The chapter therefore begins by studying the social dimensions around chemical weapons, examining the previous literature on taboos. Taboo approaches have been used to explain chemical weapons' non-use and their eventual prohibition in 1997 (Price 1997). In other words, the emphasis is on how the taboo exerts "social control" (Bentley 2022, p. 4) on states' use of chemical weapons. However, this chapter will make a case against the 'chemical weapon taboo', and instead argue that explanations for the taboo can actually be reversed in order to understand why chemical weapons are used. In fact, previous research has already established that chemical weapons do have utility in other areas. For example, Koblentz and Cross put forward convincing arguments that chemical weapons are desirable for states as a "tool of repression" and in "counter-insurgency operations" or targeted assassinations (Koblentz 2013, p. 502 and Cross 2017, p. 217–222). Palmer additionally identifies chemical weapons as "weapons of terror", and as such they have utility for terrorists (2004, p. 3). Overall, this chapter expands on this work and explains that chemical weapons serve a purpose in warfare and pose social and psychological challenges to defend against.

Therefore, the argument of this chapter is that the social ways we understand chemical weapons amplify the ability for these weapons to cause terror. This chapter uses a range of examples of chemical warfare, including the 1914–1918 First World War, the 1980–1988 Iran-Iraq war, 1970s Rhodesia, 1980s South Africa and Russia in the 2010's. This chapter then studies some social experiences with chemical warfare defence (CWD), analysing the experience of the First World War and the Iran-Iraq War. The chapter then studies contemporary CWD institutions in order to examine social perceptions of chemical weapons. This part of the chapter is based on informal discussions with practitioners within CWD in the United Kingdom (UK). In this section it appears that some of the same social perceptions are reflected within past case studies.

20.2 'Agent of Terror'

What makes chemical weapons an agent of terror? Phillip and Crouch assert that because civilians are typically the primary target of chemical weapon attacks this makes them an "agent of terror" (2023). They consider civilians are the primary target for chemical weapons because civilians cannot be adequately protected from chemical weapon attacks (Phillip and Crouch 2023). Kaszeta also reiterates similar claims that with the advent of nerve agents, civilians became disproportionately more at risk (2020, p. 259). Nerve agents according to Schmidt were perceived to be a "game changer", because they are far more lethal than other chemical weapons developed before them (Schmidt 2015, p. 176). To make matters worse, small droplets of a nerve agent can enter the body through the skin, resulting in death (Schmidt 2015, p. 175). As a result, without full protective equipment and training, civilians are vulnerable to chemical weapons, particularly nerve agents (Kaszeta 2020, p. 259).

However, what if we apply this "agent of terror" logic to other weapons? On October 1st 2017, a single gunman fired from the 32nd floor of a Las Vegas hotel onto a dense crowd of concert goers. The shooter used several automatic assault rifles and thousands of bullets to kill 60 people, and injured hundreds of others (New York Times 2023). However, national militaries and police forces have defensive capabilities to deal with active shooters. Defensive means include body armour, bullet proof helmets, smoke, and flash grenades to cover their movements, and importantly, training.[1] It is this same criterion in which Phillips and Coach make the judgment that chemical weapons are ineffective in warfare because "Most advanced militaries have protective equipment coupled with defensive training" (2023). As such this argument is not unique to chemical weapons. Civilians are always at risk, regardless of the weapon being used. We could begin to list a whole range of other weapons that would be relatively ineffective against militaries but effective at killing civilians. For example, during the Rwanda genocide the machete killed an estimated 31,117 people (Verwimp 2006, p. 13). Is the assault rifle or the machete an "agent of terror"? Why do we not label the assault rifle or the machete as a Weapon of Mass Destruction (WMD) as we do with chemical weapons (Carus 2012)?[2] Therefore, what is it about chemical weapons that makes them so scary? In order to answer this question, we need to look back at the historic social impact chemical weapons have had.

Historically, chemical weapons were not really clearly understood and defined. Instead, they were treated as poisons (Price 1997). Poisons have a long use in history. Poison arrows for example, are regularly used for hunting, both historically and in contemporary societies in the Americas (Jones 2007, p. xvi). However, poison is often associated with "barbarity", largely because of the perceived historical image of the weapon being employed by those deemed 'uncivilised' (Price 1997, p. 28). For example, the Romans would regularly accuse their enemies of employing poison, discursively labelling those who did so as without "moral right" (Bentley 2023). However, it is often difficult to distinguish this history as being the actual use of poisons, rather than just allegations (Carus 2015, p. 223). Projectiles mixed with poison are a common theme throughout history, with Ferguson labelling this as "forbidden ammunition accusations" (2019, p. 3).

The human experience with poison resulted in these weapons being associated with taboos in their use (Price 1997, p. 21). Taboos are social attitudes we give to

---

[1] During the Las Vegas shooting, off-duty British soldiers helped evacuate numerous civilians and to treat the injured (BBCNews 2018).

[2] Most states do not describe firearms as WMDs. Notable exceptions however include The Federal State of Utah, which does consider firearms to be weapons of mass destruction (Carus 2012, p. 68).

human behaviour that is condemned, meaning taboos are related to some kind of danger, imaginary or real (Bentley 2022, p. 4). Therefore, arguably there existed a taboo against chemical weapons long before their introduction to the battlefields of the First World War in 1915. Price states that the taboo is one of the main reasons for chemical weapons non-use and eventual prohibition (1997). Bentley's work on taboos (2022) offers three characteristics which expand upon Price's work. She argues that "disgust, stigmatization and fetishization" are important social ways we understand taboos with certain weapons (Bentley 2022, p. 3). However, the purpose of this chapter is not to explore justification of use of chemical weapons, or the role of the taboo in the disarmament of chemical weapons. This is because this chapter is interested in what is it about chemical weapons that makes them an 'agent of terror'. Whilst it may be the case that when the taboo is violated it may be strengthened (Price 1997), this kind of approach is particularly unhelpful in understanding why chemical weapons have utility and are employed. Despite this, it's worth applying Bentley's characteristics that make up a taboo weapon (2023) whilst also using chemical warfare psychological literature to allow us to study exactly what makes chemical weapons terrifying and as such difficult to defend against.

20.3 Fear, Disgust and Contamination

Firstly, "disgust" relates to the sicking, emotional imaginary that comes to mind with the effects that these weapons can inflict on people (Bentley 2022). With biological weapons, we all know what it is like to be ill (Bentley 2023, p. 22), and with incendiaries like Napalm we know what it is like to be burnt (Neer 2013, p. 16). With chemical weapons, we fear the effects of venomous animals like spiders and snakes, and are generally disgusted by them (Staňková et al. 2021, Gerdes et al. 2009). During the First World War, Thompson makes note of this animalistic aspect associated with chemical warfare (2017, p. 258). Thompson cites the example that; "One German soldier described the gas like an insatiable predatory animal that would 'zerfressen alle inneren Organe' or 'eat away at all of the internal organs'" (2017, p. 258).

The chemical weapon Dichloroethyl Sulphide, commonly known as 'mustard gas', has a particularly disgust-invoking impact on the human body. Mustard gas wounds people through burning their skin, lungs, and eyes, causing infection (SPIRI 1971, 50). Mustard gas could also contaminate the very ground (Schmidt 2015, pp. 35–36). It was also used by the Iraqis during the latter stages of the Iran-Iraq war, with the first documented case being in 1983 (Hiltermann 2007, p. 28). Hassan Hassani Sa'di, for example, was only eighteen when he was first exposed to mustard gas in 1985 during his time serving in the Iranian military (Time 2014). According to Sa'di, he remembers only the "smell of garlic" before "vomiting green", passing out, and awaking to painful blisters, blindness, and a lifetime of medical problems (Time 2014). During the Iran-Iraq War, tens of thousands of Iranians were injured as a result of chemical warfare.[3] Numerous Iranian veterans still suffer from long term damage, both physically and mentally (Khateri et al. 2017, pp. 1–2).

Furthermore, the Rhodesians would exploit this aspect of being poisoned to their advantage in their counter-insurgency operations. During the Rhodesian conflict of the 1960s and 1970s, the white settler Rhodesian Government in what is now known as Zimbabwe turned to chemical weapons out of desperation, over the fear that their minority ruled state could be overthrown by the black majority (Cross 2017). During the latter stages of the conflict, members of the Rhodesian security services would often send poisonous clothing out to civilian stores for guerrillas to wear (Cross 2017, pp. 103–104). The Rhodesians inadvertently realised that African tribes in Rhodesia are highly sensitive to the idea that plagues, infections, and poisoning are caused by the presence of witches (Cross 2017, pp. 121–125). As a result, many guerrillas would kill villagers in witch hunts, thinking they were behind the poisoning (Cross 2017, pp. 121–125). In reality, it was the Rhodesian security services, who were fully aware of the terror they were causing (Cross 2017, p. 125).

Therefore, as Bentley argues, disgust is associated with the real fear of contamination (2022, p. 8). Chemical weapon contamination often can spread a great deal of confusion and terror (Palmer 2004). Like Rhodesia, South Africa turned to chemical weapons during the apartheid era. The South African governments repeatedly used chemical weapons on its opponents to intimidate and instil fear (Koblentz 2013). The Soweto uprising in 1976 resulted in the massacre of 200 protesters, and the death in prison of anti-apartheid activist Steve Biko in 1977, sparking massive protests against the apartheid government. After Soweto, South Africa began employing subtler means of repression against opposition leaders (Koblentz 2013). A notable example of such an operation is the 1982 poisoning of Eastern Cape student activist Siphiwo Mtimkulu. After his release from prison, Mtimkulu, who had lost the use of his legs, was treated for two months at Groote Schuur Hospital, where thallium was found in his system. It was suspected that he had been poisoned while he was incarcerated (SAPA 1997). The case received widespread media coverage at the time, and two months later, the South African police abducted and secretly murdered Mtimkulu (SAPA 1997). South Africa would repeatedly employ poison throughout the 1980s in assassinations. In 1981, following questionable intelligence reports of chemical weapon development in neighbouring Angola, the South Africans set up a Chemical and Biological Weapon Program code-named Project Coast (Gould and Burger 2002). Project Coast was led by cardiologist Dr Wouter Basson (Gould and Burger 2002). The apartheid government reinforced the fear of poisoning and the secrecy surrounding its murder methods. Fear of contamination and the powerlessness it inflicts on oneself are important factors behind the fear of chemical weapons (Palmer 2004).

---

[3] Numerous figures have been given on the number of casualties for the Iran-Iraq War. For example, according to a report from the Iranian Veterans and Martyr Affair foundation the number was "79 335" individuals. According to the OPCW, the number was reported to be over " > 70,000" individuals still suffering from long-term injuries, but 100,000 total dead and injured (OPCW 2019).

However, being contaminated by a chemical weapon can sometimes go undiagnosed without medical attention. As such, it is worth noting that without knowing of being intentionally poisoned, fear of being contaminated in this instance would not play such a factor. However as shown by Mtimkulu's case and others like his, medical capabilities are important to countering and detecting chemical weapon poisoning. Inadvertently, this then changes the explanations for use, for example in the 2018 poisoning of Sergei Skripal, in which Russian Novichok Nerve agent was found to be behind the poisoning. However, as Kaszeta argues, the "clumsy" nature of the use Novichok was actually an intentional public message being sent by Russia (2020, p. 250). Dewey similarly makes the case that the secondary objective of Soviet and Russian chemical weapon assassinations is "intimidating domestic audiences", as well as striking fear into the dissident "émigré community" (2023, p. 1, 21). In other words, this is because once chemical weapon use has been identified by medical services, reactions of disgust and terror are to be expected. Therefore, even in this hard case of chemical assassination, disgust from the fear of contamination is still an important factor.

20.4 Stigmatisation, Social Imaginaries, Fetishisation and the Fear of the Unknown

From the disgust we have for chemical weapons, they are often discursively described as being associated with acts of "barbarity" (Price 1997, p. 33). The Germans, for example, were seen as "barbaric criminals" and an "inhuman enemy" for their use of chemical weapons against the British (Schmidt 2015, p. 27).[4] Even today, stigmatising arguments are in chemical weapon discourses. For example, British defence writer Hamish de Bretton-Gordon, in an interview with British politician Nigel Farage, reiterated a bizarre claim that the Russian desire to use chemical weapons is in their "psyche" (Youtube@GBNews 2022). This then leads on to fetishization and social imaginaries we attach to chemical weapons. In history, poisons play important dramatic roles in ancient myths and Greek tragedies (Bentley 2023, pp. 68–69). Hollywood films also depict chemical weapons use, for example the infamous bright green VX nerve agent in the 1996 film The Rock.[5] The British secret intelligence service (MI6) received reports in September 2002 that Iraq's VX was also supposedly contained in similar glass containers depicted in the film (Report of the Iraq Inquiry, 2016, p. 313). Indeed, Haber argues that "Literature, painting, film, and radio have done more for gas since 1918 than the military use of chemical weapons up to then would justify" indicating the power that social imaginaries have at amplifying fear (1986, p. 237). Moreover, scientists involved in CW programs often are

---

[4] This way we stigmatise an enemy is not entirely unique to chemical weapons. During the First World War, numerous reports, including the use of human shields and massacres of Belgium civilians, were used as examples of German atrocities (Horne and Kramer 1994, Haber 1986).

[5] Of course, VX nerve agent is not bright green but an amber liquid (CDC 2023).

associated with the social imaginaries we have with "evil or mad scientists" (Balmer 2012, p. 20). For example, Dr Basson head of South African's Project Coast, is often discursively labelled as "Dr Death" (Bale 2006, p. 51).

Finally, it is worth noting that chemical weapons are particularly feared precisely because they are "unfamiliar" or "unknown" (Stokes and Banderet 1997, p. 396, Haber 1986, p. 136). This unknown and unfamiliar characteristic of chemical weapons' effects on the human body makes it difficult to correctly and quickly identify when people have been contaminated (Palmer 2004). Indeed, a common issue is the problem of false positives, misdiagnosing symptoms of poisoning (Palmer 2004). For example, in the 1991 Gulf War Israel and Saudi Arabia were targeted by Iraqi Scuds, with numerous soldiers and civilians diagnosing themselves with conditions of nerve agent poisoning. Hundreds of civilians admitted themselves to hospital due to perceived poisoning (Palmer 2004, p. 6). Therefore, are chemical weapons truly ineffective if they are perceived to be so terrifying? As Thompson argues in his analysis of the German military experience of chemical warfare, chemical weapons arguably had little tactical impact, but instead an effect on the German military "psyche" (2017, p. 254). However, does this at all matter for chemical warfare defence?

20.5   Defence Against Terror Weapons

In April 1915, the German Empire deployed chlorine gas against the British and French at the second battle of Ypres. During the war, differing states went to great lengths within chemical warfare research. In December 1915, the Germans began using the more deadly toxic gas phosgene, which led to the demand for new and improved gas masks to be introduced (Schmidt 2015, p. 33). The British, as well, established a new research facility with the sole purpose of chemical warfare known as Porton Down in 1916. A whole evolution of British gasmasks were created throughout the war, primarily in London and Nottingham, which would replace the more rudimentary-style masks soldiers were wearing against chlorine (Carter 2000, pp. 12–14). This began with the Hypo helmet, then the PH helmet, and finally the more effective small box respirator developed in 1916 (Carter 2000, pp. 12–13). At Porton, these respirators were tested by the British military, with infantry undergoing training and exercises with them (Carter 2000, p. 12).

However, soldiers routinely lived in fear of this new equipment failing (Fitzgerald 2008, p. 617) and there was a general distrust and lack of understanding of gasmasks, particularly amongst the less experienced soldiers (Faith 2014, pp. 24–26, Fitzgerald 2008, pp. 616–617). Moreover, chemical warfare-related objects like gasmasks can sometimes provoke phobias and anxiety around their use (Stokes and Banderet 1997, p. 399). Fear of equipment failing is a common psychological problem with chemical weapons (Palmer 2004). This fear was only amplified with one's perceived weakness and vulnerability associated with the claustrophobic gasmask (Fitzgerald 2008, p. 616). Indeed, the gasmask was seen as particularly unmasculine, with them often being referred as "stink hoods, mouth drums, [and] pigs" by German soldiers (Thompson 2017, p. 253). However, developments in gasmasks hampered the ability for gas to produce the killing effect desired (SPIRI 1971, p. 56). As a result, the initial introduction of chemical weapons to the battlefield did not have the desired tactical ability commanders had hoped for (Spiers 2017, p. 155), instead they were far more effective at generally making life miserable for those in the trenches. The individual soldier during the First World War grew to fear the possibility of being killed, contaminated, or blinded by chemical weapons, contributing to the "psychological effects" of the weapons' ability to harm soldiers' morale (Schmidt 2015, p. 23).

Medical treatments also improved greatly during this period (Fitzgerald 2008, p. 618). For example, the Germans, in reaction to French mustard attacks, would issue bleaching powder that could alleviate mustard exposure to the skin (Spiers 2020, p. 35). Medical personnel went to great lengths developing a complex washing and showering procedure for decontamination of soldiers' uniforms during the First and Second World Wars (Fitzgerald 2008, p. 618 and Wiseman 1951, pp. 92–94). However, during the First World War mobile cleaning stations were few in quantity (Spiers 2020, p. 35).

Finally, tactics also changed significantly, with troops undergoing some basic training and education before being deployed to the front. For example, troops practiced putting on and correctly fitting gasmasks and conducted military exercises with protective equipment (IWM 276 1917). As we have noted, CWD training is vitally important to counter this fear caused by the unfamiliarity of operating in a chemical warfare environment (Stokes and Banderet 1997). Indeed, the Russian Empire's military, which did not undergo particularly adequate CWD training, suffered a great number of casualties during the First World War (Haber 1986, pp. 102, 179).

Chemical warfare of the First World War became an iconic and symbolic representation of horrific aspects of trench warfare, with the two becoming synonymous with each other (Schmidt 2015). Its representation in numerous mediums, including poems, artwork, and novels, provided signifiers for the horrific conditions and immorality of the First World War (Schmidt 2015, p. 43). Haber argues that this artistic process constructed a terrifying imaginary of chemical weapons (1986, p. 229), relating back to the fetishisation and disgust we have with this unfamiliar weapon.

Much like the invading German army of the First World War and the Rhodesians, invading Iraqi forces turned to chemical agents to turn the tide of the war, particularly after their initial invasion of Iran failed in 1981 (Tucker 2006, pp. 249–250). According to Saeid Sadeghi, an Iranian war correspondent, chemical weapon usage caused massive fear and panic amongst Iranian troops due to the lack of knowledge, and taking many by surprise (Tehran Peace Museum 2015). He recounted his experience of a mustard gas attack in 1984 during Operation Kheibar:

> But, you see, what started to happen after the Iraqis were defeated in an operation, was for them to immediately retaliate by dropping mustard gas bombs on our troops. From 1984 onwards, this became their modus operandi. The prospect of these gas attacks really terrified the Iranian troops. You could see the fear in their eyes. I saw for myself what chemical

weapons could do: the blisters, the burning skin just peeling off the soldiers' bodies and I watched some soldiers die a slow and agonizing death. (Tehran Peace Museum 2015)

Chemical weapons therefore proved highly effective in causing numerous casualties, particularly against an opponent like Iran, which was not ready for chemical warfare (Tucker 2006, p. 270). Furthermore, chemical weapons are perceived by Sadeghi as weapons that "terrified the Iranian troops" and invokes a strong imagery of disgust in his description of the effects of mustard gas (Tehran Peace Museum 2015). It is this element of surprise that amplifies the psychological effects of a chemical weapon, in large part because its use is rare and unfamiliar with soldiers on the battlefield (Wachtel 1941, pp. 42–43). Leading on from this, the Organisation for the Prohibition of Chemical Weapons' (OPCW) award-winning Professor Mahdi Balali-Mood, who responded to many of the chemical incidents in Iran, stated that Iran was unprepared for chemical warfare:

At first, we had a lot of problems, we were not really prepared for such a large number of gas victims. We were not ready with suitable protective clothing. As a result of this many of our own medical staff became susceptible to secondary contamination. But the real problem was that our knowledge of first aid in the event of chemical attacks was simply not good at all. (Tehran Peace Museum n.d.)

He goes on in the interview to list therapies used to treat Iranian mustard gas victims, including the careful use of showers and decontamination of clothing (see: Tehran Peace Museum n.d). What is noticeable is the striking similarity to mustard gas treatments developed during the First World War (see: Fitzgerald 2008, p. 618), indicating this improvised unpreparedness and lack of modern decontamination equipment. In fact, Professor Mahdi Balali-Mood actually used historical articles, journals and medical education from his time as a researcher in Edinburgh to re-learn treatments for mustard gas exposure (YouTube@OPCW 2015). Eventually he would improve Iran's healthcare response and training to mustard gas and other chemical agents by the end of the war (Tehran Peace Museum n.d).

The Iranians also adopted training and tactics in order to mitigate the impact of a chemical weapon attack. In particular, the Iranians increasingly used protective equipment and medical remedies such as atropine, an antidote to nerve agents (Kaszeta 2020, p. 173). Yet despite this, Iraq progressively became more effective with its use of chemical weapons. For example, it increasingly used chemical weapons for area denial to counter Iranian offensives (Tucker 2006, p. 253). Finally, also like the First World War, chemical warfare remained an iconic part of this conflict, particularly with the images from the massacre at Halabja in 1988, in which the Iraqi air force killed hundreds, if not thousands, of Kurdish civilians with a nerve agent (Hiltermann 2007). The Saddam regime would repeatedly use the horror associated with Halabja to keep the Iraqi population in line after the Iran-Iraq War and Gulf War (Koblentz 2013, pp. 516–517).

Therefore, in returning to Phillips and Couch "agent of terror" argument, it is certainly true that chemical weapons are an effective weapon of terror against civilians, particularly as they cannot defend against them. Yet, as this chapter has shown, chemical weapons' ability to kill civilians does not discount its utility as a battlefield weapon. Chemical weapons can be terrifying if defensive measures are not correctly followed.

20.6   Contemporary State CWC Institutions

To counter the threat of chemical weapons, states invest in CWD programs within differing first responder emergency services such as the military and police. These institutions train extensively to respond to chemical weapons' potential use as part of Chemical, Biological, Radiological and Nuclear (CBRN) defence, a term that now encompasses a broad range of weapons and capabilities that are typically grouped together as WMDs (Carus 2012). Incidentally, many of the same organisations responsible for chemical warfare would eventually become involved in CBRN defence. A classic example is Porton Down. Today, Porton is run by The Defence Science and Technology Laboratory (DSTL), which is a key organisation for chemical and biological defence but has its roots in Britain's offensive chemical weapon programme (DSTL, nd).

Within such institutions, both military and civilian, formal and informal rules dictate codes of conduct and can be analysed through lived or working experience (Jepperson 1991, p. 142), with social practices, and also behaviours, exhibited within them. To be clear, institutions in this case are defined as a broad range of organisations or even social phenomena (Jepperson 1991, p. 144). These CWD military and civilian institutions act as key case studies to observe social practices. Therefore, the study of institutions provides us an insight into perceptions towards chemical weapons.

As mentioned previously, this chapter uses informal discussions with experts within CBRN defence within the UK. This informal discussion took place at a workshop that was conducted within a university setting, with numerous British policy specialists and practitioners involved in CBRN defence. Because almost all of these experts are currently still working, their identities have been kept anonymous. It is worth noting that these specialists have a background in organising large-scale exercises and providing training for their respective institutions. However, gaining access to these institutions is not without its challenges. Ultimately, researchers will face numerous institutional barriers to access, such as levels of security clearance, and thus a lack of resources and data. Yet, previous research has shown that researchers can avoid some of these issues through key cooperation with "gatekeepers" often in senior positions, which is typical with most academic research in defence (Catignani and Basham 2021, p. 213). Although the workshop's purpose was to establish a forum to facilitate institutional dialogue, several questions were directed to the experts that enabled discussion on attitudes and practices towards CBRN, which included CWD. First, an overall discussion explored how they perceived CBRN defence. Although this occurred halfway through the event, it serves as an introduction into how contemporary British CBRN practitioners viewed CWD. The questions asked were around what makes chemical weapons so dangerous and to be feared. The second aspect studies what kind of ethos is needed in order to train and prepare to counter this danger.

Firstly, many of the same social and psychological pressures are felt amongst first responders regarding training. According to the experts, chemical weapons are perceived by them to be uniquely more dangerous than most other weapons throughout history (pers. comm. 26/07/2023). The explanation they gave was that the weapon is sometimes invisible, as in the case of nerve agents. Thus, if the weapon is deployed against you, it is very difficult to defend against unless you have the correct equipment and training. This reasoning links strongly to the "agent of terror" argument explored earlier in the chapter. That being said, the experts interviewed reiterated a common theme within chemical weapon history that people are far familiar with the well-known weapons of war such as blades and firearms but unfamiliar with chemical weapons (pers. comm. 26/07/2023, Brophy and Fisher 1959, p. 187 and Wachtel 1941, p. 43). Chemical weapons are rare, unknown and fetishised resulting in unfamiliarity and fear.

Furthermore, the fear of chemical exposure and contamination is taken very seriously by CWD practitioners and is emphasised in training (pers. comm. 26/07/2023). This is particularly important due to the fact that certain weapons such as nerve agent are invisible and odourless (Schmidt 2015, p. 176). This also relates back to the fear of contamination mentioned previously. Also raised was the issue of personal protective equipment. With the introduction of mustard gas, and particularly nerve agents, a gasmask would no longer be enough. As noted, mustard gas contact with skin results in burns, and nerve agent contact with skin can result in death. Therefore, full protective suits such as the US military's 'Mission Oriented Protective Posture' (MOPP) suits, are principally the safest option for CWD (Tucker 2006, pp. 307–308). However, the temperature experienced when wearing a full protective suit for a prolonged period can result in a decrease in the ability of a first responder to conduct tasks and, in some cases, risk making them a heat casualty (Yokota et al. 2014, p. 126). Indeed, this limits the amount of time first responders can reasonably be expected to wear the equipment, as in high temperatures they will need to be regularly rotated in shifts (Yokota et al. 2014, p. 132).

But what about attitudes to counter these psychological and physical challenges? When using past historical experiences of chemical weapons, for example in the case of Iran, a major factor in deaths was the unreadiness of those targeted. Being ready was important to Britain's World War Two civilian gasmask training campaign, in which the slogan was "Hitler will give no warning" (Grayzel 2022, p. 167). To provide counterbalance to the disgust and fear of the effects of chemical weapons, there is a need for good professional ethos. The general military philosophy towards chemical weapons is to have all units ready defend and operate in a contaminated environment that will not affect mission success (Brophy and Fisher 1959, pp. 187–188, Kaszeta 2023, pp. 148–150). As Kaszeta highlights, the use of the military in conjunction with civil authorities is almost always inevitable due to the fact that the military has a massive amount of technical equipment, training, and expertise (2023, pp. 148–150). However, police forces and emergency services do routinely use personal protective equipment designed for CBRN, particularly for use in crowd control situations, raids on suspected drug laboratories, as well as chemically assisted suicides (Yokota et al. 2014, p. 126, pers. comm. 26/07/2023). The challenges that emerge come down to differing levels of experience, terminology, methods, and procedures (Kaszeta 2023, p. 152). This was also raised at the workshop, yet an interesting factor that was also mentioned was the need to maintain institutional normative legacy (pers. comm. 26/07/2023). As such, due to retirement of numerous senior figures, a lot of accumulated institutional knowledge of CWD which is important in maintaining an informal social normative legacy is lost (pers. comm. 26/07/2023).

20.7 Conclusion

Current analysis on chemical weapons utility and defence does not fully consider the social psychological element around chemical weapons ability to cause terror. Although the "agent of terror" (Phillips and Crouch 2023) argument may be simplistic, it should not be dismissed out of hand. As we established, we cannot rationally explain why chemical weapons are agents of terror, and why other weapons are not. If we accept that all weapons are agents of terror, then the label fundamentally loses all meaning. However, chemical weapons are historically associated with socially constructed imaginaries of a weapon uniquely disgusting and as a result stigmatised and fetishized. This in turn makes the weapon particularly difficult to defend against partly due to the deep psychological fear around it. As this chapter has explored, a great deal of the normative attitudes, including "disgust", "unfamiliarity" and fears of "contamination" are reflected within chemical warfare even 30 years on from the CWC.

As this chapter found, close cooperation with academia and military medical services helped improve the aid response of the Iranians, as the example of Professor Mahdi Balali-Mood demonstrates. Although this example may have been an extreme case, this does highlight some of the advantages of knowing background historical knowledge of chemical warfare. Professor Mahdi Balali-Mood effectively modelled the Iranian medical response to Iraq's mustard gas on the British experience during the First World War. In other words, academics should be aware that first responder communities are interested and open to having more historical background to various chemical agents. According to some experts, they are also open to having more critical analysis of past historical usages of chemical warfare and chemical attacks (pers. comm. 26/07/2023). Overall, the Iranian experience with chemical weapons is key in allowing us to understand why first responders still take the threat of chemical warfare seriously despite the low threat of the weapons being employed. As shown, the cost of being unprepared was devastating for the Iranians.

To conclude, we should remain sceptical to the idea that chemical weapons are a weapon of the past and lack utility. Chemical weapons still have utility in that they are perceived to be deadly weapons of mass destruction that cause terror. As such, they are an attractive weapon for states, and particularly for conflicts in which the opposing side lacks advanced medical and defensive capability. Therefore, although it is now thirty years since the establishment of the CWC, the challenges faced from chemical weapons remain today.

Bibliography

Bale M (2006) South Africa's project coast: "death squads", covert state-sponsored poisonings, and the dangers of CBW proliferation. Democr Secur 2(1):27–59

Balmer B (2012) Secrecy and science: a historical sociology of biological and chemical warfare. Routledge, London

BBCNews (2018) Las Vegas shootings: British soldier awarded for bravery. https://www.bbc.co.uk/news/uk-england-nottinghamshire-46334900. Accessed 11 July 2023

Bentley M (2016) Syria and the chemical weapons taboo: exploiting the forbidden. Manchester University Press, Manchester

Bentley M (2023) The biological weapons taboo. Oxford University Press, Oxford

Bentley M (2022) A new model of "Taboo": disgust, stigmatization, and fetishization. Int Stud Rev 24(3)

Brophy LP, Fisher GJ (1959) United States Army in World War II: the technical services; the chemical warfare service. Organizing for War. Office of the Chief of Military History, Department of the Army. Centre of Military History United States Army, Washington DC

Carter GB (2000) Chemical and Biological Defence at Porton Down, 1916 2000. Stationery Office Books; Rev. and updated ed. Edn, London

Carus S (2012) Defining 'weapons of mass destruction. CSWMD Occasional Paper, No. 8, January 2012.

Carus WS (2015) The history of biological weapons use: what we know and what we don't. Health Sec 13(4):219–255

Catignani S, Basham VM (2021) The gendered politics of researching military policy in the age of the 'knowledge economy.' Rev Int Stud 47(2):211–230

CDC (2023) VX: exposure, decontamination, treatment. CDC, chemical emergencies, chemical fact sheets. (Online). https://www.cdc.gov/chemicalemergencies/factsheets/vx.html#:~:text=VX%20is%20an%20oily%20liquid,as%20slowly%20as%20motor%20oil. Accessed 14 Feb 2024

Cross G (2017) Dirty war: rhodesia and chemical biological warfare 1975–1980. Helion and Company, Warwick

Dewey K (2023) Poisonous affairs: Russia's evolving use of poison in covert operations. Nonproliferat Rev 29(4–6):155–176

Dorsey MG (2023) Holding their breath: how the allies confronted the threat of chemical warfare in World War II. Cornell University Press, Ithaca

DSTL (n.d) Dstl's history. About us the defence science and technology laboratory (Dstl) is the science inside UK defence and security. Gov.uk (online). https://www.gov.uk/government/organisations/defence-science-and-technology-laboratory/about. Accessed 15 Feb 2024

Faith TI (2014) Behind the gas mask: the US chemical warfare service in war and peace. University of Illinois Press, Champaign

Ferguson J (2019) You never dreamt of a poysoned bullet': 'Forbidden' ammunition from the 16th century to the present. In: Tucker J, Hacker BC, Vining M (eds) A right to bear arms? The contested role of history in contemporary debates on the second amendment. Smithsonian Institution, Washington

Fitzgerald GJ (2008) Chemical warfare and medical response during World War I. Am J Public Health 98(4):611–625

Gerdes AB, Uhl G, Alpers GW (2009) Spiders are special: fear and disgust evoked by pictures of arthropods. Evol Hum Behav 30(1):66–73

Gould C, Burger M (2002) Secrets and lies: Wouter Basson and South Africa's chemical and biological warfare programme. Struik Publishers, Cape Town

Grayzel SR (2022) The age of the gas mask: how british civilians faced the terrors of total war. Cambridge University Pres, Cambridge

Haber LF (1986) The poisonous cloud: chemical warfare in the First World War. Oxford University Press, Oxford

Hiltermann JR (2007) A poisonous affair: America, Iraq, and the gassing of Halabja. Cambridge University Press, Cambridge

Horne J, Kramer A (1994) German" atrocities" and Franco-German opinion, 1914: the evidence of german soldiers' diaries. J Mod Hist 66(1):1–33

IWM 276 (1917) Gasmasks of the first world war [Allocated Title]. War Office. Imperial War Museums. https://www.iwm.org.uk/collections/item/object/1060022760. Accessed 20 Sep 2023

Jepperson D (1991) Institutions, institutional effects, and institutionalism. In: DiMaggio PJ, Powell WW (Eds), The new institutionalism in organizational analysis. The University of Chicago Press, Chicago, pp 143–164

Jones DE (2007) Poison arrows: North American Indian hunting and warfare. University of Texas Press, Austin

Kaszeta D (2020) Toxic: a history of nerve agents, from Nazi Germany to Putin's Russia. Oxford University Press, Oxford

Kaszeta J (2023) CBRN and hazmat incidents at major public events: planning and response, 2nd edn. John Wiley & Sons, Hoboken

Khateri S, et al (2017) Mental health status following severe sulfur mustard exposure: a long-term study of Iranian war survivors. Asia-Pacific Psychiatry 9(2)

King W (2021) Nerve agents in postwar britain: deterrence, publicity and disarmament, 1945–1976. Springer International Publishing, Cham

Koblentz GD (2013) Regime security: a new theory for understanding the proliferation of chemical and biological weapons. Contemp Sec Policy 34(3):501–525

Mansour Razavi S, et al (2012) A review on delayed toxic effects of sulfur mustard in Iranian veterans. DARU J Pharmaceut Sci 20(1):51

Neer RM (2013) Napalm: an American biography. Harvard University Press, Harvard

New York Times (2023) Gunman in 2017 Las Vegas shooting was angry at casinos, new F.B.I. files show. (Online). https://www.nytimes.com/2023/03/30/us/las-vegas-shooting-gunman.html. Accessed 14 Feb 2024

OPCW (2019) Practical guide for medical management of chemical warfare casualties. opcw.org. (Online). https://www.opcw.org/sites/default/files/documents/2019/05/Full%20version%202019_Medical%20Guide_WEB.pdf. Accessed 09 Mar 2024

Palmer I (2004) The psychological dimension of chemical, biological, radiological and nuclear (CBRN) terrorism. BMJ Milit Health 150(1):3–9

Phillips L, Crouch D (2023) Have chemical weapons been used in Ukraine? The Royal United Services Institute (RUSI). https://www.rusi.org/explore-our-research/publications/commentary/have-chemical-weapons-been-used-ukraine. Accessed 21 Jan 2024

Price RM (1997) The chemical weapons taboo. Cornell University Press, Ithaca

Report of the Iraq Inquiry (2016) Section 4.3 Iraq WMD assessments, October 2002 to March 2003. The National Archive https://webarchive.nationalarchives.gov.uk/ukgwa/20171123122743/http:/www.iraqin.quiry.org.uk/the-report/. Accessed 21 Jan 2024

SAPA (1997) Cops deny poisoning but admit murder in amnesty bid. (Online). https://www.justice.gov.za/trc/media/1997/9709/s970918a.htm. Accessed 11 July 2023

Schmidt U (2015) Secret science: a century of poison warfare and human experiments. Oxford University Press, Oxford

Schmidt U (2017) Preparing for poison warfare: the ethics and politics of britain's chemical weapons program, 1915–1945. In: Friedrich B, et al (Eds.) One hundred years of chemical warfare: research, deployment, consequences. Springer International Publishing, Cham, pp 77–101

Spiers E (2017) The gas war, 1915–1918: if not a war winner, hardly a failure. In: Friedrich B, et al (Eds.) One hundred years of chemical warfare: research, deployment, consequences. Springer International Publishing, Cham, pp 153–168

Spiers EM (2020) Agents of war: a history of chemical and biological weapons. Reaktion books. 2nd edn. London

SPIRI (1971) Volume I, The rise of CB weapons. (Online). https://www.sipri.org/sites/default/files/CBW_VOL1.PDF. Accessed 14 Feb 2024

Staňková H, et al. (2021) The ultimate list of the most frightening and disgusting animals: negative emotions elicited by animals in central European respondents. Animals 11(3):747

Stokes JW, Banderet LE (1997) Psychological aspects of chemical defense and warfare. Mil Psychol 9(4):395–415

Tehran Peace Museum (2015) Saeid Sadeghi. Interview with Mr Saeid Sadeghi at the Tehran Peace Museum. Translator: Elaheh Pooyandeh. https://www.tehranpeacemuseum.org/index.php/en/en-news/1353-saeid-sadeghi-oral-history-en.html. Accessed 20 Sep 2023

Tehran Peace Museum (n.d.) Professor Mahdi Balali-Mood. Terhran Peace museum. Translator: Elaheh Pooyandeh. https://www.tehranpeacemuseum.org/index.php/en/component/content/article/194-english/oral-history-en/1063-professor-mahdi-balali-mood-en.html. Accessed 20 Sep 2023

Thompson P (2017) The chemical subject: phenomenology and German encounters with the gas mask in the World War I. Hist Technol 33(3):249–271

Time (2014) Iran still haunted and influenced by chemical weapons attacks. Time.com. (Online). https://world.time.com/2014/01/20/iran-still-haunted-and-influenced-by-chemical-weapons-attacks/. Accessed 14 Feb 2002

Tucker J (2006) War of nerves: chemical warfare from World War I to Al-Qaeda. Anchor

Van Courtland Moon JE (1984) Chemical weapons and deterrence: the World War II experience. Int Sec 8(4):3–35

Verwimp P (2006) Machetes and firearms: the organization of massacres in Rwanda. J Peace Res 43(1):5–22

Wachtel C (1941) Chemical warfare. Chapman & Hall LTD, London

Wiseman LCDJC (1951) War Office. The second world war 1939–1945 army special weapons and types of warfare. Volume I. gas warfare. The Naval & Military Press Ltd.

Yokota M, Karis AJ, Tharion WJ (2014) Thermal-work strain in law enforcement personnel during chemical, biological, radiological, and nuclear (CBRN) training. Int J Occup Environ Health 20(2):126–133

Youtube@GBNews (2022) Talking pints with Nigel Farage and chemical weapons expert Hamish De Bretton-Gordon. Youtube.com (Online). https://youtu.be/2ASjaqt7_SQ?t=478 (Time stamped: 8:05). Accessed 14 Feb 2024

YouTube@OPCW (2015) Interview with 2015 OPCW-The Hague award winners.Youtube.com. (Online) https://youtu.be/EoXd2EIrvFI?t=349 (Time stapped: 5:50). Accessed 11 July 2023

Lincoln Sheff is a PhD student at the University of Bath under the supervision of Brett Edwards. His current research is working with Brett Edwards to establish a forum to bring stake-holders involved in CBW defense together in an informal setting to work on ways to provide better education.

Part VI
Information Warfare: The Power of Images

Chapter 21
The Photographic Document: Aesthetics and Politics of Militarized Photographs

Katja Schmidt-Mai

Abstract The OPCW establishes in Fact-Finding Missions (FFM), if alerted to the usage of toxic chemicals as weapons, whether such an incident took place and what substances were used. In order to do so, the FFM obtains a variety of information from multiple sources including photographs. This article addresses the role of photography as a tool to record and document subject matters of war and conflict, with a particular focus on the deployment of chemical weapons. Firstly, the origins and abilities of war photographs are summarized, followed by an introduction to photographs as pictorial messengers. This section is contextualized by a short history of war photographs. Secondly, the ethics of war photographs are investigated, leading to the complex issues of the coverage of chemical warfare terrorism and the advent of AI-generated photographs.

Keywords War photography · Chemical warfare · Chemical warfare terrorism · History of photography · Documentary photography · Contextuality of war photographs · Ethics of photographic illustration · AI-generated images

21.1 Introduction

At the time of writing, the internet is awash with photographs of war and conflict given the ongoing Ukraine war as well as the Gaza conflict. Images portraying warfare and conflict play not only a significant role in the history of photography, but also influence people's perception of war and conflict today. Most people's knowledge and visual perception of war derives from photographic images or film stills rather than written accounts or personal experience, therefore photographs of war play a significant role in narrating and documenting these events. There have been numerous

K. Schmidt-Mai (✉)
University of Hamburg, Hamburg, Germany
e-mail: Katja.schmidtmai@uni-hamburg.de

Research Associate, Institute for Liberal Arts and Sciences, Monetastrasse 4, 20146 Hamburg, Germany

publications that address the role of war and conflict photographs and images in terms of representational ethics (Sontag 2003; Campbell 2004; Barthes 1977, 2006), their influence on politics and security policies (Möller 2007; Goalwin 2013; Dodds 2007; Dauphinée 2007), and their strategies of visual rhetoric (Foss 2005; Barthes 1978a, b). The aim of this article is to build upon these publications by addressing the role of photography to record and document subject matters of war and conflict, with a particular focus on photographs depicting the deployment of chemical weapons. The OPCW establishes in Fact-Finding Missions (FFM), if alerted to the usage of toxic chemicals as weapons, whether such an incident took place and what substances were used. Fact-Finding Missions obtain a variety of information from multiple sources including photographs. The art historian Ernst Gombrich once advised: "Any picture, by its very nature, remains an appeal to the visual imagination; it must be supplemented in order to be understood" (Gombrich 2002, p. 204). Firstly, therefore, this chapter will summarize the origins, categories and abilities of war photographs, followed by an introduction to photographs as key pictorial messengers and the idea and usage of documentary photographs. This section is contextualized by a short history of war photographs. This is followed by an investigation of the ethics of war photographs, which leads to the complex issues of the coverage of chemical warfare terrorism via CTTV and private mobile phone footage. Finally, the advent of AI-generated photographs is discussed, along with what we can expect from the photographic document in the age of AI.

Alexander Gardner, The Dead of Antietam, 1862

Already in 1862, only 23 years after photography was invented,[1] the Scottish photographer Andrew Gardner (1821–1882)[2] who immigrated to the U.S. in 1856, took one of the first war photographs: a close-up displaying dead soldiers during the American Civil War. Photographs of war and conflict often portray subject matter of human tragedy, scenes of victory and environmental destruction. Well-known war photographs often depict the decisive moment of human pain or suffering, for example the image of a nine-year-old Jewish boy with his hands up in the air in the Warsaw Ghetto in 1943 probably taken by SS-Hauptsturmführer Franz Konrad (1906–1952) or the photograph by Nick Ut (*1951) capturing the moment of children fleeing from a napalm attack during the Vietnam War.[3] Yet other war photographs have become icons of victory, depicting the triumphant moment in which a flag is raised, as seen in Joe Rosenthal's (1911–2006) photograph "Raising the Flag on Iwo Jima" (1945)[4] or Yevgeny Khaldei's (1917–1997) shoot "Raising a Flag over the Reichstag" (1945).

War photography plays a pivotal role in depicting the impact of armed conflicts and can visualize their consequences. Before the invention of photography, written accounts, drawings and paintings would provide war coverage, but none of these possessed the immediacy to document the events like a photograph. Photographs are able to capture an apparent reality of conflict and provide a visual record which might further our understanding of the consequences of violence and disruption. War photographs can generate awareness of the human cost and consequences of armed conflict by highlighting the suffering of civilians and military personnel. They also possess the ability to raise awareness of humanitarian crises and therefore can sway public opinion and shape public perception. War photographs can motivate the international community to take action to mitigate suffering and negotiate peace. In the aftermath of war and conflict some photographs might become "icons",[5] which can make them useful to advocate for peace, human rights or humanitarian

---

[1] If we consider the daguerreotype by Louis-Jacques-Mandé Daguerre (1787–1851), which was presented to the wider public in Paris in 1839, as the first proper photographic process.

[2] Alexander Gardner, Confederate dead at Antietam north of the Dunker Church on the west side of the Hagerstown Pike, September 1862, Wet collodion negatives, the United States Library of Congress, Prints and Photographs division, Digital ID cwpb.01097. Photograph of the field at Antietam, American Civil War. Confederate dead by a fence at the Hagerstown Turnpike, looking north; the Turnpike is to the right of the fence, the dirt lane on the left leads to the farm of David Miller (September 1862). All rights reserved © the United States Library of Congress, Prints & Photographs Division, Civil War Photographs, reproduction number cwpb.01079.

[3] Nick Ut, The Terror of War June 8, 1972, Gelatin Silver Print, Associated Press, National Gallery of Art, Accession Number 2006.133.131.

[4] Rosenthal's iconic photograph was used by the sculptor Felix de Weldon (1907–2003) and the architect Horace W. Peaslee (1884–1959) as a model for the U.S. Marine Corps War Memorial (1954), which is also known as "Iwo Jima Memorial" at Arlington, Virginia.

[5] In order for a photograph to become "iconic", it needs to possess narrative properties that transcend the visual surface and might establish a personal connection with the viewer in terms of emotion and personal pictorial memory. These photographs have the ability long after the moment they were shot to affect, inspire and evoke a reaction by the viewer. However, they have often left their original context behind.

intervention, and yet other photographs will provide testimony that might serve as evidence in legal proceedings to hold perpetrators of war crimes and human rights abuses accountable.

Photographs are often interpreted as spontaneous and authentic representations of real events and real human experience that played out in front of a camera lens. Photography certainly has the ability to represent reality, but it does not represent reality in its entirety. Photographic content is shaped by humans using different techniques, compositions and perspectives, thus changing or adapting their narrative. In addition, selecting a particular photograph for publishing can change and influence the reading of its pictorial content. Furthermore, the presentation context might have a bearing on how it is perceived and interpreted by the viewer. Numerous options of photo manipulation, especially in the digital age, also influence the way how a photographic message is conveyed. Therefore, it is essential to analyze photographs and verify their sources to obtain a comprehensive knowledge of their creation.

Most photographs depicting the deployment of chemical weapons cannot be illustrations of the actual incident. The representation of substances that are lethal to human beings produces certain types of images. These images can be roughly put in to three categories: they either represent troops or military personal in hazmat gear training with chemical weapons for war deployment, depict distant aerial views of the actual usage of chemical weapons, or they show the graphic aftermaths and consequences of their usage on humans and their environment. With the world's first multilateral disarmament agreement to provide for the elimination of chemical weapons in 1997, photographs of the usage of such weapons often provides evidence of the violation of the agreed ban. These photographs enter the public domain either through publication by government officials or journalists (two sources which can be verified), or they are taken by eyewitnesses (a source that is much more difficult to verify).

Together with information stemming from official government and intelligence sources, these photographs can galvanize an international response, prompting public condemnation, the imposition of government sanctions, or even trigger military intervention that can quickly have serious consequences. For example, on 7th April 2018 a Syrian military offensive took place in Douma on the outskirts of Damascus. Troops of President Bashar al-Assad gained the upper hand in the area with the help of chemical weapons. International news outlets quickly broadcast film footage and photographs that depicted the incident, focusing strongly on affected children and their hospital treatments. The pictorial footage referred in its comments or subtitles to the symptoms of the victims who suffered from breathing problems, convulsions, and foaming at the mouth. The footage, in combination with information obtained by intelligence sources, prompted the United States administration and its allies to issue a statement that the Syrian regime had used some kind of chemical weapon, possibly chlorine.[6] Seven days later, on 14th April 2018 the United States, France and

---

[6] https://news.sky.com/story/douma-chemical-attack-that-killed-43-in-syria-was-likely-work-of-regimes-air-force-international-watchdog-finds-12796578, last accessed 2024/4/28; https://www.cbc.ca/news/world/chemical-weapons-inspectors-syria-reach-douma-1.4630142, last accessed

Britain launched missile strikes on Syrian military installations.[7] However, in order to confirm the usage of chemical agents and to establish the facts of the incident, it needed experts to visit the site in a timely manner to identify remaining traces of chemicals. Seven months later an OPCW team visited the site.[8] Three months after the attack, the research agency Forensic Architecture, based at Goldsmiths, University of London, also published their findings on the Douma attack via their website, including 3-d modelling, video footage, and reconstructed photographic footage.[9] On 27th January 2023, the news outlet Reuters reported that an investigation by the global chemical watchdog found that the Syrian military had dropped gas cylinders on civilian infrastructure.[10] On the same day the OPCW Investigation and its Identification Team (IIT) released a third report "that there are reasonable grounds to believe that the Syrian Arab Air Forces were the perpetrators of the chemical weapons attack on 7 April 2018 in Douma, Syrian Arab Republic".[11]

This example shows that photographs can be a starting point or supplementary source for an investigation, but they cannot be the sole evidence. In addition, it highlights the importance of the visual narrative of a press photograph, which can shape public opinion and might even be able to influence the perception of events, which in some cases can be a helpful outcome for a government.[12] The digital age inextricably links photography and social media with visual narratives, being the dominant language of the internet. Platforms like Instagram, Pinterest or Telegram have become dynamic hubs for visual storytelling, captivating audiences and leaving lasting impressions, far apart from any code of conduct and ethics for photography.

2024/4/28;   https://abcnews.go.com/International/site-suspected-chemical-attack-syria-10-days-reporters/story?id=54526689, last accessed 2024/4/28.

[7] https://www.nbcnews.com/news/world/trump-announces-strikes-syria-following-suspected-chemical-weapons-attack-assad-n865966, last accessed 2024/4/28.

[8] In November 2018, a team was established by member states at the Hague-based OPCW to investigate the chemical attacks in Syria.

[9] https://forensic-architecture.org/investigation/chemical-attacks-in-douma, last accessed 2024/4/28.

[10] See   Reuters.com: https://www.reuters.com/world/middle-east/chemical-weapons-watchdog-blames-syrian-air-force-douma-attack-2023-01-27/, last accessed 2024/1/12; https://www.reuters.com/world/middle-east/chemical-weapons-watchdog-blames-syrian-air-force-douma-attack-2023-01-27/ last accessed 2024/1/12; https://www.opcw.org/media-centre/news/2023/01/opcw-releases-third-report-investigation-and-identification-team, last accessed 2024/5/2.

[11] https://www.opcw.org/media-centre/news/2023/01/opcw-releases-third-report-investigation-and-identification-team, last accessed 2024/5/2.

[12] See for example the article by Griffin analyzing the coverage of the Vietnam War and its influence on print and television photojournalism (Griffin 2010).

21.2 Photographs as Key Pictorial Messengers

Photographs are outstanding pictorial messengers. A team of neuroscientists from the Massachusetts Institute of Technology has found that the human brain can process images in the rapid speed of 13 ms, with participants asked to identify a specified subject matter by using keywords for example "picnic" or "smiling couple". They were then exposed to what is called "rapid serial visual presentation (RSVP)", presenting "a series of six or twelve pictures at between 13 and 80 ms per picture" (Potter et al 2014). The moment the participants identified the sought image, a visual reading took place processing information regarding composition (shape, color, contrast). The study showed that the human brain is able to correctly identify subject matter in as little as 13 ms; after that, humans will guess what they might have seen and try to quickly connect the visual information in front of them to information already stored in their memory (Potter et al 2014).

In the humanities, the French literary theorist, philosopher, and semiotician Roland Barthes (1915–1980) also tried to offer an explanation how humans read an image by studying the semiotics of representations, focusing primarily on Western cultural context (Caves 2004; Crow 2010). Semiotics investigate modes of representation with a syntax (structure), with semantics (meaning) and with pragmatics (usage). Photographs, according to Barthes, therefore, have a grammatical structure and are polysemic. Barthes also specifies how the viewer deciphers "meaning" in an image: "the literal meaning" of an image (denotation) is what the viewer can see in the image; "the symbolic meaning" of the image (connotation) can be based on the viewer's cultural and societal context. Barthes goes on to distinguish between "the non-coded iconic message" of an image, which literally represents what the image depicts, and which is accessible without any learned cultural and societal knowledge, for example an apple can be identified as just a piece of fruit. "The coded iconic message" of an image, on the other hand, needs the application of any learned cultural and societal knowledge the viewer has acquired in his lifetime, in case of the apple, it could be read as healthy and fresh.

Photographs are exceptional pictorial messengers given that they are perceived by the viewer as records of an encounter by a photographer, which seem to enable the viewer to experience the depicted scene and acquire knowledge about it. Photographs are thought to present an indexical link, a term established within the field of semiotics by Charles Sanders Peirce (1839–1914).[13] The indexical link refers to a photograph as a preserved trace or imprint of the photographed object (Peirce 1865/1982). The photographic process therefor establishes an invisible bond between the photographed object with the photographic print, which leads viewers to believe in the existence of the representation (Barthes 1985; Well 1997).

---

[13] The term "indexicality" refers originally to a concept in semiotics which was utilized for the visual arts, including photography.

The idea and usage of documentary photography since the mid-twentieth century is based on this idea that a photograph is able to record and depict reality. Documentary photography emerged around the 1920s and 1930s parallel to the appearance of the popular press, including photo magazines like Life or Picture Post (Williams 1992). Here photographs were strongly tied to terms of objectivity and immediacy. Photographs promised to provide viewers with direct access to truthful depictions, which were achieved by certain aesthetic and compositional characteristics: no cropping, no retouching, no posing, no staging, no additional lighting. As a fine print it should be added that none of these characteristics can be verified by the viewer in the moment of viewing the image.

Why is a photograph seen as a documentary tool? A document, by definition, is a recorded or written representation of information. It can take various forms, including written texts, images or a combination of both (i.e., journalistic photographs). Documents serve to convey, store or communicate information and they can range from simple notes to complex reports, legal agreement or official records. The definition of what a document is has widened with technological progress, and now also includes digital files, images and other electronic formats. A photograph might be treated as a document if it serves as evidence or contains important information. War photographs in themselves are not legal documents. Legal documents typically involve written or recorded information that hold legal significance, such as contracts, treaties or official records. However, war photographs can be used as evidence in legal contexts, especially in documenting events, war crimes, or as part of investigative processes. These photographs may be considered important visual records, aiding in the understanding and documentation of historical events or legal cases, and while not legal documents per se they can play a crucial role in providing evidence of context. Context is another important factor in decoding a photograph. Photographs are too easily seen as transparent and unmediated mechanical records of reality. The meaning of photograph rests to a large extend on the uses to which it is put. Its meaning can be altered by generally adjusting the context in which it is shown. Barrett distinguishes between internal, original and external context (Barrett 1985). Internal context refers to the visible surface of the photograph; original context refers to invisible knowledge of when, where and why the photograph was made; external context refers to the photograph's presentational environment and the perception (Barrett 1985).

To explain how a photograph can be a pictorial messenger, let me start off with a playful example to explain the concept of message, source and time segment, which can be used to analyze and decipher a photograph. On 21st April 1934, the British newspaper Daily Mail published an enlarged photograph of the Loch Ness Monster on its front page, the so-called surgeon's photograph, taken by the London physician Robert Kenneth Wilson (1899–1969).

Cropped photograph of Nessie depicted on front page of the Daily Mail, April 21st, 1934[14]

The source of a photographs i.e. the photographer has, in case his identity is known, the strongest influence whether we perceive an image as an objective document. The art historian John Tagg (*1949) asked in his book "The Burden of Representation", "under what conditions would a photograph of the Loch Ness Monster (of which there are many) be acceptable?" (Tagg 1988) For Tagg the source of a photograph changes our perception of the content; in his view "it matters whether it [the Loch Ness photograph] was made by a private enthusiast or by the military" (Tagg 1988). To decipher a photograph, it needs to be seen as one section of different time segments. Similar to film stills, there is a moment before and after the shot was taken, which also need to be considered. In addition, there is also the act of preparation for taking the photograph as well as the act of publishing the photograph. In case information about these steps are at hand they prove to be helpful for an extensive analysis. In the case of the Nessie photograph, it is known what happened before and after the shooting: The "monster" was produced out of plastic, clockwork, a tinplate and a toy submarine and had to be launched into the water. The original photograph was cropped before it entered the public domain through the Daily Mail in 1934, giving the impression through the cropping process that the photographer was able to get up close to Nessie. It took sixty years, until 1994, to solve the riddle of Nessie's photographic emergence.[15]

---

[14] So-called "surgeon's photograph" of 1934. The photograph is cropped, to give the appearance of a close-up, of the Loch Ness monster swimming, which is made out of plastic, a clockwork, a tinplate and a toy submarine. The photographer is disputed, but the photograph was originally published under the claimed authorship of the physician Robert Kenneth Wilson, who died in 1969. Later Christian Spurling, the stepson of Wetherell, who created the hoax monster, suggested that Marmaduke Wetherell, who died in 1939, took the photograph and used Wilsons name. Public Domain.

[15] The production process of the Nessie photograph is discussed in detail in Martin 1999.

21.3   A Short History of Photographs Portraying War and Conflict

To provide a better understanding about photographs portraying war and conflict, I would like to briefly summarize the developments of this photographic genre from the First World War until the 1990s, the arrival of the digital turn in photography. Photographs of war and conflict can draw public attention, and by doing so are able to influence public opinion, thereby potentially reinforcing or eroding public support for the depicted scene. The photographic surface can be a representation of the traditional, cultural, political, religious, and ethnical upbringing of the person behind the camera. For example, Morley's photograph[16] summarized the British Government's "Keep Calm and Carry On" motto during the Second World War (Slocombe 2010) and emphasized the perseverance of the British working class, who continued to do their jobs despite the war. The photograph is staged. It was taken the morning after a bomb raid of London by the German Luftwaffe. In terms what happened before and after the shooting it is known that on that morning, Morley encountered a milkman, borrowed his jacket and crate, and his assistant dressed up and posed as a milkman. For the British government, the Morley photograph was a thoroughly successful illustration of the officially stated slogan to persist in a calm manner. In times of conflict, crisis and war, governments work to control, limit, or even delay photographic production and circulation. This includes occasionally shielding the public from particular coverage, pushing the distribution of preferred images to get their message across and preferring photographs which seem to take the notion of objectivity, documentary and balance seriously. Given that war is "a high-stake enterprise", as Michael Griffin has put it, "public opinion and public perception are never left to chance" (Griffin 2010).

During the First World War, photographic compositions and aesthetics were shaped by the use of available camera techniques: view cameras were bulky, heavy and immobile. Exposure times were too long to take precise photographs in combat. Small format cameras suffered from the lack of telephoto lenses and needed perfect light conditions to create acceptable photographs. Nevertheless, these photographs made it possible for people to keep up to date with the unfolding events of war. In 1916, Geoffrey Malins and John McDowell even shot the British documentary and propaganda war film "The Battle of the Somme", which included for the first-time sequences depicting dead soldiers. The film was shot in the trenches and on the battle sites, only a few scenes were staged. At the time, an advertisement in the Yorkshire Evening Post for the film read: "If the exhibition of this picture all over the world does not end war, god help civilization!".[17]

---

[16] A milkman delivering milk in a street, devastated in a German bombing raid, in the Holborn area of London, 9th October 1940. Firemen are extinguishing the ruins behind him. For reference see photograph by Fred Morley/Fox Photos/Hulton Archive/Getty Images.

[17] Yorkshire Evening Post advertisement for "The Battle of the Somme" (1916 British film), 28th August 1916, in Cecil and Liddle (1996).

Left: C type aerial reconnaissance camera fixed to the fuselage of an aircraft, 1916[18] Right: The Pigeon Spy and his work in war published in Popular Science monthly, 1872[19]

Another way to provide war coverage for the public were aerial photographs. There were two options to obtain photographs of the battlefields, either with a reconnaissance camera attached to the fuselage or by pigeons that had a camera attached to their bodies with a small leather harness; either way these photographic processes did not involve much human intervention. The button of the reconnaissance camera, to expose the negative, was pressed by the pilot along the way while flying the plane. In case of the pigeon, photographs were shoot with the help of a timer and the only human intervention was to collect the film after the bird had landed.

Aerial photographs also made it possible in 1916 to photograph the opening of German gas containers from a safe distance on the Eastern Front, producing an early example of a chemical warfare photograph during combat.

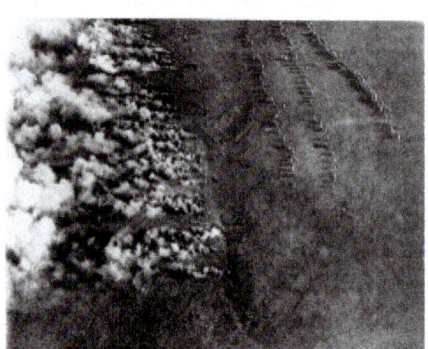

---

[18] A sergeant of the Royal Flying Corps demonstrates a C type aerial reconnaissance camera fixed to the fuselage of a BE2c aircraft, 1916, All rights reserved © Imperial War Museums, IWM (Q 33850).

[19] During the First World War, the German army used pigeons with attached cameras and inbuilt timers for aerial photography. All rights reserved © Internet Archive, digital upload: D. Appleton, Popular science monthly, 1872; see also: https://archive.org/details/popularsciencemo88newyuoft/page/n3/mode/2up?q=pigeon.

Left: First World War German gas attack on the eastern front[20]Right: British 55th Division troops temporarily blinded by a gas attack[21]

Other chemical warfare photographs depicted mainly static and arranged subject matters given that sometimes single plated cameras with longer exposure times were used. Therefore, certain pictorial themes were repeated, for example soldiers posing in battlefield, wounded soldiers in military camps, dead bodies mainly of hostile armies, guns and canons, vehicles, and from 1915/16 also gas masks and tanks.

To shoot these subject matters, access and permission by the army was required, and consequently official interpretations were depicted. Censorship of photographs was implemented by cropping images, leaving details out of the image that would help to identify time, date, location or the scale of destruction. To provide more vivid combat images, some photographers used composite photographs combining two negatives in the darkroom into a single print.[22] The Australian photographer Frank Hurley (1885–1962) argued that as war was conducted on a vast scale, it was impossible to capture the essence of war in a single negative. In 1919, he wrote in The Australasian Photo-Review: "None but those who have endeavored can realize the insurmountable difficulties of portraying a modern battle by the camera. To include the event on a single negative, I have tried and tried, but the results are hopeless. […] Now, if negatives are taken of all the separate incidents in the action and combined, some idea may then be gained of what a modern battle looks like" (Hurley 2002).

During the Second World War, photographic compositions became more dynamic, given that small format cameras and accompanying equipment were smaller, more portable and lightweight, which allowed photographers to get closer to their subjects. Cameras did not necessarily depend on tripods and were equipped with wide-angle lenses which aided the appearance of close ups and photographs capturing motion. In addition, photographers were either embedded with troops, or military personnel included trained photographers and motion picture cameramen. The British Ministry of Information commissioned well-established photographers like Cecil Beaton (1904–1980) and Bill Brandt (1904–1983) to photograph on the home front, in bomb shelters and abroad (Haworth-Booth and Mellor 1985). In 1941, the British forces formed the Army Film and Photographic Unit (AFPU), which consisted of trained photographers and cameramen (Chapman 2000).[23] Not only did photographic technology progress during the second half of the twentieth century, but also the usage of

---

[20] Aerial view of First World War German gas attack on the eastern front by a Russian airman titled: "German Frightfulness from the Air", 1916, All rights reserved © Bundesarchiv, Bild 183-F0313-0208-007/CC-BY-SA 3.0.

[21] Thomas Keith Aitken (Second Lieutenant) photographed the British 55th Division troops temporarily blinded by a tear gas attack at an Advanced Dressing Station at Béthune, France during the Battle of Estaires on April 10, 1918, as part of the German offensive in Flanders. All rights reserved © Imperial War Museum, Part of Ministry of Information First World War Official Collection, Signature: Q 11586.

[22] Composite photographs were invented by Sir Francis Galton (1822–1911) around 1880. See Galton (1878).

[23] Chapman (2000).

photographs changed, given that picture magazines like Life in the U.S and Picture Post in Great Britain gained popularity during the war years, which were in constant need of photographic coverage.

21.4 Ethics of War Photographs

After the invention of photography in 1839, one could argue that the abilities of the photographic medium to document scenes of war and conflict were still limited, and the inevitable question of what is permitted to document would not arise with the same intensity that is does nowadays. Practitioners were more concerned with what the technical conditions of the photographic medium would allow them to document. Daguerreotypes and calotypes (also known as talbotype),[24] the most common photographic procedures during the nineteenth century, were labor-intensive and in need of skillful preparation and chemical knowledge before the actual shooting could take place. If we look, for example, at the coverage of the Crimean war by Roger Fenton (1819–1888), who was sent by the British government in 1855 to document the war efforts, his photographs exclude the dead, the maimed or sick soldiers and troops. Instead, he often depicted static scenes behind the front lines. Fenton's use of a wet-collodion process meant he had to be mindful of the involved photographic procedure to bring his pictures to light (Baldwin 1991). The wet-collodion process required the material quickly to be coated, sensitized, exposed, and developed in around fifteen minutes. This was done in his portable darkroom, located in his horse-drawn carriage (Werner-Marien 2002). This example illustrates that the technical standards of the time dictated the subject matter, which often might appear to capture war scenes in a more ethical way, given that scenes were static, sometimes staged and often represented the aftermath. This mode of war photography would continue almost through the First World War. All of this would change with the arrival of the first small format camera using a 35-mm film with the option to expose thirty-six images.[25] These lightweight and handheld cameras allowed photographers to point and shoot in a quick manner. In terms of war photography, pictures could now capture combat in action and their severe and horrendous consequences. Quickly, the question arose what can be photographed, and with it came a debate about the ethics of seeing, most famously championed by Susan Sontag in her book Regarding the Pain of Others, by examining photography's ability to represent war and violence from the mid-nineteenth century to more recent conflict zones in Rwanda, Israel and Palestine

---

[24] "The daguerreotype consists of cleaned silver-plated copper plate with a polished surface. The plate was sensitized over iodine. After exposure to light, the plate is developed over hot mercury. To fix the photograph, the plate is dipped in sodium thiosulfate or salt and toned with gold chloride. The calotype consists of a sheet of paper, which is coated with silver chloride. The prepared paper was exposed to light in a camera obscura. Those areas hit by light became dark in tone, yielding a negative image." Baldwin (1991), pp. 15–16, 35.

[25] The first commercially available small format camera was a Leica 35mm camera invented by Oscar Barnack and available to the public from 1925.

(Sontag 2003). Sontag raises questions of responsibility and authority: why should we look at photographs depicting the horrors of war if we are not able to do anything about it? What is the purpose of these images? Looking at photographs of war and conflict, becoming a distanced witness of atrocities is a double-edged sword: on the one hand it can generate empathy, but on the other, it can also result in voyeurism, including feelings of guilt and shame. However, these photographs can also act as evidence and, as Sontag has pointed out, they are traces of the world, which answer the human desire to confirm reality via a pictorial representation (Sontag 1977).

In the closing days of the Iran–Iraq War, on 16th March 1988, a chemical attack took place in Halabja, Kurdistan, Iraq. A United Nations investigation concluded that the attack was carried out with mustard gas and other unknown nerve agents (Hilterman 2007, p. 195).

The website of the U.S. State Department lists 5000 civilians as immediate fatalities of the attack and 10,000 civilians who suffered from immediate medical effects including skin burns, impaired vision, breathing difficulties, respiratory shutdown, vomiting, diarrhea. In the years after the attack, the numbers of casualties as well as victims with long term illnesses rose further, including chronic illnesses and birth defects.[26] Western press coverage of the attack was mainly provided by Agence France Presse in collaboration with the Iranian News Agency. Several photographs portrayed the aftermaths of the attack, with wide angle shoots of dead bodies without visible facial features.

The photographic coverage by AFP abided by the illustration standards of the agency and therefore did not include close ups of injuries, wounds, dismembered limbs or corpses.[27] AFP's code of editorial standards and best practices are voluntary guidelines, which try to guard and uphold the documentary value on any pictorial material obtained. A particular focus lies on any "staging and re-enactments" which have to be avoided under any circumstances in order not to undermine the trustworthiness of photographs. One could assume that an ethics code for media illustration would have been a necessity in the digital era only, but a voluntary code was already created in 1947 by the Hutchins Commission, named after Robert Maynard Hutchins, the president of the University of Chicago.[28] The commission was established after the publisher of Time, Life, Fortune and Sports Illustrated, Henry Luce (1898–1967), contacted Hutchins asking him to form a commission, which would look into the role of the print media and press in a modern democracy (Pickard 2015, p. 144; Bates 2020, pp. 10, 23–24). A main concern of Luce and Hutchinson was a potential loss of trust in press coverage by the general public and its impact on democracy. When the findings of the commission were published it was hoped that the press was able "to meeting society's needs."[29] In addition, the commission declared: "The time has

---

[26] U.S. Department of State, Web archive, Saddam's Chemical Weapons Campaign: Halabja, 16th March, 1988 in: https://2001-2009.state.gov/r/pa/ei/rls/18714.htm, last accessed 2024/04/10.

[27] Source: AFP editorial standards and best practices, in: https://www.afp.com/sites/default/files/afp_ethic_septembre_2023_0.pdf, last accessed 2024/02/12.

[28] The Hutchins Commission's official name was the Commission of Freedom of the Press.

[29] Blanchard (1977).

come for the press to assume a new public responsibility" in order "to equip the people for the complex world of the late 1940s."[30] In 1947, the Editor & Publisher Magazine[31] commented that the commission report had identified that press freedom was in danger, given that press and media outlets were owned only by a few people, thereby endangering access of the general public to broad spectrum of information.[32] It is astounding how much the commission's findings resonate with the role of today's media and social platforms.

In terms of the Halabja attack, other news outlets like Hawar News Agency (ANHA)[33] and Firat News Agency (ANF), a Kurdish outlet, have released images with more graphic details of the victims, including close ups of children and babies with *en face* shots.[34] Firat News Agency also released a detailed photograph of a dead man still holding an infant, both lying close to a doorstep. This particular photograph has been duplicated by multiple news outlets and print media. Nowadays the photograph is still used on social media platforms. A sculpture for the Halabja memorial site, which opened in 2003 located in the Kurdistan Region of Iraq, was modelled after the photograph and the museum of the memorial site features a section in which life size figures of the victims are on display, including the dead man with a baby.

The photograph of the man and infant has become an icon, representing the essence of the pain and loss of the Halabja attack. As the saying goes: "A picture is worth a thousand words", and this photograph compresses the horror of the attack by depicting an innocent infant in the arms of a man who tried to flee the attack, maybe by trying to get indoors. One can hardly imagine that a viewer of this image can escape the subject matter, given that such a visual rhetoric triggers multiple emotions in humans regardless of culture, class or religion. Most photographs depicting chemical warfare are photographed from a distance; they might be able to cause a strong reaction in the viewer, but one could argue that this is not because of the depicted attack but due to triggering of viewers' feelings and memories that are associated with the consequences of such an attack, which might include loss, death and pain. As the photographer Stephen Shore has argued, a photograph has a depictive level for which the photographer has imposed his order (by choosing a vantage point, a frame, a moment of exposure, a plane of focus) and a photograph inheres a mental level which "elaborates, refines and embellishes our perception of the depictive level" (Shore 2007, pp. 37, 97).

---

[30] Blanchard (1977).

[31] Blanchard (1947).

[32] ERIC Document from the database of Education Resources Information Center by the United States Department of Education. The Hutchins Commission, The Press and the Responsibility Concept. The Association for Education in Journalism (1977), in: https://files.eric.ed.gov/fulltext/ED139017.pdf, last accessed 2023/12/10.

[33] ANHA is associated with the Syrian Democratic Forces.

[34] Hawar News Agency (ANHA) (2023).

Left: Father holding an infant, photographed and filmed by journalists in Halabja[35] Right: Memorial of Halabja located in the memorial garden of the Organization for the Prohibition of Chemical Weapons headquarters in The Hague[36]

The Halabja photograph became iconic because it was able to transport the pain and horror of the attack by reinforcing the emotions of this particular moment across potential cultural, social and political boarders. Photographic icons stand out from the flood of images and take a representative position for the actual event as a whole, and possess the ability to touch the collective pictorial memory regardless of culture, class or religion. An icon goes beyond a mere visual representation; it becomes the initiator of a narrative that evokes memories in the viewer and might define cultural and historical events.

Nowadays, the Halabja photograph is still widely circulated on the internet, social media outlets and websites as a defining parameter for the attack. Over the course of time this photograph acquired additional meaning and narrative elements mainly by using it in different contexts. The political impact of such a photograph stems from its symbolic character and its degree of usefulness and importance for political objectives. It is important to note that images are per se polysemic; their interpretation always depending on the perception of the viewer looking at them.

21.5 The Shifting Nature of Pictorial Coverage of Recent Chemical Warfare Terrorism

In recent years we have dealt with a different type of photographic documentation: photographs and film stills of chemical warfare attacks originating from CTTV footage and film footage from private mobile phones. While a photograph is a segment

---

[35] The Father, Omer Khawar, holding his infant filmed by journalists in the Kurdish village of Halabja on 20th March 1988, Source AFP/ IRNA, ARP2179338 Courtesy of Getty Images, https://www.gettyimages.de/detail/nachrichtenfoto/photograph-and-cameraman-take-pictures-on-march-20-nachrichtenfoto/154211839?adppopup=true, All rights reserved © Getty Images.

[36] Memorial of Halabja located in OPCW's memorial garden in The Hague, Courtesy of Organization for the Prohibition of Chemical Weapons, All rights reserved © OPCW 2024.

of reality (a moment in time that is photographically frozen, which needs to be contextualized in terms of message, source and time segment), film footage from private mobile phones is more difficult to contextualize, and suddenly enables us to see chemical attacks unfold and even witness their terrible consequences.

Previously, a photograph was the result of a picture taken by a photographer of an object in front of the camera; nowadays with mobile phone footage, these clearly defined roles have grown hazy. The person who produces the images is often part of the portrayed scene, and the intention to film or photograph is different from a professional photographer. This user-generated content might enter the public domain via social media platforms without any contextualizing captions or fact checks.

In the case of footage and film stills of CTTV, we are looking at images made for surveillance purposes by security firms, police forces or intelligence services. These images were not intended for public dissemination. CTTV can record continuously or if motion is detected, often filming a sequence of a time set. For press coverage, often only single film stills from different cameras will be published, which need to be accompanied by written (print press) or spoken (TV coverage) information. The assassination of Kim Jong Nam, the older half-brother of the leader of North Korea Kim Jong Un is a typical example for such a practice. On 13th February 2017 at about 9am Kim Jong Nam was attacked at the Kuala Lumpur International Airport in Malaysia by two women with a VX nerve agent near an airport self-check-in in the departure hall at Kuala Lumpur International Airport while waiting for his flight to Macau.[37]

Kuala Lumpur International Airport CTTV[38]

---

[37] McCurry (2023); BBC News (2017).

The CTTV footage obtained from the airport security cameras illustrated television news either with film footage or, in the case of print news, offered numerous film stills. Different CTTV cameras captured the attack from different angles bit by bit: firstly, Kim Jong-nam can be seen carrying a backpack and looking at the flight information screen at the airport. Secondly, a woman walks towards him and then turns away. Thirdly, he can be seen talking to airport security officials, pointing with his hands towards his face seemingly trying to explain what had happened. He is then led away by two men. The last film still depicts a medical center at the airport.[39] Rolling coverage of the incident proves to be inconclusive for the spectator; therefore most editors directed the viewers' eyes by encircling or highlighting single events that in the end led to Kim Jong-nam's death. Above it, word captions accompanied rolling coverage as well as film stills to explain the unfolding plot.[40] Using the film stills in combination with text bears similarities to the narrative style of a picture story: in chronological order the viewer/reader is led through the unfolding story, while a picture editor has carefully chosen the right illustration to narrate the introduction, main part and end of the story. Here, viewers might get the impression that they have witnessed the complete incident, leaving aside the fact that there are numerous stills in between the chosen ones. Besides, the spectator depends strongly on the word captions and visual directions, otherwise the footage would not make sense at all.

Almost a year later the public witnessed another chemical attack, this time in the United Kingdom. On 4th March 2018, Sergei Skripal, a former Russian officer and his daughter, Yulia Skripal, were poisoned by Novichok, a nerve agent, in the city of Salisbury, England.[41]

Left: Boshirov and Petrov at the Salisbury train station[42]Right: Petrov and Boshirov in London[43]

---

[39] Wingfield-Hayes (2017).

[40] For example, the Wall Street Journal used the CTTV Airport Film footage on its online outlet accompanied by word captions. https://www.wsj.com/video/the-moment-kim-jong-nam-was-att acked-cctv-footage/D4AAC6B7-60BE-4457-8B20-093F3F063197, last accessed 2023/10/21.

[41] Harding (2020); Salisbury Novichok poisoning (2018).

[42] All rights reserved © Public Domain, CTTV Image, London Metropolitan Police.

[43] All rights reserved © Public Domain, Scotland Yard.

According to British authorities, two men using the pseudonyms Alexander Petrov and Ruslan Boshirov traveled from a hotel in East London to Salisbury in Wiltshire.[44] They sprayed Novichok with a perfume bottle on the door handle of the Skripal's' front door. This caused the poising of Sergei and Yulia Skripal who survived the attack. The attack also caused the death of a member of the public, Dawn Sturgess, who sprayed herself with the perfume bottle together with her partner Charlie Rowley, who was poisoned and survived. In addition, the police officer Nick Bailey was poisoned while investigating the Skripal's house, despite wearing a hazmat suit.[45]

CTTV footage for media purposes included stills from cameras in Gatwick (the arrival airport of the suspects), images taken in Waterloo Station (from which the suspects boarded a train to Salisbury), and images taken at Heathrow Airport (the departure airport of the suspects). Other photographic coverage offered depictions after the attack took place. British media outlets circulated photographs of the cordoned off area of the incident and the bench on which the Skripals were found, as well as police photographs of the counterfeit perfume box and bottle that contained the nerve agent.

Again, most of these photographs and film stills needed word captions and were only used for illustration purposes of the attack, especially given that they stem from numerous CTTV cameras operating in different locations of London.

On 20th August 2020, the now deceased anti-corruption activist and Russian opposition figure Alexei Navalny was poisoned with a Novichok nerve agent. The poisonous attack was represented through film stills of surveillance videos, private mobile phone photographs and film stills, and social media reports. During a flight from Tomsk to Moscow Navalny fell ill. After an emergency landing in Omsk, he was taken to a hospital and two days later he was flown to the Charité hospital in Berlin, Germany. The German Government later confirmed that Navalny was poisoned by Novichok.

The Washington Post, for example, covered the attack with a short film containing a mixture of open-source material from social media accounts, private mobile phone film footage and other web-based sources to reconstruct the incident.[46] For media outlets to report about attacks of this nature, visual illustrations are of the utmost importance in an image-dominated news environment and there is no shortage of such visuals originating from web-based sources. However, to comply with press standards including trustworthy sources, unmanipulated photographs or film footage, media outlets run visual forensic units to verify images. All pictorial representations

---

[44] The Netherlands-based investigative journalism group Bellingcat identified three assassins in the Skripal case: Anatoliy Chepiga, Denis Vyacheslavovich Sergeev, Alexander Mishkin. See also: Moritz Rakuszitzky, Third Suspect in Skripal Poisoning Identified as Denis Sergeev, High-Ranking GRU Officer, in: https://www.bellingcat.com/news/uk-and-europe/2019/02/14/third-suspect-in-skripal-poisoning-identified-as-denis-sergeev-high-ranking-gru-officer/, last accessed 2023/11/14; Bellingcat Investigation Team, Full report: Skripal Poisoning Suspect Dr. Alexander Mishkin, Hero of Russia in: https://www.bellingcat.com/news/uk-and-europe/2018/10/09/full-report-skripal-poisoning-suspect-dr-alexander-mishkin-hero-russia/, last accessed 2023/11/14.

[45] Lennon (2023).

[46] Visual Forensics (2024).

are evaluated in terms of authenticity and origins. Images are geolocated and their sources are identified to provide truthful coverage of the event. However, these images in most instances are not illustrating the words, which are too inconclusive; it is the words, the word captions or spoken reportages that illustrate the images and lead the spectator's eyes.

21.6   AI-Generated Photographs

In recent years, discussions around artificial intelligence (AI) and AI image generators have widened as the technology becomes more sophisticated and widespread. As early as the late 1960s, the artist Harold Cohen (1928–2016) started to use computers to create artworks. By the mid-1970s, he had created the first AI image generating software, named AARON, at the University of California, Berkeley (Héder 2021 pp. 1–2, 378). Cohen trained the program with data of art techniques and styles, as well as data of different art works. AARON could then utilize this knowledge to create art works from scratch.[47] The Whitney Museum of American Art, New York dedicated an exhibition in 2024 to Cohen's collaborative work with the AARON software.[48] Cohen's approach to make AARON learn is very different from AI generators today, which are mainly dependent on databases.

But as technology advances, so does its impact on our society and coexistence. What are the opportunities and dangers of AI-generated images, in particular photographs of war and conflict? How can we ensure that these are used in an ethically responsible manner? AI in general is constructed around neural networks, which are based on algorithms similar to structures of the human brain. These algorithms can detect patterns in enormous amounts of data. The neural networks are based on neurons, which are interlocked with each other. Every neuron receives input, evaluates it, and then forwards the information to the next layer of neurons. By doing so, neurons in these neural networks notice patterns.[49] Neural networks can utilize data from a variety of sources, for example text, audio recordings, images and so on. However, these amounts of data need to be enormous to filter the most relevant and suitable data patterns. This process is called deep learning, which establishes a profound foundation from which AI can draw its knowledge and can result in AI-generated images that can hardly be identified as such (Lecun et al 2015).

How does an AI image generator learn? AI image generators are also known as Generative Adversarial Networks (GANs). GANs consist of two neural networks that communicate with each other. One network, the generator, creates an image which mimics the forms, shapes, color and texture of real subject matter. The other network, the discriminator, keeps watch and detects differences between the image

---

[47] AARON can be seen at work. Harold Cohen Drawing Machine Livestream in: https://www.you tube.com/watch?v=m_tR3gIdWxc. Accessed 23 Apr 2024.

[48] The Whitney Museum of American Art, New York, Exhibition Harold Cohen (2024).

[49] Hardesty (2024).

created by the generator and the real images. The generator and the discriminator critically collate their pictorial compositions; a process that helps the system to train its skills and produce images that are indistinguishable from real images (Goodfellow et al. 2014). AI image generators (like DALL-E from Open AI, MidJourney or Stable Diffusion) create images by a written text instruction or a fitting catchword, called prompts. Image generators offer an enormous variety of editing modes and styles in combination with models to achieve the desired composition, style and aesthetic. AI systems do not possess morals and ethics, and their actions are based on data and algorithms. AI image generators are based on prompts and therefore rely on the morals and ethics of their users.

Whether AI systems are able to learn what acceptable moral and ethical standards are will be seen in the future. Currently, it proves a challenge to detect and distinguish AI-generated images from real images, unless they are clearly marked as such. This process might not be necessary for AI images utilized for private use, but images for public use to illustrate press publications need to be tagged. AI could create images of war and conflict that never happened and are presented as photographs, making use of an aura of truthfulness and objectivity to mislead the viewer, with potentially serious consequences in an era in which conspiracy theories are booming. We might see these generated images as the artificial worm in the digital photographic apple, which is in desperate need of more research. These types of image will pose enormous challenges in the future and need a new set of rules for usage, publication and source visibility.

21.7   Conclusion

The aim of this article has been to address the role of photography to record and document, inform and disseminate subject matter relating to war and conflict, with a particular focus on photographs depicting the deployment of chemical weapons. War photography plays an important role in representing the impact and consequences of armed conflicts. These images provide visual records that can sway public opinion and shape public perception. In the aftermath of war and conflict, some photographs might become iconic, enabling them to be used to advocate for peace, human rights or humanitarian intervention, while other photographs might serve as evidence in legal proceedings to hold perpetrators of war crimes and human rights abuses accountable. Photographs depicting the deployment of chemical weapons often show troops or military personal in hazmat gear training with chemical weapons for war deployment, distant aerial views of the actual usage of chemical weapons, or the graphic aftermaths and consequences of their usage on humans and their environment.

Photographs are key pictorial messengers, given the human brain's ability to process them quickly. On a depictive level, the photographic surface provides an undisputed immediacy of the subject matter that viewers can decipher easily. However, to read a photograph correctly the origins, context and purpose of the image need to be clarified. In general, photographic compositions and aesthetics are shaped by the available camera techniques at the time. During the First World War, view cameras with tripods and longer exposure times were used in the beginnings, which were later replaces by small format cameras. These technical developments influence the shift from static scenes to scenes portraying motion. Representing chemical warfare was confined to aerial reconnaissance cameras or static and arranged subject matters given that sometimes single plated cameras with longer exposure times were used. Photographs of the Second World War became more dynamic because lightweight small format cameras were used, which resulted in images of troops as well as snap shots capturing civilians and the destruction of landscapes and cityscapes. Not only did photographic technology progress during the second half of the twentieth century, but also the usage of photographs changed as picture magazines gained popularity during the war years, which were in constant need of photographic coverage.

War and conflict photographs can act as evidence, and as Sontag has pointed out they are traces of the world, which answer the human desire to confirm reality via a pictorial representation, as exemplified by the photograph of a man holding an infant who died in the chemical attack of Halabja.

In recent years, another type of photographic documentation evolved: photographs and film still of chemical warfare attacks originating from CTTV footage and film footage emerging from private mobile phones. Most of these images on their own prove to be inconclusive for the spectator. Therefore, focal points are highlighted or encircled by picture editors, with journalists providing text captions to narrate a picture story.

The usage of AI image generators proves to be another challenge: AI can create a new type of pictorial presentation which appears in disguise as a photograph. AI systems could potentially create images of war and conflict that never happened, misleading the viewer with serious consequences. AI photographs therefore need to be clearly marked.

In summary, the technical developments and photographic examples illustrate that the overall nature and character of war photographs have changed. They must be examined in regard of their production process and origins, especially if they are to be considered as evidence of the usage of chemical weapons in a Fact-Finding Mission.

References

Baldwin G (1991) Looking at photographs. A guide to technical terms. British Museums Press, London

Barrett T (1985) Photographs and contexts. J Aesthet Educ 19(3):51-64

Barthes R (1977) Elements of semiology, , farrar, straus and giroux, reissue edition. New York

Barthes R (1978a) Death of the author. In: Image-music-text, trans. Stephen Heath, Hill and Wang, New York, pp 142–148

Barthes R (1978b) Rhetoric of the image. In: Image-music-text, trans. stephen heath. Hill and Wang, New York, pp 32–51

Barthes R (1985) Die helle Kammer. Bemerkungen zur Photographie, Suhr-kamp, Frankfurt am Main

Barthes R (2006) Schockfotos, pp 105–108. In: Kemp W (ed) Theorie der Fotografie, Bd. III, 1945–1980. Schirmer/Mosel, München

Bates S (2020) An aristocracy of critics: luce, hutchins, niebuhr, and the committee that redefined freedom of the press. Yale University Press, New Haven, Connecticut

Campbell D (2004) Horrific blindness: images of death in contemporary media. J Cult Res 8(1):55–74

Caves RW (2004) Encyclopedia of the city. Routledge, London

Cecil H, Liddle PH (eds) (1996) Facing armageddon: the first world war experienced. Leo Cooper Ltd., London

Chapman J (2000) The British at War. Cinema, state and propaganda, 1939–1945, Reprint, I.B. Tauris, London/New York

Crow D (2010) Visible signs: an introduction to semiotics in the visual arts. Bloomsbury Publishing PLC, London

Dauphinée E (2007) The politics of the body in pain: reading the ethics of imagery. Secur Dialog 38(2):139–155

Dodds K (2007) Steve Bell's eye: cartoons, geopolitics and the visualization of the 'war of terror. Secur Dialog 38(2):157–177

Foss SK (2005) Theory of visual rhetoric. In: Smith KL, Moriarty S, Kenney K, Barbatsis G (eds.) Handbook of visual communication: theory, methods, and media. Routledge, New York, pp 141–152

Galton F (1878) Composite portraits. J Anthropol Inst Great Britain Ireland 8:132–142

Goalwin A (2013) The Art of War: Instability, insecurity, and ideological imagery in northern Ireland's political murals, 1979–1998. Int J Polit Cult Soc 26(3):189–215

Goodfellow I, Pouget-Abadie J, Mirza M, Xu B, Warde-Farley D, Ozair S, Cour-ville A, Bengio Y (2014) Generative adversarial nets. In: Proceedings of the International Conference on Neural Information Processing Systems (NIPS 2014). pp 2672–2680. https://proceedings.neurips.cc/paper_files/paper/2014/file/5ca3e9b122f61f8f06494c97b1afccf3-Paper.pdf. Accessed 2 May 2024

Gombrich EH (2002) Art and Illusion. A study in the history of pictorial representation. Phaidon, New York

Griffin M (2010) Media images of war. Media, War & Conflict 3(1):7–41

Hardesty L (2024) Explained: neural networks. Ballyhooed artificial-intelligence technique known as "deep learning" revives 70-year-old idea, in: MIT News Office 14.4.2017. https://news.mit.edu/2017/explained-neural-networks-deep-learning-0414. Accessed 2 May 2024

Harding L (2020) A chain of stupidity: the skripal case and the decline of Russia's spy agencies 23rd June 2020. https://www.theguardian.com/world/2020/jun/23/skripal-salisbury-poisoning-decline-of-russia-spy-agencies-gru. Accessed 21 Oct 2023

Haworth-Booth M, Mellor D (1985) Bill Brandt. Behind the Camera, Aperture, New York

Hilterman JR (2007) A poisonous affair: America, Iraq, and the Gassing of Halabja. Cambridge University Press, Cambridge

Héder M (2021) AARON. In: Frana PL, Klein MJ (eds.) Encyclopedia of artificial intelligence: the past, present, and future of AI. ABC-Clio Publishing, San-ta Barbara, California, pp 1–2, p 378

Hurley F (2002) War photography, Australasian photo-review, 15th February 1919. In: Ennis H (ed.) Man with a camera. Frank Hurley Overseas, National Library of Australia, Canberra, p 164

Lecun Y, Bengio Y, Hinton G (2015) Deep learning. Nature, HAL Open Science, 2015, 521 (7553), pp.436–444, ff10.1038/nature14539ff. ffhal-04206682f. https://hal.science/hal-04206682/file/Lecun2015.pdf. Accessed 2 May 2024

Martin D, Boyd A (1999) Nessie. The surgeon's photograph exposed, Martin and Boyd, East Barnet, London

Mc Curry J (2023) Kim Jong-un's half-brother dies after 'attack' at airport in Malaysia. https://www.theguardian.com/world/2017/feb/14/kim-jong-un-half-brother-reportedly-killed-malaysia-north-korea. Accessed 11 Dec 2023

Möller F (2007) Photographic Interventions in post-9/11 security policy. Secur Dialog 38(2):179–196

Peirce CS (1865) Logic of the sciences. In: Fisch MH (Eds.) Writings of Charles S. Peirce: a chronological edition, vol 1. Indiana University Press, Bloomington, Indiana, pp 162–302

Pickard V (2015) America's battle for media democracy: the triumph of corporate libertarianism and the future of media reform (communication, society and politics. Cambridge University Press, Cambridge

Potter MC, Wyble B, Hagmann CE, McCourt ES (2014) Detecting meaning in RSVP at 13 ms per picture. Attent Percept Psychophys 76(2):270–279

Shore S (2007) The nature of photographs, 2nd edn. Phaidon Press Limited, London

Slocombe R (2010) British posters of the second world war. Exhibition catalogue Imperial War Museum, London

Sontag S (2003) Regarding the pain of others. Picador, New York

Sontag S (1977) On photography. Farrar, Straus and Giroux, New York

Tagg J (1988) The burden of representation. University of Minnesota Press, Minneapolis, Essays on Photographies and Histories

Well L (1997) Photography: a critical introduction. Routledge, London, London

Williams R (1992) The growth of the popular press. In: Williams R (ed.) The long revolution, London. Hogarth Press, London, pp 195–236

Warner-Marien M (2002) Photography. A cultural history. Laurence King Publishing, London

Web Based Sources

ABC News

https://abcnews.go.com/International/site-suspected-chemical-attack-syria-10-days-reporters/story?id=54526689. Accessed 28 Apr 2024

Agence France Press com.

https://www.afp.com/sites/default/files/afp_ethic_septembre_2023_0.pdf. Accessed 12 Feb 2024

BBC News

BBC News (2017) Kim Jong-nam death: North Korea asks for return of body. https://www.bbc.com/news/world-asia-38980474. Accessed 12 Jan 2023

Salisbury Novichok poisoning (2018) Russian nationals named as suspects. https://www.bbc.com/news/uk-45421445. Accessed 21 Oct 2023

Wingfield-Hayes R (2017) Kim Jong-nam killing: 'VX nerve agent' found on his face, including video and CTTV footage. https://www.bbc.com/news/world-asia-39073389. Accessed 21 Oct 2023

BBC Radio Wiltshire

Lennon M (2023) Salisbury Novichok poisoning medics 'still traumatized'. BBC Radio Wiltshire. https://www.bbc.com/news/uk-england-wiltshire-64742249. Accessed 20 Oct 2023

Bellingcat com.

Bellingcat Investigation Team (2023) Full report: skripal poisoning suspect Dr. Alexander Mishkin, Hero of Russia. https://www.bellingcat.com/news/uk-and-europe/2018/10/09/full-rep ort-skripal-poisoning-suspect-dr-alexander-mishkin-hero-russia/. Accessed 14 Nov 2023
Rakuszitzky M (2023)Third suspect in skripal poisoning identified as denis sergeev, high-ranking GRU officer. https://www.bellingcat.com/news/uk-and-europe/2019/02/14/third-suspect-in-skr ipal-poisoning-identified-as-denis-sergeev-high-ranking-gru-officer/. Accessed 14 Nov 2023

CBC Canada News

https://www.cbc.ca/news/world/chemical-weapons-inspectors-syria-reach-douma-1.4630142. Accessed 28 Apr 2024

Database of Education Resources Information Center by the U.S. Department of Education

Blanchard MA (1977) The REPORT is issued in The Hutchins Commission, the press and the responsibility concept. In: The association for education in journalism, p. 30. https://files.eric. ed.gov/fulltext/ED139017.pdf. Accessed 10 Dec 2023
Blanchard MA (1947) Commission asserts press menacés itself. The Editor & Publish-er Magazine, Hendersonville, Tennessee, p 62. https://files.eric.ed.gov/fulltext/ED139017.pdf. Accessed 10 Dec 2023
The Hutchins Commission (1977) The press and the responsibility concept. The Association for Education in Journalism. https://files.eric.ed.gov/fulltext/ED139017.pdf. Accessed 10 Dec 2023

Hawar News Agency

Hawar News Agency (ANHA) (2023) Halabja, a massacre in memory, similar danger Kurds live today. https://hawarnews.com/en/164741250029583. Accessed 24 Nov 2023

Forensic Architecture org

https://forensic-architecture.org/investigation/chemical-attacks-in-douma. Accessed 28 Apr 2024

NBC News

https://www.nbcnews.com/news/world/trump-announces-strikes-syria-following-suspected-che mical-weapons-attack-assad-n865966. Accessed 28 Apr 2024

OPCW.org

https://www.opcw.org/media-centre/news/2023/01/opcw-releases-third-report-investigation-and-identification-team. Accessed 2 May 2024

Reuters.com

https://www.reuters.com/world/middle-east/chemical-weapons-watchdog-blames-syrian-air-force-douma-attack-2023-01-27/. Accessed 12 Jan 2024.

Sky News

https://news.sky.com/story/douma-chemical-attack-that-killed-43-in-syria-was-likely-work-of-reg imes-air-force-international-watchdog-finds-12796578. Accessed 28 Apr 2024

Wall Street Journal

Wall Street Journal (2023) https://www.wsj.com/video/the-moment-kim-jong-nam-was-attacked-cctv-footage/D4AAC6B7-60BE-4457-8B20-093F3F063197. Accessed 21 Oct 2023

Washington Post

Visual Forensics, Washington Post (2020) Tracking Alexei Navalny's last movements before he was allegedly poisoned. https://www.youtube.com/watch?v=QDSfZGzZofY. Accessed 12 Mar 2024

Web Archive U.S. Department of State

U.S. Department of State (1988) Web archive. Saddam's Chemical Weapons Campaign, Halabja. https://2001-2009.state.gov/r/pa/ei/rls/18714.htm. Accessed 10 Apr 2024

Whitney Museum New York

The Whitney Museum of American Art, New York, Exhibition Harold Cohen (2024) AARON
   https://whitney.org/exhibitions/harold-cohen-aaron. Accessed 2 May 2024

Youtube Live Stream AARON

AARON can be seen at work. Harold Cohen Drawing Machine Livestream. https://www.youtube.
   com/watch?v=m_tR3gIdWxc. Accessed 23 Apr 2024

Katja Schmidt-Mai is a trained art historian (M. A. Art History, Pre-and Ancient History, Classical Archaeology; Ph.D. History of Photography) specializing in theories, methodologies and historiographies of photography. She has worked as a Research Assistant at Birkbeck College at the University of London and the Graves Art Gallery in Sheffield and has held posts as Lecturer at Kent University. In 2021, she joined the University of Hamburg as a Research Associate at the Department of Art History. Since March 2023, she started work as a Research Associate at the Department of Philosophy, Institute of Liberal Arts and Sciences, Hamburg. She organised a series knowledge transfer projects which resulted, among others, in the online exhibition "Composed: An Introduction into Photographic Fiction" and "Photography in Context" at the Institute for Contemporary Art and Transfer. She has published on 'Art under Dictatorship: Propaganda, Plunder, and Provenance' (with U. Schmidt, 2019), and on 'A wartime medical experiment as propaganda: the malaria case" (with J. M. Moreno, 2019) in an edited volume on Propaganda and Conflict, and is currently preparing a book chapter on 'Capturing the Everyday: The Transnational Aesthetics of Social Documen-tary Photography in the GDR' (forthcoming 2025/26).

The manufacturer's authorised representative in the EU is Springer
Nature Customer Service Centre GmbH, Europaplatz 3, 69115 Heidelberg,
Germany. If you have any concerns regarding our products, please
contact ProductSafety@springernature.com

Printed and bound by CPI Group (UK) Ltd, Croydon, CR0 4YY
23/04/2026
02095596-0002